Teas, Cocoa and Coffee

This book is dedicated to Professor Atsushi Komamine (1929–2011), kind mentor, inspirational lecturer, and outstanding researcher.

Teas, Cocoa and Coffee

Plant Secondary Metabolites and Health

Edited by

Professor Alan Crozier
School of Medicine,
College of Medical, Veterinary and Life Sciences,
University of Glasgow, Glasgow G12 8QQ, UK

Professor Hiroshi Ashihara
Department of Biological Sciences,
Graduate School of Humanities and Sciences,
Ochanomizu University, Otsuka, Bunkyo-ku,
Tokyo 112-8610, Japan

Professor Francisco Tomás-Barbéran
CEBAS CSIC, PO Box 164,
Espinardo 30100 Murcia, Spain

WITHDRAWN

A John Wiley & Sons, Ltd., Publication

This edition first published 2012 © 2012, 2006 by Blackwell Publishing Ltd.

Blackwell Publishing was acquired by John Wiley & Sons in February 2007. Blackwell's publishing program has been merged with Wiley's global Scientific, Technical and Medical business to form Wiley-Blackwell.

Registered office: John Wiley & Sons, Ltd, The Atrium, Southern Gate, Chichester, West Sussex, PO19 8SQ, UK

Editorial offices: 9600 Garsington Road, Oxford, OX4 2DQ, UK
The Atrium, Southern Gate, Chichester, West Sussex, PO19 8SQ, UK
2121 State Avenue, Ames, Iowa 50014-8300, USA
111 River Street, Hoboken, NJ 07030-5774, USA

For details of our global editorial offices, for customer services and for information about how to apply for permission to reuse the copyright material in this book please see our website at www.wiley.com/wiley-blackwell.

The right of the author to be identified as the author of this work has been asserted in accordance with the UK Copyright, Designs and Patents Act 1988.

All rights reserved. No part of this publication may be reproduced, stored in a retrieval system, or transmitted, in any form or by any means, electronic, mechanical, photocopying, recording or otherwise, except as permitted by the UK Copyright, Designs and Patents Act 1988, without the prior permission of the publisher.

Designations used by companies to distinguish their products are often claimed as trademarks. All brand names and product names used in this book are trade names, service marks, trademarks or registered trademarks of their respective owners. The publisher is not associated with any product or vendor mentioned in this book. This publication is designed to provide accurate and authoritative information in regard to the subject matter covered. It is sold on the understanding that the publisher is not engaged in rendering professional services. If professional advice or other expert assistance is required, the services of a competent professional should be sought.

Library of Congress Cataloging-in-Publication Data

Teas, cocoa and coffee : plant secondary metabolites and health / edited by Professor Alan Crozier, Professor Hiroshi Ashihara, Professor F. Tomás-Barbéran.
     p. cm.
   Includes bibliographical references and index.
   ISBN-13: 978-1-4443-3441-8 (hardcover : alk. paper)
   ISBN-10: 1-4443-3441-7
  1.Plant metabolites.   2. Metabolism, Secondary.   I. Crozier, Alan.   II. Ashihara, Hiroshi.
III. Tomás-Barbéran, F. A. (Francisco A.)
   QK881.T43 2011
   613.2′8–dc23
                                                                                              2011015215

A catalogue record for this book is available from the British Library.

This book is published in the following electronic formats: ePDF 9781444347067; Wiley Online Library 9781444347098; ePub 9781444347074; Mobi 9781444347081

Set in 10/12pt Minion by Aptara® Inc., New Delhi, India
Printed and bound in Singapore by Markono Print Media Pte Ltd

1   2012

Contents

Contributors	ix

1 The Origins of Tea, Coffee and Cocoa as Beverages — 1
Timothy J. Bond

    1.1 Introduction — 1
    1.2 The beverages in question — 1
    1.3 Discoveries – myth and legend — 2
        1.3.1 Tea — 3
        1.3.2 Coffee — 4
        1.3.3 Cacao products — 5
    1.4 Global domination begins — 8
        1.4.1 Tea – overland and a race by sea — 9
        1.4.2 Coffee – from persecution to epitomising the protestant work ethic — 13
        1.4.3 Chocolate – from lying down . . . to sitting up — 14
    1.5 From foreign fancies to the drinks of the masses — 15
    1.6 Tea, coffee and chocolate 'go public' — 18
    1.7 Opinion is divided on the merits of the three beverages — 19
    1.8 Tea, coffee and chocolate – the future — 22
    References — 22

2 Purine Alkaloids: A Focus on Caffeine and Related Compounds in Beverages — 25
Michael E.J. Lean, Hiroshi Ashihara, Michael N. Clifford and Alan Crozier

    2.1 Introduction — 25
    2.2 Occurrence of purine alkaloids — 26
    2.3 Biosynthesis of purine alkaloids — 27
    2.4 Degradation of purine alkaloids — 27
    2.5 Decaffeinated tea and coffee — 29
    2.6 Metabolism of caffeine by humans — 31

	2.7	Effects of caffeine consumption on human health		33
		2.7.1	Biochemical and biological actions of caffeine	34
		2.7.2	Mental performance enhancement	37
		2.7.3	Physical performance enhancement	37
		2.7.4	Caffeine toxicity	38
		2.7.5	Tolerance, withdrawal and dependence	39
		2.7.6	Caffeine in pregnancy	39
		2.7.7	Toxicity in other species	40
	2.8	Summary	40
		References	40
3	Phytochemicals in Teas and Tisanes and Their Bioavailability	45
	Michael N. Clifford and Alan Crozier	
	3.1	Introduction	45
	3.2	Phytochemical content of teas and tisanes	45
		3.2.1	*Camellia* teas	45
		3.2.2	Yerba maté tea	54
		3.2.3	Itadori tea	58
		3.2.4	Rooibos tea	59
		3.2.5	Honeybush tea	59
		3.2.6	Chamomile tea	62
		3.2.7	Hibiscus tea	62
		3.2.8	Fennel tea	63
		3.2.9	*Anastatica* tea	63
		3.2.10	*Ficus* tea	66
	3.3	Bioavailability – absorption, distribution, metabolism and excretion	66
		3.3.1	Green tea	68
		3.3.2	Black tea	77
		3.3.3	Itadori tea	80
		3.3.4	Rooibos tea	81
		3.3.5	Honeybush tea	84
		3.3.6	Hibiscus tea	85
		3.3.7	Fennel tea	85
		3.3.8	Other teas	87
	3.4	Summary	87
		References	88
4	Teas, Tisanes and Health	99
	Diane L. McKay, Marshall G. Miller and Jeffrey B. Blumberg	
	4.1	Introduction	99
	4.2	Black, oolong and green tea (*C. sinensis*)	100
		4.2.1	Black tea	100
		4.2.2	Oolong tea	107
		4.2.3	Green tea	109
	4.3	Other teas and tisanes	116
		4.3.1	Yerba maté (*Ilex paraguariensis*)	116

		4.3.2	Itadori (*Polygonum cuspidatum*)	118
		4.3.3	Chamomile (*Chamomilla recutita* L.)	119
		4.3.4	Hibiscus (*Hibiscus sabdariffa* L.)	120
		4.3.5	Rooibos (*Aspalathus linearis*)	126
		4.3.6	Honeybush (*Cyclopia intermedia*)	128
	4.4	Summary and conclusions		130
	References			131

5	Phytochemicals in Coffee and the Bioavailability of Chlorogenic Acids	143

Angelique Stalmach, Michael N. Clifford, Gary Williamson and Alan Crozier

	5.1	Introduction		143
	5.2	Harvesting coffee beans, roasting and blending		144
	5.3	Phytochemicals in coffee		144
		5.3.1	Effects of roasting on the phytochemical content of coffee beans	149
		5.3.2	Chlorogenic acid intake and coffee consumption	154
	5.4	Bioavailability of coffee chlorogenic acids in humans		155
		5.4.1	Studies involving volunteers with and without a functioning colon	156
	5.5	Conclusions		164
	References			164

6	Coffee and Health	169

Gary Williamson

	6.1	Introduction		169
	6.2	Antioxidant status		170
		6.2.1	Effect of coffee consumption on antioxidant status: epidemiological and cohort studies	179
		6.2.2	Effect of coffee consumption on antioxidant status: intervention studies	179
	6.3	Diabetes		180
		6.3.1	Effect of coffee consumption on diabetes risk: epidemiological and cohort studies	180
		6.3.2	Effect of coffee consumption on diabetes risk: intervention studies	182
	6.4	Cardiovascular disease		183
		6.4.1	Effect of coffee consumption on cardiovascular risk: epidemiological and cohort studies	183
		6.4.2	Effect of coffee consumption on cardiovascular risk: intervention studies	184
	6.5	Effect of coffee on inflammation		186
	6.6	Effect of coffee consumption on cancer risk		186
		6.6.1	Effect of coffee consumption on cancer risk: epidemiological and cohort studies	186

		6.6.2	Effect of coffee consumption on cancer risk: intervention studies	188
	6.7	Summary		188
	References			188

7	Phytochemicals in Cocoa and Flavan-3-ol Bioavailability			193
	Francisco Tomás-Barbérán, Gina Borges and Alan Crozier			
	7.1	Introduction		193
	7.2	Phytochemicals in cocoa		194
		7.2.1	Purine alkaloids, theobromine and caffeine	194
		7.2.2	Flavan-3-ols	194
		7.2.3	Phenolic acid derivatives	196
		7.2.4	Minor phytochemicals	197
	7.3	Bioavailability of cocoa flavan-3-ols		198
		7.3.1	Background	198
		7.3.2	Flavan-3-ol monomers	200
		7.3.3	Procyanidins	210
	7.4	Conclusions		212
	References			213

8	Cocoa and Health		219
	Jennifer L. Donovan, Kelly A. Holes-Lewis, Kenneth D. Chavin and Brent M. Egan		
	8.1	Introduction	219
	8.2	Composition of cocoa products	220
	8.3	Worldwide consumption of cocoa and its contribution to flavonoid intake	222
	8.4	Epidemiological and ecological studies of cocoa	222
	8.5	Cocoa effects on vascular endothelial function and platelet activity	224
	8.6	Cocoa and hypertension	227
	8.7	Antioxidant and anti-inflammatory effects of cocoa	229
	8.8	Effects of cocoa consumption on lipid and lipoprotein metabolism	232
	8.9	Cocoa effects on insulin sensitivity	233
	8.10	Cocoa effects on cerebral blood flow and neurocognitive functioning	234
	8.11	Potential negative health effects of cocoa consumption	237
		8.11.1 Obesity	237
		8.11.2 Testicular health	237
		8.11.3 Acne	238
		8.11.4 Dental caries	238
	8.12	Effects of consumption of cocoa with milk or other foods	238
	8.13	Conclusions	239
	References		240

Index 247

A color plate section falls between pages 6 and 7

Contributors

Hiroshi Ashihara	Department of Biological Sciences, Graduate School of Humanities and Science, Ochanomizu University, Otsuka, Bunkyo-ku, Tokyo 112-8610, Japan
Jeffrey B. Blumberg	Jean Mayer USDA Human Nutrition Research Center on Aging, Tufts University, 711 Washington Street, Boston, MA 02111, USA
Timothy J. Bond	Finlay Tea Solutions, Swire House, 59 Buckingham Gate, London SW1E 6AJ, UK
Gina Borges	School of Medicine, College of Medical, Veterinary and Life Sciences, University of Glasgow, Glasgow G12 8QQ, UK
Kenneth D. Chavin	Department of Surgery, Medical University of South Carolina, Charleston, SC 29401, USA
Michael N. Clifford	Food Safety Research Group, Centre for Nutrition and Food Safety, Faculty of Health and Medical Sciences, University of Surrey, Guildford, Surrey GU2 7XH, UK
Alan Crozier	School of Medicine, College of Medical, Veterinary and Life Sciences, University of Glasgow, Glasgow G12 8QQ, UK
Jennifer L. Donovan	Departments of Psychiatry and Behavioral Sciences, Medical University of South Carolina, Charleston, SC 29401, USA
Brent M. Egan	Department of Medicine, Medical University of South Carolina, Charleston, SC 29401, USA
Kelly A. Holes-Lewis	Departments of Psychiatry and Behavioral Sciences, Medical University of South Carolina, Charleston, SC 29401, USA

Michael E.J. Lean	University of Glasgow College of Medical, Veterinary and Life Sciences, Walton Building, Royal Infirmary, 84 Castle Street, Glasgow G4 0SF, UK
Diane L. McKay	Jean Mayer USDA Human Nutrition Research Center on Aging, Tufts University, 711 Washington Street, Boston, MA 02111, USA
Marshall G. Miller	Department of Psychology, Tufts University, 490 Boston Avenue, Medford, MA 02155, USA
Angelique Stalmach	School of Medicine, College of Medical, Veterinary and Life Sciences, University of Glasgow, Glasgow G12 8QQ, UK
Francisco Tomás-Barbéran	CEBAS CSIC, PO Box 164, Espinardo 30100 Murcia, Spain
Gary Williamson	School of Food Science and Nutrition, University of Leeds, Leeds LS2 9JT, UK

Chapter 1
The Origins of Tea, Coffee and Cocoa as Beverages

Timothy J. Bond

Finlay Tea Solutions, Swire House, 59 Buckingham Gate, London SW1E 6AJ, UK

1.1 Introduction

What are the origins of tea, coffee and cocoa? How were they discovered and how did they become such important items in the everyday lives of billions of people? To answer these questions we need to cover a lot of ground from geography and social anthropology to plant biochemistry and human physiology. The so-called Western cultures began to consume these beverages only very late in their history, 'discovering' tea, coffee and cocoa through the forces of international commerce and trade and the expansion of their empires in the seventeenth century. Prior to that, the drinks had been consumed for a millennium, and as we shall see, the real 'discoveries' were that while many plants were toxic, tea, coffee and cocoa could yield products that could be consumed and have a beneficial impact on human physiology, emotional state and culture. Today, for many consumers it is difficult to imagine life without the first cup of coffee or tea of the day or without a chocolate bar.

1.2 The beverages in question

Tea, *Camellia sinensis*, is a small tree native to the Assam area of North India where North Burma and South China meet. This region has had a tumultuous history, linked to the Opium Wars between the United Kingdom and China, and the War of Independence between the United States and the United Kingdom, the growth of urban centres, the expansion of the British Empire, the birth of the British Industrial Revolution and many more events. After water, tea is now the most consumed beverage in the world (UK Tea Council 2010), drunk for both pleasure and health.

Coffee is produced from the seeds of another small tree, originating this time from Africa, spread via the slave trade to the Arabic empires where it gained pre-eminence

---

Teas, Cocoa and Coffee: Plant Secondary Metabolites and Health, First Edition.
Edited by Professor Alan Crozier, Professor Hiroshi Ashihara and Professor F. Tomás Barberán.
© 2012 Blackwell Publishing Ltd. Published 2012 by Blackwell Publishing Ltd.

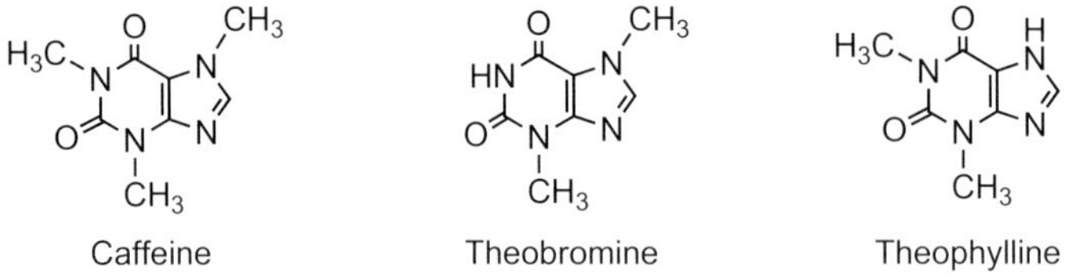

Figure 1.1 Structures of caffeine, theobromine and theophylline.

due to the Muslim ban on fermented alcoholic beverages. *Coffea arabica*, *Coffea robusta* and *Coffea liberica* were discovered later and were transplanted across the world to establish plantations/estates in as far away places as Hawaii, Brazil and Vietnam. From this position, coffee has become, after oil, the second most valuable traded commodity on the global stock market.

Cocoa (*Theobroma cacao*) originates from the rainforests of Central America. As a drink, it was once the preserve of the Mayan and Aztec elite and a form of currency. It became intimately linked to Cortez's pursuit of gold in Southern America, and was much loved by the Spanish court. Cocoa made the move from drink to chocolate to become the favourite snack of the world's children and numerous 'chocoholic' adults.

These products created change not only in our societies and habits but also in our overall health and drove major changes in transport, especially shipping, contributing to the development of major ports such as Hamburg and Hong Kong, and indirectly creating markets for European pottery and silverware.

Chemically, cocoa, tea and coffee all comprise complex mixtures of plant secondary metabolites (see Chapters 2, 3, 5 and 7). Key amongst these are the purine alkaloids caffeine, theophylline and theobromine (Figure 1.1), which have an impact on alertness and are the basis of the 'pick-up' effects of all three beverages. In addition, simple and complex polyphenols are present, which provide taste and colour characteristics. The exact make-up of these constituents is influenced by the type of raw material, its origin, processing that is driven by manufacturing technology, and whole empires for the growth, trading, blend and distribution of these plant materials as they migrate from bush to cup.

The three beverages were unknown to Western civilisations prior to the seventeenth century. Today, they are items for everyday consumption, having been converted from very different cultural uses. Tea was originally a medicine, coffee was as a religious aid and symbol, and cocoa was the ultimate status symbol consumed by gods, priests and royalty. But where did they come from, who 'discovered' and introduced them to Western cultures and how did they assume their current paramount importance?

1.3 Discoveries – myth and legend

The discovery and initial utilisation of tea, coffee and cocoa as beverages are very much the stuff of myth and legend, and some of the earliest stories presented here should be treated as just that – fascinating and insightful, but with limited credibility. These mythical tales are chiefly a consequence of how the histories of the discovering cultures were recorded. In the case of tea, the character representing tea seems to have evolved along with language in general, so it is difficult to be sure that in the earliest texts, the 'tea' being referred to was not some other bitter herbal infusion. In Ethiopia, where coffee has its origins, history was mainly verbal rather than written, and in Mesoamerica, the pictographic language of the early civilisations has only recently been deciphered and even then is open to different interpretation. This makes verifying the information difficult but does give a flavour as to the origins of these beverages.

1.3.1 Tea

Tea is intimately linked with Asia, herbal medicine and Buddhism. Almost 4000 years ago, in 2737 BC, the great herbalist, 'Divine Healer' and Chinese emperor Shen Nung discovered tea (Ukers 1935). The story is that he observed leaves that had fallen from a nearby and unassuming tree (Plate 1.1) being boiled in water by his servant. Upon tasting the brew, the emperor found it to his liking and thus green tea was 'discovered'. This also indicates that the Chinese already knew the value of making water drinkable by boiling it to kill microbial contamination. This habit still remains in China to this day where boiled drinking water is preferred, being served still and warm, rather than the glass of cold iced water demanded by other cultures. Tea, as luck would have it, has antimicrobial properties from the inherent catechin and caffeine contents, so it has a double benefit when consumed hot! In Chinese herbal medicine, 'bitter' is seen as a desirable trait; it would be interesting to see how the story might have evolved if tea had been discovered by a society that did not value this taste to the same extent.

This part of the story of the discovery of tea is well-established mythology. What is less well known is what became of the Shen Nung, father of tea. Shen Nung, tea and Chinese herbal medicine are all intimately linked. The emperor went around tasting innumerable herbs and plants, cataloguing them and documenting their health-promoting properties, and published his findings in the *Pen ts'ao* or *Medicinal Book*. It was during the course of this work that he reputably tasted a herb that was so poisonous that it could kill you before you had taken ten steps. Only tea leaves could save him, but alas none was available and he died after taking seven steps. How an author could publish a book having died during the course of his research is an interesting question, but not the only one surrounding the historical accuracy of Shen Nung's book. In the discovery of tea, much hinges on the following passage in the good book:

> Bitter t'u is called cha, hsun and yu. It grows in winter in the valleys by the streams, and the hills of Ichow, in the province of Szechwan, and does not perish in severe winter. It is gathered on the third day of the third month, April, and dried. (Ukers 1935)

It seems that again we have some controversy as both Szechwan and Fujian provinces claim to be the historical home of tea. This aside, there is a historical linguistic issue with this quotation as the term 'cha' came into use only after the seventh century. The first Chinese book of tea the *Ch'a Ching*, published in 780 AD by Lu Yu, is widely recognised as an authoritative text covering harvesting, manufacturing and preparation of the beverage, as well as its consumption and history (Ukers 1935).

Despite the conjecture about the precise origins of tea, it is widely acknowledged that it was discovered in China and spread from there to surrounding countries and then onto Japan, sometime during the eighth century. Not to be outdone though, Japanese have their own myths about the birth of tea in China. The Japanese version of events is that Bodhidarma, a Buddhist saint, fell asleep during meditation. Furious with himself, a not very Buddhist sentiment, and to ensure that this did not happen again he cut off his eyelids and threw them to the ground. A bush sprang from the ground where they landed, with leaves curiously of the same shape as his eyelids, which when brewed produced a drink that banished fatigue (Ukers 1935).

1.3.2 Coffee

Coffee's discovery is just as emotive, of which we have hints in the naming conventions of the types of coffee that are now consumed around the world. Coffee is thought to be a native of the Kefa region of old, which was also known as Harar, in Northern Africa (Allen 2000). Decimating the ancient kingdoms, Western empires divided Africa in the nineteenth and twentieth centuries to artificially create territories of control. The Kefa region is now in Ethiopia. Legend has it that sometime around 850 AD in Abyssinia, current-day Northern Ethiopia, a goat herder named Kaldi noticed that his goats were acting in a rather strange way, with even the oldest of them running and gambolling in the heat of the midday sun. Paying attention to this, he noticed that they were especially frantic after eating red berries from a small, broad-leaved shrub (Roden 1981). These red berries were, of course, coffee cherries – the red, fleshy 'berry' covering the green coffee bean, the seed of the coffee tree (Plate 1.2). From here the legend diverts from Kaldi either to the abbot of a local monastery or, via trade, to the abbot of a monastery in Yemen. Either way, the resident abbot seems to have been concerned with a lack of alertness of monks during early morning prayers (Ukers 1922). Hearing of the effects the beans had on goats, he tried them on the acolytes, and again, we have a link between an invigorating beverage and keeping awake during religious ceremonies with, in due course, coffee trees becoming common fixtures of monastery gardens across the Arab world.

Coffee was not initially consumed as a beverage; instead, the beans were often mixed with fat and eaten as a high-energy snack during long journeys (Allen 2000). Between the first and fifth centuries AD, a lucrative trade route between Arabia and East Africa grew up, with slaves as one of the most valuable commodities (Segal 2002). Coffee seems to have been brought by the slave trains overland from Ethiopia and across to the port of Mocha, a name that still has meaning for the modern-day coffee connoisseur. It is said that coffee trees line the old trade routes as a result of discarded snacks along the way, germinating to produce the berry-yielding trees. The Mocha region, which is in current-day Yemen, became both a major area for cultivation and the ancient port from which coffee was exported to the rest of the world.

How coffee was converted from a food to a beverage is less clear, though there are accounts of a 'fermented beverage' being produced from the red coffee cherries (Ukers 1922). Like other food and beverage items, the modern-day incarnation of coffee is very different from its origins. In the case of modern coffee, the cherries are picked and the outer pulpy 'cherry' removed either by soaking in water (wet coffee) or alternatively the cherries are laid out to dry in the sun (dry coffee) and the dried outer coating is removed by abrasion. With both methods the removal of the covering reveals the green coffee beans that are the cotyledons of the seed. But this is not coffee as we know it; the beans must be roasted to a dark-brown/black colour, the effects of which develop the rich aroma and characteristic taste of the coffee. The roasted beans are ground to a coarse/fine powder and finally infused in boiling water to produce what we would consider a 'cup of coffee'. What brought about the move from beans in a ball of fat to the 'roast and ground' coffee drink we recognise today is far from clear, but another tale of the discovery of coffee may throw some light on the matter. This time the tale is of:

> ... the dervish Hadji Omar was driven by his enemies out of Mocha into the desert, where they expected him to die of starvation. This undoubtedly would have occurred if he had not plucked up the courage to taste some strange berries which he found growing on a shrub. While they seemed to be edible, they were very bitter; and he tried to improve the taste by roasting them. He found, however, that they had become very hard, so he attempted to soften them with water. The berries seemed to remain as hard as before, but the liquid turned brown, and Omar drank it on the chance that it contained some of the nourishment from the berries. He was amazed at how it refreshed him, enlivened his sluggishness, and raised his drooping spirits. Later, when he returned to Mocha, his salvation was considered a miracle. The beverage to which it was due sprang into high favour, and Omar was made a saint. (Ukers 1922)

Coffee initially gained acceptance in the Muslim world as an acceptable, non-alcoholic drink that also conveniently removed fatigue before prayer.

1.3.3 *Cacao products*

Cacao is more sensitive to agronomic conditions than are tea and coffee, the tree being broad leafed and thriving in narrow climatic conditions defined by shaded areas in latitudes within 20 degrees of the equator (Coe and Coe 2003). Cacao is *Theobroma cacao* according to the Linnaeus classification system. Theobroma means 'food of the gods', which shows either Linnaeus great reverence for the product or his understanding of the cultural relevance of cacao. Cacao originates from the rainforests of Central America, sometimes termed 'Mesoamerica', where on modern maps southern Mexico's Yucatan peninsula, Guatemala, Belize, El Salvador and Honduras are located. The discovery and early use of cocoa are intimately linked with the ancient cultures of these areas – the Olmecs (1500–400 BC), Mayans culture (250–900 AD) and later the Aztecs (*ca.* 800 AD to mid-sixteenth century). These three cultures had advanced pictographic written languages, and real advances in deciphering them have been made only in the last 50 years. The Olmecs and Mayans are all the more mysterious as their cultures seemed to die out in great 'collapses', with one passing and the other flourishing in its place. The reason for the collapse of the Olmec culture is unknown, while that of the Mayans is thought to be linked to environmental degradation and political infighting. The collapse of the Aztecs is better recorded and intimately linked with their conquest and colonisation by the Spanish. We have limited information about the Olmecs, with some remaining pictographs, but more is known about the Mayans. It is possible though that the word 'cacao' has its origins with the Olmecs as '*Kakawa*', was a term reserved for what we now know as *T. cacao* (Coe and Coe 2003). Many Mayan pictographs remain with depictions of the cacoa tree. Here, we are very lucky that the cacao tree is so distinctive, unlike the physically unremarkable tea tree and coffee bush, with the flowers and, after pollination, the seed pods actually growing directly from the trunk of the tree rather than at the ends of branches and shoots amongst the leaves. This makes identification even on ancient pottery relatively easy (Plate 1.3).

For the Mayans and Aztecs, the origin of cocoa was less a discovery than divine interventions in their creation myths. For the Mayans, cacao was provided to the newly created humans by the gods on the 'Mountain of Sustenance' (translated from the now-lost original pictographic *Popol Vuh* – originally written on perishable bark paper):

> and so they were happy over the provisions of the good mountain, filled with sweet things, thick with yellow corn, white corn, and thick with pataxte (i.e. *Theobroma bicolour*) and cacao.

For the Aztecs, all necessary foods had their source deep within a mountain and were brought to the surface by the intervention of gods such as Quetzalcoatl (Coe and Coe 2003). Cacao's divine nature is further reinforced through ancient records. The hot and humid climate of the mid-Americas was not kind to the bark-paper books, and only a handful survive in European museums and libraries, including the *Dresden* (http://www.dresdencodex.com/) and *Madrid Codex* (http://www.mayacodices.org/). In these depictions, cacao was always present with the gods. It was equally important when beginning the journey into the afterlife to have pottery containing chocolate-based beverages placed in the tomb. We know this because current-day analysis of the vessels has revealed that they contain residual theobromine and caffeine, evidence of the original presence of cocoa (Coe and Coe 2003).

For the Mayans, cocoa started as a drink for the elite where it was utilised as a spiritual beverage and valued to such an extent that cocoa pods and nibs were used as currency. Trade was very important in ancient Mesoamerica, especially as the seats of power were frequently not in the lowlands where cacao flourished. Long-distance trade routes developed, and the merchants and bearers were surely happy that the most valuable commodities, cacao and brightly coloured bird feathers, were not as heavy as the minerals and elements favoured by other ancient cultures.

Depictions on excavated pottery vessels help us to further understand how cacao was consumed – as a hot and cold beverage with copious amounts of esteemed 'froth' on the surface. The earliest known depiction of this traditional method is depicted on a vessel from the eighth century held in the Princeton Art Museum. The cacao liquid was poured from vessel to vessel, often at a distance, in order to raise the foam.

Plate 1.1 A flowering shoot of *Camellia sinensis*. (*Curtis Botanical Magazine*, Plate 998, Volume 25, 1801. Reproduced with permission of The Linnean Society.)

Plate 1.2 Flowers and ripe cherries of *Coffea arabica*. (*Curtis Botanical Magazine*, Plate 1303, Volume 32, 1810. Reproduced with permission of The Linnean Society.)

Teas, Cocoa and Coffee: Plant Secondary Metabolites and Health, First Edition.
Edited by Professor Alan Crozier, Professor Hiroshi Ashihara and Professor F. Tomás Barberán.
© 2012 Blackwell Publishing Ltd. Published 2012 by Blackwell Publishing Ltd.

Plate 1.3 Mayan pottery depicting cacao tree, Princeton Museum. Kerr number: K631. Palace scene with cacao tree. The individual using the metate is probably grinding chocolate pods into powder or paste. (Reproduced with permission of Barbara Kerr.)

Plate 1.4 Chinese characters depicting the changing face of tea. (Reproduced with permission of H. Wang, Finlay Tea Solutions.)

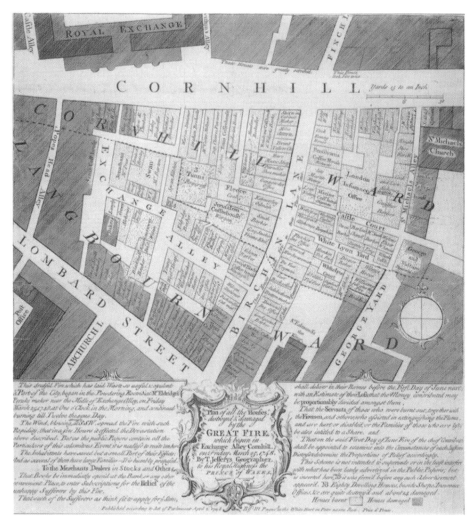

Plate 1.5 M. Payne, A plan of all the houses destroyed and damaged by the great fire that began in Exchange Alley, Cornhill, 25 March 1748, published at the White Hart in Paternoster Row, London, 1748. (© The British Library Board (Maps cc.5.a.293).)

Before this, the cacao 'nuts' were ground with other additives as commented on by the Spanish Bishop Landa:

> They make of ground maize and cacao a kind of foaming drink which is very savory, and with which they celebrate their feasts. And they get from their cacao a grease which resembles butter [cocoa butter?], and from this and maize they make another beverage which is very savoury and highly thought of .... (Coe and Coe 2003)

The elite classes of the Aztec empire had their ruling city Tenochtitlan, which is close to modern-day Mexico City, far from cacao production, but through conquest, they gained cacao-growing territories. The nobles and warriors were important 'classes' of citizen in the top ranks. A third class of elites was that of the 'long-distant merchant' who were important as they were responsible for overland expeditions to source cacao and bring it back to the nobles and priests.

The Aztecs favoured cacao both as a beverage and as currency as noted by many Western travellers to that region. The first Westerner to 'discover' cacao was Christopher Columbus in 1502 off the coast of Guanaja Island, which is 50 km north of what is now Honduras. Because of translation and exaggeration what actually transpired has been summarised as an encounter between Columbus's crew and a war/trade dugout canoe containing amongst other things:

> [M]any of those small almonds which in New Spain are used for money. They seem to hold these almonds at a great price, for when they were brought on board ship together with their goods, I observed that when any of these almonds fell, they all stooped to pick it up as if an eye had fallen! (Morrison 1963)

Here, we must recognise that the additional terms and phrases used in the description of cacao products – including cocoa, chocolate, chocolatte, etc. – are all derivations or mispronunciations of native American words:

> [T]he species name is cacao, and one can understand that Englishmen, finding it difficult to get their insular lips round this outlandish word, lazily called it cocoa. (Knapp 1920)

Sixteenth-century dictionaries translating the local dialects into Spanish give '*cacahuatl*' (*huatl* meaning water) for drinking chocolate (Coe and Coe 2003). Francisco Hernández used the word *chocolatl*, which one assumes later became chocolatte and indeed finally chocolate as the name for the beverage, and in modern parlance, the confectionary.

Not all the people of 'New Spain', the conquered territories of Latin America or indeed 'Old' Spain took to this new drink:

> The chief use of Cocoa is in a drincke which they call Chocolaté, whereof they make great account foolishly and without reason; for it is loathsome to such as are not acquainted with it, having a skumme or frothe that is very unpleasant to taste .... (De Acosta 1604)

Nonetheless, there were those in New Spain who came to depend on their cacao beverage – especially the ladies of New Spain. Written in 1648 and published in Gaze's *New Survey of the West Indies*, the following account shows just how much this beverage was esteemed:

> The women of that city [Chiapa], it seems, pretend much weakness and squeamishness of stomacke, which they say is so great that they are not able to continue in church while the mass is briefly hurried over, much lesse while a solemn high mass is being sung and a sermon preached, unless they drinke a cup of hot chocolattte [*sic*] and eat a bit of sweetmeats to strengthen their stomackes. For this purpose it was much used by them to make their maids bring them to church, in the middle of mass or sermon, a cup of chocolate, which could not be done to all without a great confusion and interpreting both mass and sermon. The Bishop, perceiving this abuse, and having given faire warning for the omitting of it, but all without amendment, thought fit to fix in writing upon the church dores an excommunication against all such as should presume at the time of service to eate or drinke within the church. This excommunication was not taken by all, but especially by the gentlewomen, much to heart, who protested, if they might not eat or drinke in the church, they could not continue in it to hear what otherwise they were bound unto. But none of these reasons would move the Bishop. The women, seeing him so hard to be entreated, began to slight him with scornefull and reproachfull words; others slighted his excommunication, drinking in iniquity in the church, as the fish doth water, which caused one day such an uproar in the Cathedrall that many swordes were drawn against the Priests, who attempted to take away from the maids the cups of chocolate which they bought unto their mistresses, who at last, seeing that neither faire nor foule means would prevail with the Bishop, resolved to forsake the Cathedrall: and so from that time most of the city betooke themselves to the Cloister Churches, where by the Nuns and Fryers they were not troubled .... The Bishop fell dangerously sick. Physicians were sent for afar and neere, who all with a joint opinion agreed that the Bishop was poisoned. A gentlewoman, with whom I was well acquainted, was commonly censured to have prescribed such a cup of chocolatte to be ministered by the Page, which poisoned him who rigorously had forbidden chocolatte to be drunk in the church .... And it became afterwards a Proverbe in that country: 'Beware of the Chocolatte of Chiapa! (Gaze 1648)

The first commercial shipment of cocao arrived in Seville from Veracruz in 1585 (Jamieson 2001), but it would be naive to assume that it had not found its way to Spain earlier, probably through the Jesuit religious networks whose senior influential priests introduced it to their social circles. Indeed, Baroque Spain with its flamboyant style would have been the ideal vehicle for the dissemination of ideals and standards to other Catholic regions.

1.4 Global domination begins

So, three plants in three different parts of the world have been discovered and utilised by the peoples of these regions. All have very different flavours and modes of consumption, but all are linked in local psyche with alertness, vitality and enhanced physical ability. How did these drinks spread and why did people from far-away lands such as Europe, with very different tastes, choose to sample and ultimately embrace these beverages? The key to this is global trade and the desire of European powers for expansion overseas and the control of trade routes both overland and by sea.

The old trade routes including the 'Silk Road' and the 'Spice Route' were the key overland routes for goods to travel from China, across India and the Middle East through Persia (the old name for the Iran and Iraq of today) to Venice and from there to the rest of Europe. Spices, Far Eastern cloth, tea and coffee all came along these routes, on camel and horseback, from their native lands to European capitals. The British and Dutch decided that this convoluted route was both slow and inefficient, and probably more to the point, out of their control and with too many middlemen. They decided to go direct by sea. The Spanish and Portuguese conquests of South and Latin America brought them into intimate contact with cocoa, and later, their colonies were also linked to coffee.

1.4.1 Tea – overland and a race by sea

The etymology of tea, the derivation of the word gives some insights, on which we shall be brief, but they are informative. As discussed previously, the changing characters of the ancient Chinese language have added to the challenge of tracking tea through history earlier than 780 AD. The character *t'u* (Plate 1.4b) has a number of meanings, including sow thistle, bitter cabbage, grass/rush and, finally, tea. Only in the seventh century AD was the second horizontal stroke on the lower vertical of *t'u* eliminated, giving *Ch'a* (Plate 1.4a) – a character that has only ever had one meaning, tea.

Insights from the language of tea are not limited to ancient etymology; they allow us to track how tea spread from ancient China to the rest of the world. The term *Ch'a* was spread and assimilated by word of mouth through overland and short sea journeys to Japan (*Cha*), Persia (*Cha*), Turkey (*Chay*) and Russia (*Chai*). Most of these journeys probably began from old Canton (modern-day Guangzhou). The other ancient language that appears most prominently in the history of tea is that of Amoy (current-day Xiamen) on the southern coast of China in Fujian province, which is still an important tea-producing region. The Amoy dialect pronounced tea as '*Tay*', which can be traced to those regions who received their tea by sea routes such as the Netherlands (*Thee*), Italy (*Te*), Sri Lanka (for the Singhalese *Thay* and for the Tamils *Tey*), British (Tea), French (*Thé*) and, of course, the Latin *Thea* (Ukers 1935). Clearly, after the initial designations, languages continued to evolve, and in many circles and sea ports in Europe, '*Cha*' became linked more with the working classes. The educated citizens choosing to be associated with more refined languages, such as French, incorporated '*Thé*' in their everyday language (Hobhouse 1999).

Word of tea drinking spread, preceding the actual tasting of tea, and indeed, various Europeans repeated tales of far-flung lands, recounted to them by visitors to their cities, almost as gospel. Most notable was that written by Giambattista Ramusio (1485–1559) in the second volume of his *Navigatione et Viaggi* written in 1559. In this, the secretary of the powerful and influential Venetian 'Council of Ten' collected and published travellers' tales including that of a Persian subject called Hajji Mohammed (or Chaggi Mehmet) who told of a herb '*Chai Catai*':

> [T]ake of that herb whether dry or fresh and boil it well in water. One or two cups of this decoction taken on an empty stomach removes fever, headache, stomach ache, pain in the side or in the joints and it should be taken as hot as you can bear it. (Ukers 1935)

Interestingly, doubt has been cast on some of the travels of Marco Polo (1254–1324) who told of his journeys over 24 years to the court of Kublai Khan and across the whole of China. Why, if he travelled so widely across the vast lands of China, did he never come across tea, or if he did, why did he never comment upon it? Omission or fraud, we shall never know.

The first printed account of tea in English (*Chaa*, indicating a short sea trip or an initial overland exit from China) was in a 1598 translation of the adventures and travels of a Dutchman Jan Hugo van Linschoten (1563–1663) in his *Itinerio*, originally published 1596. He wrote that the Japanese:

> [A]fter their meat use a certain drinke, which is a pot of hote water, which they drinke as hote as ever they may endure, whether it be Winter or Summer ... the afore said warme water is made with the powder of a certain herbe called Chaa, which is much esteemed .... (Griffiths 1967)

It is little wonder then that as the Europeans came into contact with such a highly esteemed beverage they wanted to sample it. The earliest mention of tea by an Englishman comes from a Mr Wickham – agent of the East India Company in Hirando, Japan, dated 27 June 1615 (Forrest 1973). No account of trade in the seventeenth century in general and regarding tea in particular would be complete without a mention of the East India Company, colloquially known as 'John Company' (Ukers 1935). The East India Company was formed in 1600 and traded until 1833, controlling all trade to 'The Indies' from England. The Dutch version and great rival to 'John Company' was the Dutch East India Company, established in 1602 with a 21-year monopoly on trade with Asia (Ferguson 2004).

The East India Company grew from a group of merchants, with joint stock options and a royal warrant for trade, to a global force with armies at its command, one of the greatest navies of its time, with administrative control over whole countries and the ability to mint its own money. In all, 'John Company' acted like a separate country or state, similar to Venice in its day. The East India Company lost its monopoly on all trade with the Indies in 1833 when trade was opened to all, with the obvious consequences of more competition and falling prices.

John Wickham wrote in 1615 to a Mr Eaton, another officer in the company in Macao, China, requesting a 'pot of the best Chaw'. One book that helps us date the absence of tea, coffee and chocolate is the *Treatise on Warm Beer – The Ufe and Neceffity of Drinke* by Martin Grindel in 1741. In olde English, the letter 'S' was depicted as 'F', which we shall keep here for authenticity. In this article, Mr Grindel, whose whole treatise was based on the fact that drinks consumed warm were better for you, mentioned tea only in passing, with no mention of coffee or chocolate, indicating that they were relatively unknown at this time. He stated right at the end of his article that:

> I will prove by Giovani Petro Maffei, the Jesuit who in his 6th book of Hiftories, related that the people of China do for the moft part drink hot the ftrained liquor of an Herb, called Chai*.

There is a footnote to this article:

> 'Tea I fuppose, this treatife being firft wrote when that plant was fcarce known in England'. (Grindal 1741)

Tea first arrived in Europe in 1610, after chocolate (1528) but before coffee (1615) (Ukers 1922). This in itself gives an insight into the animosity and distances travelled between these nations in the fifteenth century (and beyond). Much is made of the Chinese drink 'tea', and in early references, it seems to be of a single type, rather than the whites, greens, blacks, oolongs, etc., of today. If Lu Yu's book of tea (the *Ch'a Ching*) is consulted, in the section on preparation, he refers to seven processes from plucking to packing. Tea is (i) plucked, (ii) steamed, (iii) pounded, (iv) patted, (v) baked, (vi) packed and (vii) finally, re-packed (Lu Yu 780). This process, due to the initial steaming, would lead to green tea. Looking through the old literature there are multiple mentions of Bohea (black) and Viridis (green). With little else to go on and no knowledge of the manufacture it was assumed that two separate bushes gave green and black tea. Indeed, until the forays into China by people like Robert Fortune, Ex-director of the Chelsea Physic Garden in London, little was known about tea. Fortune arrived in China looking for tea in 1848. He 'went native' in disguise complete with fake 'quill' (i.e. ponytail) to penetrate into the interior of China and reach the tea-growing areas (Griffiths 2007). This to modern eyes may seem ill advised, especially the dressing up, but in the seventeenth century, it was illegal, punishable by death, for a foreigner to leave the 'cantonment' of Canton, the thin strip of land with warehouses, to which the 'foreign barbarians' were restricted in order to protect both tea and the Chinese way of life. Interestingly, despite these and other precautions, tea, coffee and cocoa bushes have all been stolen by the spies of Western nations and transplanted to other nations and regions controlled by them for commercial reasons.

Carl Linnaeus took the original Amoy pronunciation (*Tay*), and applying the country of discovery to his two-part naming system (genera first and then species), he came up with *Thea sinensis*. Subsequently, he names the two varieties *Thea viridis* and *Thea bohea*. An excellent overview of the evolution of the knowledge that all tea comes from the same bush and finalisation of the current nomenclature *C. sinensis L. (O) Kuntz* is given by Griffiths (2007) in his excellent book *Tea – The Drink That Changed the World*.

It is interesting that in 2010 ~98% of the tea drunk in the United Kingdom was black and consumed with milk, and sometimes sugar (UK Tea Council 2010). Originally, the tea introduced into Europe was green and it was green tea that gave the British their taste for this now national beverage as indicated by the Scots physician Thomas Short, who wrote:

> [T]he Europeans contracted their first acquaintance with the green tea: then Bohea took its place. (Short 1750)

It is probable that 'black tea' (also known as 'red tea' by the Chinese as opposed to 'dark tea' meaning oxidised after heat inactivation, such as puer) was actually created on the long journey from China to the UK ports. The journey from China by East Indiaman around the Cape of Good Hope would have taken 6 months in the seventeenth century. During this journey, there would have been plenty of time for the leaves to oxidise and gradually turn more like modern black tea in character than green tea. Polyphenol oxidase – one of the key enzymes involved in the transformation of green to black tea – has been found to remain active at low levels in green and black tea even with efficient modern-day processing (Mahanta *et al.* 1993). Old-style processing with ineffective inactivation of the enzymes, packaging with low barrier properties to oxygen, and a long and arduous journey would have not been favourable in delivering green tea of the best quality. As Bohea was later thought to provide the best-quality black tea, maybe we could view this from a different angle and suggest that initially Bohea actually produced the lowest quality green tea with the best chance to oxide during its journey and so it arrived as black tea. After overseas peoples found black tea to their liking, perhaps the Chinese responded to demand and through experimentation produced a product that modern eyes would recognise as black tea.

The idea that tea was originally only of the green variety is also supported by a 1685 pamphlet:

> The Author of the book intitled, the embaffy of the United Provinces to the Emperor of China, Printed at Leyden in the years 1655. In the Description which he makes of Empire, speaks thus of tea. 'The moft excellent leaves of Cha or Tea, and found in the Provinces of KIangnon, and fpecially near the city of Hoeicheu; this leaf is little, and the tree thereof is very linke the fhrub called by Pliny, Rhus Coriarus, or Curriers Sumack' . . . . To make this drink of Cha fo much efteemed by the Indians, they only look for the firft leaf which comes forthe in Spring, which alfo is the moft foft and delicate, they gathr it with great care and one after the to'ther, and feperately, afterwards they prefently heat it a little while, and foftly on a gentle fire, and warp it in a very fine, thin and fmooth piece of Calico, often ftirring and rubbing it with the hands, then they fet it on the fire again, being alfo wrapped up, and turn'd, and rub it the fecund time till it curls up together and becomes quite dry, after which they pour it into tin boxes, fealing and ftopping it very clofe, for fear the Spirits and the too fubtil quality, fhould evaporate, for after you have kept it a long while, if you put it into boyling water it will retake its former verdure extending and fpreading forth it felf; if it be good it leaves behind it in the Water a fmell and tafte very agreeable to the palate, and withal a greeifh colour. (Crook 1685)

1.4.2 Coffee – from persecution to epitomising the protestant work ethic

Coffee drinking, once discovered, seemed to have had a troubled journey compared to tea. Tea was seen as a healthy drink and, in a secular society focused on herbal remedies, rapidly grew from medicinal use to daily consumption. The word for coffee originally was through its Turkish form *kahveh* (Ukers 1922). Coffee, in the region in which it was discovered, appealed to the Muslim majority as being non-alcoholic and neither from grape nor from grain, received the Koran 'seal of approval'. Coffee drinking spread to Aden and from there to Mecca (The British Magazine 1750). This was initially thought as a good thing, but as the faithful of Mecca spent more and more time in the social coffee house – even preceding morning prayer – a backlash began. One idea is that an invigorating beverage consumed by the 'chattering classes' led to discussion of politics that was less than complimentary to the state and eventually dissent and revolution. The traditional sharing of the cup, also associated with alcoholic drinks, enabled the religious advocates not content with a more alert congregation to challenge coffee. In *ca.* 1515, when Kair Bey, governor of Mecca, found a group of worshippers drinking what he thought was liquor within the mosque, he was initially outraged. When he discovered that they were in fact drinking coffee, and learning of the properties of the drink, he concluded that it may incline people to 'extravagances prohibited by law', and he decided to suppress it. Later in 1570, Imams called for Mohammedans to renounce the drink as coffee houses were full and the mosques were almost empty (Ukers 1922). As coffee spread so did the restrictions, with the Turks being among the strictest. When coffee was banned around 1656, the Ottoman Grand Vizier Koprili imposed a 'two strikes and you are out' system: the first offence was met with a cudgelling (beating) and on the second offence the offender was sewn up in a leather bag that was thrown in the Bosphorous for the offender to drown (Roden 1981).

Europe's first experience of coffee probably came from travellers from the Levant – an area now occupied by modern-day Israel, Palestine, Jordan and Syria. The first European to go into print about coffee (or '*chaube*' as it was printed) was Leonhard Rauwolf, the town physician of Augsburg who, in 1582, published Rauwolf's travels, the tales of his tours in the Far East. He stated:

> [T]hey have a very good drink, by them called Chaube that is almost as black as ink, and very good in illness, chiefly that of the stomach; of this they drink in the morning early in open places before everybody. (Rauwolf 1582)

Drinking coffee with sugar was also noted by Johann Vesling (1598–1649) during a visit to Cairo:

> [S]ome did begin to put sugar in their coffee to correct the bitterness of it, and others made sugar-plums of the berries. (Ukers 1922)

In Christian countries one of the biggest problems seems to have been the colour of the beverage – black. Black was seen as an overly negative colour with connotations of Satan – the 'Devil'; indeed, priests petitioned Pope Clement VIII (1535–1605) to ban the 'devils brew'. As coffee was approved by Muslims, who banned Christ's sanctified beverage – wine, it was clearly a brew to tempt Christians to the Devil's side. This attempt backfired and Pope Clement, inspired by the aroma, drank the brew and proclaimed that it was too delicious to allow Satan exclusive use of it. So coffee was 'baptised' for the use of Christians. (Kilham 2001)

Also, a problem was the fact that coffee was a bean:

Although the use and eating of beans were heretofore forbidden by Pythagoras . . . 'bacaufe that their flowers being fpotted with a black colour, did represent a melancholy fhape, and the souls that did dwell therein; And though there be others that reject them affirming that ufe of them dulls the fenfes, and causes troublefome dreams: Yet because they serve us in the nature of Victuals and Phyfick, I fhall not think my time and labour mifimployed, if I communicate to the publick, fomething on the subject of beans'. (Chamberlayne 1685)

The author goes on to discuss coffee further:

I will fpeak for the prefent of a certain bean of Arabia called Bon from which they make a Drink termed coffee which was heretofore in ufe amongst Arabians and Egyptians; and which is now days in very great requeft amongft the Englifh, French and Germanes . . . . The Firft that makes mention of the Property of the Bean, under then name Bunchum in the 9th Century after the birth of our saviour, was Zachary Mabomet Rafes, commonly called Rhafio, a very famous Arabian Phyfician . . . he was the firft, that did explain what was the meaning of unchum, affuring that it is hot and dry, very good for the Stomach, it hinders the unpleafant fmell of Sweat, and of Depilatory Oyntments. (Chamberlayne 1685)

Slowly but surely coffee spread throughout the Middle East to Europe and became generally available.

1.4.3 *Chocolate – from lying down . . . to sitting up*

As we have seen with the spread of tea and coffee, the controlling country of overseas territories, their empires if you like, very much influenced the nature of the new beverages. In the case of chocolate, the lynchpin was 'His Most Catholic Majesty – the King of Spain', and as such, chocolate was very much considered a 'catholic' beverage. Chocolate was thought to have been brought back to Spain by the conquistadors in the early 1600s, within the reign of King Philip III.

The consumption of chocolate as a drink continued in vogue, although with the addition of another import, sugar, originally from India, but subsequently grown in Spanish territories, it became increasingly popular – but more on this later. Drinking chocolate was initially very much the drink of the Catholic aristocracy and became associated in the minds of the Protestant North with the idle rich. Indeed, chocolate was often what the *Ancien Régime* drank before getting out of bed in the moments between lying down and sitting up! (Schivelbusch 1992). The drink, although without the immediate invigorating properties of coffee or tea, was very nourishing and exactly what a weary aristocrat needed to prepare for the hectic social engagements of the day. Chocolate soon became a drink for the clergy and the wealthy on fast days as the devout could not allow food to pass their lips but liquids were allowed (*Liquidium non frangit jejunum*) (Schivelbusch 1992). Chocolate became an allowable substance being a drink rather than a food and was:

> faid to be fo nutritive, that an ounce of them [nibs] contains more real nourifhment than a pound of beef. (The British Magazine 1750)

Chocolate spread easily to the other great Catholic realms of Europe, most notable Italy, where the additives including musk, jasmine flowers and ambergris became even more opulent (Young 1994). Such expensive ingredients were more commonly associated with perfumery, indicating that the elite status of chocolate remained into the latter half of the seventeenth century.

The process to defat cocoa nibs to remove cocoa butter and create cocoa powder was developed in 1820 and patented in 1828 by Coenraad Johannes Van Houten, a Dutch chemist (Roden 1981). As we have seen, this was more of a redefining of the process that was known by the Mayans. From Van Houten's process, the cocoa drinks of modern day were created and the division between solid and liquid chocolate consumption was created.

1.5 From foreign fancies to the drinks of the masses

So, the three beverages having been discovered and transported to Europe at vast expense were consumed by society's elite. But why did they change from an interesting fashion item to the favoured drink of the masses? To answer this question we need to first look at what beverages were available in Europe before tea, coffee and chocolate arrived. Water was untreated and usually dangerous to health – this was before water was identified as a vector for typhus and other diseases, although it was common experience that drinking water frequently led to health problems. Milk soon spoiled and so was consumed only in rural areas, so what was left? The answer seems to be alcoholic drinks, and as wine consumption really became popular only in the nineteenth century, beer was the favoured drink. It seems strange to the modern eye, but beer was consumed at breakfast and throughout the day as 'beer soup' – a mix of beer and eggs served warm. Europe's relationship with beer has been long and complex both as a nutritional beverage (Fredrick the Great was a firm supporter) (Roden 1981) and as a form of social control (the Finnish army were given 7 litres of beer per day as part of their rations) (Allen 2000). Caffeinated beverages have been linked with the birth of the Industrial Revolution (Macfarlane and Macfarlane 2009). This, of course, is understandable, as even though ancient beer was rather weak, containing less than 3% alcohol, it would have been challenging to operate complex machinery whilst consuming 8–9 litres of beer per day. Beer was consumed from time immemorial to modern day, but in the mid-1700s, it was almost supplanted in the United Kingdom by gin, which was originally introduced by the Dutch monarch 'William of Orange' in 1689 (Hallgarten 1983). What we often forget is that gin was not consumed in modern-day quantities and style, which is diluted with a mixer such as tonic. In the 1700s, gin was drunk neat and served in the same measures as beer (i.e. quarts and pints!) (Warner 2002). Beer was regarded as wholesome, while gin had the opposite reputation, hence the adage 'mothers ruin'. So, we find tea and coffee being linked with the temperance movements driven by the middle classes as a way of saving the 'inferior classes' from their alcoholic ways.

Beer contributed to the diets of the urban and rural poor of Great Britain as a ready form of energy. Tea in the form commonly consumed in the twenty-first century in Europe, with milk or lemon or a little sugar, is a good source of water for hydration, health-protective flavonoids, fluoride and caffeine (Dufresne and Farnworth 2001) and is relatively light on calories. A further reason tea became popular amongst the working classes is another import, this time from India – sugar. Sugar followed trade routes through Persia and came to Europe, like coffee, via Venice. It arrived in England in 1318, Denmark in 1374 and Sweden in 1390 (Hobhouse 1999), but it was not until the eighteenth century that sugar became available to the general population at a viable price. Unlike Far Eastern cultures, the West did not view bitterness as a positive feature of food and beverage items, so sugar played a definite role in the increasingly widespread acceptance of tea and coffee. Sugar is a very energy-dense food and beverage ingredient, and tea and jam were important sources of energy for factory workers. Indeed, the bitterness of tea made it a convenient vehicle for the consumption of large quantities of sugar. Whilst a cup of tea with three to four spoonfuls of sugar is very sweet to those unused to sweetened tea, a cup of water with the same amount of sugar would be unbearably sweet. In modern-day India, tea is often consumed as street food – highly sweetened with unrefined sugar or jaggery, with milk and spices – and is used as a food substitute for relatively little cost. Sugar was a very important additive for the consumption of tea especially in the United Kingdom (Smith 1992). Coffee was also noted in Turkey to be drunk with some sugar, although milk was rarely added as it was linked in the Middle East with causing leprosy (Pocoke 1659). In Europe, chocolate was consumed most commonly with sugar and other adjuncts, one recipe being:

> Of Cacoa 700[pods], of white sugar, one pound and a halfe; Cinnamon 2, ounces of long red pepper 14 [chillies], of Cloves, halfe an ounce: Three Cods of logwood or Campeche tree; or instead of that, the weight of 2 reals or a shilling of Anniffeeds, as much of Achiote, as will give it colour, which is about the quantity of a Hafell-nut. Some put in Almonds, kernels of Nuts, and Orange flower-water. (Colmenero 1640)

Tea, coffee and chocolate were consumed in the late 1600s in public places quite separate from the familiar bars and taverns. The first coffee house was actually in Oxford in 1650 at 'the Angel at the parish of St. Peter in the East', and in London, the first coffee house was opened in St. Michael's Alley Cornhill in 1652 (Ukers 1922). Usually, all three beverages were served in these houses as a 1665 newspaper advertisement showed:

One Constantine, a Grecian, living in Threadneedle street, overagainst St-Christopher's church, London being licenced to sell Coffee, Chocolate, Cherbert, and Tea.

The phenomenon of these houses was amazing, as by 1700, there were more than 3000 coffee, tea and chocolate houses in London. A small section of London of around Exchange Alley is shown in Plate 1.5, with ~12 coffee houses within an area of <0.5 square miles. These establishments became the open meeting places for business, politics and literature. Coffee houses were open to all who could afford a cup and so became egalitarian centres where class had no meaning as the following 'Rules and Orders of the Coffee House' (*ca.* 1674) were shown in many coffee houses (Schivelbusch 1992):

> Enter sirs freely, but first if you please,
> Peruse our Civil-Orders, which are these,
> First, Gentry Tradesman and all are welcome hither,
> And may without affront sit down together,
> Pre-eminence of place, none here should mind,
> But take the next fit seat he can find;
> Nor need any, if Finer Persons come,
> Rise up to assigne to him their room

So, despite the class system being alive and well, all men could sit together. The coffee houses were not, however, places for women. Women were confined to socialising at home, which soon led to the advent of the tea party and coffee party where women would entertain their friends. However, whereas coffee and tea were great levellers of class in the public eye, this same openness did not translate to entertaining those outside your station in life within your own house. Thus, the tea or coffee party were select events, highly ritualised to maintain social order and exclude those who did not have the benefit of high birth. It was not until the arrival of the tea gardens in London that men and women of all classes could meet in public for the consumption of non-alcoholic beverages.

Going back to the tea and coffee houses, people gathered in substantial numbers to discuss a great many things and some houses even began to record these musings and print them in what was the forerunner of modern-day newspapers and magazines. Indeed, individual houses became known for different areas of knowledge: Ozinda's chocolate house for news of the aristocracy; foreign and domestic news was the speciality of St. James' coffee house; and for items about learning you went to the Grecian.

Doctors used booths in coffee houses to meet patients and businessmen often gathered in other booths for commerce. After a while they started renting booths from the proprietors on a more permanent basis. Eventually, this even influenced the future of these coffee houses. Edward Lloyd's coffee house in Tower Street became the meeting place of seafaring men, and underwriters came to listen for news of ships and the sea. Merchants and ship owners came to insure their ships and so Lloyds of London – the insurance firm – was born. Until well into the 1950s, Lloyds offices were based around the booth design and runners referred to as 'waiters'. The oldest fire insurance firm in

London, the 'Hand-in-Hand' was formed at Tom's coffee house and later transformed into the 'Commercial Union Bank' (Roden 1981). In Change Alley, Jonathan Castaing – the proprietor of Jonathan's coffee house – began posting stock and commodity prices and it soon became the gathering place for stockbrokers. This, after the loss of the original shop in the great fire and rebuilding, is the basis of the modern-day London Stock Exchange (Werner 2008).

1.6 Tea, coffee and chocolate 'go public'

Having helped establish printed journalism, the three beverages now used it. The first advertisement for coffee appeared in 1657. In an issue of the *Publick Adviser* on 26 May the London public were advised:

> In Bartholomew Lane on the back side of the old Exchange, the drink called Coffee (which is a very Wholsome and Physical drink, having many excellent vertues, clofes the Orifice of the Stomack, fortifies the heat within, helpeth Digeftion, quickneth the Spirits, maketh the heart lightform, is good againft Eye-fores, Coughs or Colds, Rhumes, Confumptions, Head-ach, Dropfie, Gout, Scurvy, Kings Evil, and many others is to be fold both in the mornings, and at three of the clock in the afternoon. (Ukers 1922)

Chocolate was advertised 2 weeks later in the *Publik Adviser* on 16 June 1657:

> In Bishopgate street, in Queens Head Alley, at a Frenchman's house is an excellent West India drink called chocolate, to be sold, where you may have it ready at any time, and also unmade at reasonable rates.

Tea was first publically sold that same year at Garraway's, also known as Garway's, in Exchange Alley, but the first advertisement appeared only a year later in the *Mercurius Politicus*, No. 435, for September 1658:

> That excellent and by all Physitians approved China Drink, called by the Chineans Tcha, by other nations Tay, alias Tee, is sold at the Sultaness Head, a cophee-house in Sweetings Rents, by the Royal Exchange, London. (Forrest 1973)

One idea is that Garraway was also the author of the *Mercurius Politicus* article:

> Samuel Pepys tried all three beverages, referring to coffee as 'The bitter black drink'. On tea '. . . and afterwards did send for a Cupp of Tee (a China drink) of which I had never drank before' (Taylor 1955) – although we have no positive comment, there is no negative one either. On chocolate Pepys was more positive, commenting 'To a coffee house to drink jocolatte, very good'. (Pepys 1664)

1.7 Opinion is divided on the merits of the three beverages

In the modern day, health claims are a key area on which consumers try to make informed choices on what to eat and drink. The sixteenth- and seventeenth-century things were no different, and health claims were being made in pamphlets, books and advertisements. Claims were, as today, sometimes contradictory and there were many on both the positive and negative views.

Much is made in the literature of phrases, such as 'hot and drying' or 'cold and moist', and to understand these we need to consider the medical system in vogue at the time, the Hipocrates/Galen 'humerol' system. This system linked four bodily fluids, or humors, namely blood, yellow bile, black bile and phlegm with certain properties, organs and temperaments, respectively. Thus, blood was warm and moist, and associated with the liver and sanguine or extrovert temperaments; yellow bile was warm and dry, associated with the gall bladder and choleric or dominating temperaments; black bile was cold and dry, associated with the spleen and melancholic or thoughtful temperaments; phlegm was cold and moist, associated with the kidneys and phlegmatic or accepting temperaments. Good health was determined by balance, so medicine was about identifying the properties of patients and foods/remedies and putting them together.

Thomas Garraway, gathering medical knowledge of the time, claimed the particular virtues of tea, the nature of which being moderately warm included:

> It maketh the body active and lusty; It helpeth the Headache, giddiness and heaviness therof; It removeth the obstructions of the spleen; It is very good against the stone and gravel cleaning the Kidneys and Uriters, being drunk with Virgins Honey instead of Sugar. It taketh away the difficulty of breathing, opening obstructions .... It is good against Crudities, strengthening the weakness of the Ventricle or Stomack, causing good appetite and Digestion, and particularly for men of a Corpulant Body, and such are great eaters of Flesh. (Ukers 1935)

It seems strange that coffee, initially gaining its success as a drink that promotes alertness, was in Europe more associated with melancholic personalities and indeed impotence. Indeed, in the Women's Petition Against Coffee (1674), the anonymous author(s) claimed:

> The occasion of which Insufferable Disaster, after a serious Enquiry, and Discussion of the Point by the learned of the faculty, we can Attribute to nothing more that the Excessive use of that Newfangles, Abominable, Heathenish Liquor called COFFEE, which Riffling Nature of her Choicest Treasures, and Drying up the Radical Moisture, has so Eunuchs our Husbands, and Crippled our more kind Gallants, and they become as Impotent, as Age, and as unfruitful as those Deserts whence that unhappy Berry is said to be brought. (Anonymous 1674)

Chocolate had a slightly easier time of it in the press, although it was seen as 'cold and dry' in the 'humerol' system and the custom to drink this with a variety of additives, with their own properties, allowed these to balance cacao's inherent melancholic properties and probably be seen as a 'balanced' drink as and of itself. Dr Stubbs in the *Philosophical Tranfactions* affirms:

> ... that well prepared chocolate is an excellent diet for thofe who are fcorbutic [have scurvy], afflicted with the [kidney] ftone or arthritic pains, and to prevent convulsions; and yet fome of the beft phyficians have obferved that drinking chocolate to exfes contributes to the formation of ftones, especially in the gall bladder. (The British Magazine 1750)

Indeed, even in 1640 there were health cautions:

> Any many do fpeake diverfly of it, accordingly to the benefit, or hurt, they receive from it: Some faying, tjhat it is ftopping: Others, and thofe the greater part that it makes one fat. (Colmenero 1640)

Taking these claims within its stride, chocolate's reputation as an aphrodisiac (Dillinger *et al.* 2000) cannot have hurt sales and are understandable if we consider the sensuous mouth feel and its phenylethylamine contents, but in the seventeenth century these attributes were unknown.

If all three beverages were present in the markets and advertised strongly, why did the United Kingdom become a tea-drinking culture and much of Europe and the United States coffee-drinking cultures? What happened to chocolate? These issues, like any societal change, were a complex matter and probably involved issues such as availability of product and economic, cultural and public opinion. The East India Company, and indeed, the Dutch East India Company, that controlled most of the legal tea entering Europe obviously had a vested interest in the success of tea over coffee. Economically and politically, they had many allies, and indeed, they even used the press of the day. The Dutch physician Dr Nikolas Dirx writing under the pseudonym Nikolas Tulp (also known as Tulpii, or Tulpius) published his treatise on Medical Observations in 1652 (Ukers 1935), with positive views on tea, not surprising as he was a director of the Dutch East India Company. Tea was seen as a patriotic drink in England and was positioned as such to the public who supported the drink that supported their colonies. Great Britain's desire to gain further control over access routes to tea is intimately linked with China. The First Opium War (1839–1842) has its basis in a triangular trade of tea, silver and opium. The Qing Dynasty, not being interested in any of England's exports such as coarse cotton cloth, demanded silver in exchange for tea. The English soon ran out of available silver and looked for a product that they could sell to the Chinese to get their currency back. The answer they settled on was opium, which the East India Company produced in Bengal. Hindsight is 20:20 and we can see that if the Chinese were less restrictive in their currency demands or the British had chosen a more consumer-friendly trade item, the colossal problem that was created afterwards could have been avoided. The British did not invent the opium issues within China, but they clearly exploited them. When the emperor banned opium and attempted to stop British ships entering China, a small skirmish turned into a full-blown war where China was forced to cede in perpetuity access to five ports and the deep-water harbour of Hong Kong. Despite the preponderance of Starbuck's and other high-street cafés, tea in 2010 remains the dominant hot beverage in the United Kingdom.

Germany was never really on the world stage in the seventeenth and eighteenth centuries and had to buy all its coffee from the Dutch. Coffee was, hence, a great drain on German currency and with substantial amounts of money leaving the country. This led to coffee being labelled un-German and 'chicory-coffee' being invented (Smith 1985). Chicory being grown in Germany stopped the flow of German currency to foreign countries and indeed to trading nations such as Britain and Holland – sometime bitter rivals. Indeed, 'coffee sniffers' roamed the streets of Germany tracking the distinctive smell of roasting 'bean coffee' and imposing heavy fines (Smith 1985). Although restrictions on coffee in Germany were lifted soon after the use of chicory to adulterate 'bean' coffee spread to France and beyond. An interesting anecdote regarding Prince Bismarck goes that upon entering a French country inn he asked if the innkeeper had any chicory. He had, so the count asked for it all to be bought to him and soon had it all. 'Now', he said, 'go and make me a pot of coffee!' (Roden 1981). Bean coffee though was soon back and became established as the hot drink of choice.

King Gustav III of Sweden, who is said to have been interested in the different merits of tea and coffee, undertook an interesting, if ethically dubious, experiment. Two identical twins were tried for their crimes and condemned to death. The sentence was commuted to life imprisonment on the proviso that one twin drank tea and the other coffee for the rest of their life. Unfortunately, the doctors supervising the experiment were the first to die, presumably, of natural causes. The twin who drank tea died first at the age of 83, and ever since coffee has been the dominant beverage in Sweden (Smith 1985)

One of the greatest coffee-drinking cultures, based on volume anyway – the United States – has tea, or rather taxes placed upon tea, as the root cause for coffee's success. It is most likely that tea first arrived in the Americas either via official British exports begun in 1711 or via smuggled Dutch tea (Walsh 1892). The Tea Act of 1773 was, some say, the root of the American Revolution and the birth of the United States. Taxes on everyday objects such as tea were all about raising revenue for war efforts. The Stamp Act of 1765 tried to impose taxes on overseas colonies to get them to pay, at least in part, for the British troops that protected them. This was repealed in 1766 as it was supremely unpopular with the colonists and a political pawn for the government opposition at the time led by William Pitt. The tax was reinstated in 1767 amongst other items such as lead and glass and the American colonies chose not to import from Britain. This did not do down well with the merchants of London and so all taxes apart from a 3p per pound (lb) tax on tea were repealed. With the colonists turning to Holland for tea supplies, the Tea Act of 1773 sought to give the East India Company sole rights to trade tea to the Americas direct, cutting out other British and American merchants. This was met with the predictable dissention by the colonists, culminating in 'The Boston Tea Party' of 16 December 1773, when a party of colonists dressed as American Indians boarded British ships moored at the Griffins wharf (Ferguson 2004). They brought the tea onto the deck, broke open the chests and emptied the contents into Boston harbour. This night was followed by other 'tea parties', such as the Greenwich Tea Party, when tea was burned in the Market Square on 22 December 1773 by colonists dressed as American Indians, and the Philadelphia Tea Party (Ukers 1935).

General dissent led to tea being thought of an un-American, despite chocolate being present in the colonies and coffee becoming the national beverage. Chocolate's long-term success has been as confectionary rather than drink and is another story.

1.8 Tea, coffee and chocolate – the future

So, with such vibrant, colourful and formative pasts and their current status as world-dominating beverages, apart from drinkable water, tea- and coffee-based beverages are the big two. Basically, they contain caffeine, polyphenols (see Chapters 2, 3 and 5) and, depending on final make-up, sugar. As we go forward, we are gaining more insight into the health-promoting properties and indeed mechanisms of how the active principles are having the beneficial effects on health and well-being (see Chapters 4 and 6). More information allows the consumer to make more informed choices. Nutraceuticals, food as medicine, is an established and growing area that is well promoted by the 'pseudo-nutritionists' of the popular press, and as a result, tea, coffee and chocolate are not the only plant-based beverages out in the market. What of rooibos and maté, acai and yumberry, and other products which year on year make their assault on supermarket and health-store shelves through the medium of marketing and new product innovation. Will tea, coffee and chocolate hold their own? The author for one thinks, and hopes, that they will, for despite their undoubted health-promoting benefits, they are what the human race wants on a daily basis, great tasting, invigorating beverages to be consumed on innumerable occasions. Truly, we cannot imagine everyday life without tea, coffee and chocolate. Over 6 billion people cannot be wrong!

References

Allen, S.L. (2000) *The Devil's Cup, Coffee, The Driving Force in History*. Cannongate Books Ltd., Edinburgh.
Anonymous (1674) Pamphlet entitled "*The Women's Petition Against Coffee*".
Chamberlayne, J. (1685) *The Manner of Making Coffee, Tea, and Chocolate. As It Is Used in Most Parts of Europe, Asia, Africa, and America. With Their Vertues. Newly Done out of French and Spanish*. William Crook, London.
Coe, S.D. and Coe, M.D. (2003) *The True History of Chocolate*, Thames and Hudson, London.
Colmenero, A. (1640) *A Curious Treatife of the Nature and Quality of Chocolate*. Written in Spanifh by Antonio Colmenero, Doctor in Phyficke and Chirurgery. And put into Englifh by Don Diego de Vades-forte; Imprinted at London by F. Okes, dwelling in Little St. Bartholmews.
Crook, W. (1685) *The New Relation of the Use and Virtue of Tea*. Printed for W. Crook at the sign of the Green Dragon without Temple Bar, London.
De Acosta, J. (1604) Cited in Knapp, A.W. (1920) *Cocoa and Chocolate. Their History from Plantation to Consumer*. Chapman and Hall, London.
Dillinger, T.L., Barriga, P., Escarcega, S. et al. (2000) Food of the Gods: cure for humanity? A cultural history of the medicinal and ritual use of chocolate. *Am. Soc. Nutr. Sci.*, **130**, 2057S–2072S.

Dufresne, C.J. and Farnworth, E.R. (2001) A review of the latest findings on the health promoting properties of tea. *J. Nutr. Biochem.*, **12**, 404–421.
Ferguson, N. (2004) *Empire – How Britain Made the Modern World*. Penguin Books, London.
Forrest, D. (1973) *Tea for the British*. Chatto and Windus, London.
Gaze, T. (1648) New survey of the West Indies. In B. Head (ed.) *The Food of the Gods*. George Routledge and Sons, London, pp. 81–83.
Griffiths, J. (2007) *Tea – The Drink That Changed the World*. Andre Deutsch, London.
Griffiths, P. (1967) *The History of the Indian Tea Industry*. Weidenfeld and Nicolson, London.
Grindal, M. (1741) *Warm Beer, A Treatise Proving That Beer So Qualify'd Is Far More Wholesome Than That Which Is Drank Cold*. T. Read, London.
Hallgarten, P. (1983) *Spirits and Liqueurs*. Faber and Faber, London.
Hobhouse, H. (1999) *Seeds of Change – Six Plants That Transformed Mankind*. Papermac, London.
Jamieson, R.W. (2001) The essence of commodification: caffeine dependencies in the early modern world. *J. Soc. History*, **35**(2), 269–294.
Kilham, C. (2001) *Psyche delicacies – coffee, chocolate, chiles, kava and cannabis, and why they're good for you*. Rodale Inc, Emmaus, PA, USA, p. 29.
Knapp, A.W. (1920) *Cocoa and Chocolate. Their History from Plantation to Consumer*. Chapman and Hall, London.
Lu Yu (780) *Ch'a Ching*.
Macfarlane, A. and Macfarlane, I. (2009) *The Empire of Tea*. Overlook Press, New York.
Mahanta, P.K., Boruah, S.K., Boruah, H.K. *et al.* (1993) Changes of polyphenol oxidase and peroxidase activities and pigment composition of some manufactured black teas (*Camellia sinensis* L.). *J. Agric. Food Chem.*, **41**, 272–276.
Morrison, S.E. (1963). *Journals and Other Documents on the Life and Voyages of Christopher Columbus*. Heritage Press, New York. Cited in Coe, S.D., Coe, M.D. (2003).
Pepys, S. (1664) *The Diary of Samuel Pepys*. Diary entry 24 November 1664. http://www.pepysdiary.com/archive/1664/11/24.
Pocoke, E. (1659) *The Nature of the Drink Kauhi, or Coffee, and the Berry of Which It Is Made, Described by an Arabian Phisitian*. Henry Hall, Oxford.
Rauwolf, L. (1582) *Rauwolfs Travels – Journeys to the Lands of the Orient*.
Roden, C. (1981) *Coffee*. Penguin Books, London.
Schivelbusch, W. (1992) *Tastes of Paradise*. Vintage Books, Random House, London.
Segal, R. (2002) *Islam's Black Slaves – The Other Black Disapora*. Farrar, Straus and Giroux, New York.
Short, T. (1750) *Discourses on Tea, Sugar, Milk, Made Wines, Spirits, Punch, Tobacco etc. With Plain Rules for Gouty People*, T. Longman & A. Millar, London, p. 72.
Smith, R.F. (1985) A history of coffee. In M.N. Clifford and K.C. Wilson (eds.), *Coffee: Botany, Biochemistry and Production of Beans and Beverage*. Croom Helm, London, pp. 1–12.
Smith, W.D. (1992) Complications of the commonplace: tea, sugar and imperialism. *J. Interdisciplinary History*, **23**, 259–278.
Taylor, A.E. (1955) *And so to Dine, S. A. E. Strom Vignettes: A Brief Account of the Food and Drink of Mr Pepys Based on His diary*. Fredrick Books, George Allen & Unwin Ltd., London, p. 11.
UK Tea Council (2010).
Ukers, W.H. (1922) *All About Coffee*. The Tea and Coffee Trade Journal Co., New York.
Ukers, W.H. (1935) *All About Tea*. The Tea & Coffee Trade Journal Co., New York.
Walsh, J.M. (1892) *Tea: It's History and Mystery*. Henry T. Coates & Co., London.

Warner, J. (2002) *Craze: Gin and Debauchery in an Age of Reason.* Four Walls, Eight Windows, New York.

Werner, A. (2008) cited in C. Ross and J. Clark (eds), *London: The Illustrated History*, Museum of London, London.

Young, A.M. (1994) *The Chocolate Tree – A Natural History of Cacao.* Smithsonian Institutional Press, Washington, DC.

Chapter 2
Purine Alkaloids: A Focus on Caffeine and Related Compounds in Beverages

Michael E.J. Lean[1], Hiroshi Ashihara[2], Michael N. Clifford[3] and Alan Crozier[4]

[1] University of Glasgow College of Medical, Veterinary and Life Sciences, Walton Building, Royal Infirmary, 84 Castle Street, Glasgow G4 0SF, UK
[2] Department of Biological Sciences, Graduate School of Humanities and Science, Ochanomizu University, Otsuka, Bunkyo-ku, Tokyo 112-8610, Japan
[3] Food Safety Research Group, Centre for Nutrition and Food Safety, Faculty of Health and Medical Sciences, University of Surrey, Guildford, Surrey GU2 7XH, UK
[4] School of Medicine, College of Medical, Veterinary and Life Sciences, University of Glasgow, Glasgow G12 8QQ, UK

2.1 Introduction

Methylxanthines and methyluric acids are purine alkaloids (Figure 2.1). Caffeine (1,3,7-trimethylxanthine) and theobromine (3,7-dimethylxanthine) are the best-known purine alkaloids being found in popular non-alcoholic beverages, such as tea, coffee, maté and cocoa. Synthetic caffeine is added to many soft drinks and some foods. Purine alkaloids are endogenous constituents of at least 80 plant species, although usually only in trace quantities. The distribution is wider than other alkaloids such as nicotine, morphine and strychnine, but narrower than trigonelline (*N*-methylnicotinic acid), which occurs in coffee beans along with caffeine (see Chapter 5; Ashihara 2006). Typically, caffeine is the major purine alkaloid, but in a few plant species theobromine or methyluric acids such as theacrine (1,3,7,9-tetramethyluric acid) (Figure 2.1) accumulate rather than caffeine. By far, the majority of research has been carried out on the purine alkaloid-rich plants used to prepare beverages belonging to the genera *Camellia*, *Coffea*, *Ilex* and *Theobroma* with the findings reviewed by Ashihara et al. (2008, 2011). In this chapter, we first describe distribution, biosynthesis and degradation of purine alkaloids in *Camellia*, coffee, cacao and maté, after which methods for producing decaffeinated coffee and tea beverages are evaluated. Finally,

---

Teas, Cocoa and Coffee: Plant Secondary Metabolites and Health, First Edition.
Edited by Professor Alan Crozier, Professor Hiroshi Ashihara and Professor F. Tomás Barberán.
© 2012 Blackwell Publishing Ltd. Published 2012 by Blackwell Publishing Ltd.

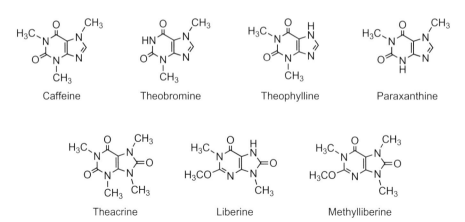

Figure 2.1 Structures of key purine alkaloids include the methylxanthines caffeine, theobromine, theophylline and paraxanthine and the methyluric acids theacrine, liberine and methylliberine.

the metabolism of caffeine by humans is discussed along with its pharmacological effects and impact on health.

2.2 Occurrence of purine alkaloids

Nagata and Sakai (1984) examined the distribution of caffeine in 23 species of *Camellia*. The purine alkaloid was found in young leaves of first flush shoots of *Camellia sinensis* var. *sinensis* (2.8% of dry weight), *C. sinensis* var. *assamica* (2.4%), *Camellia taliensis* (2.5%) and *Camellia kissi* (<0.02%), but was not detected in the other 20 species. In addition to caffeine, the young leaves of *C. sinensis* that are used for commercial tea production synthesize theobromine, albeit in smaller amounts. However, they do not accumulate other purine alkaloids, such as theophylline (1,3-dimethylxanthine) and paraxanthine (1,7-dimethylxanthine) (Figure 2.1), in detectable quantities.

Most cultivars of *Coffea arabica* contain caffeine at a concentration of 1.0–1.1% of dry weight. Some *Coffea* species, including *Coffea canephora* cv. Robusta (1.7%), contain higher concentrations of caffeine than *C. arabica*. Species with beans containing less caffeine than *C. arabica* include *Coffea eugenioides* (0.4%), *Coffea salvatrix* (0.7%) and *Coffea racemosa* (0.8%) (Mazzafera and Carvalho 1992). Unlike *C. arabica*, leaves of *Coffea liberica*, *Coffea dewevrei* and *Coffea abeokutae* contain the methyluric acids, theacrine, liberine (O(2),1,9-trimethyluric acid) and methylliberine (O(2),1,7,9-tetramethyluric acid) (Figure 2.1) (Baumann *et al.* 1976; Petermann and Baumann 1983). There is also evidence that these compounds occur in beans of *C. liberica* and *C. dewevrei* (Clifford *et al.* 1989).

The cocoa bean is the dried, fully fermented seed of cacao (*Theobroma cacao*) from which cocoa solids and cocoa butter are extracted. Theobromine is the dominant purine alkaloid; mature beans contain 2.2–2.7% theobromine and 0.6–0.8% caffeine (Senanayake and Wijesekera 1971). Cupuaçu (*Theobroma grandiflorum*), also known as Cupuassu and Copoasu, is a tree that grows in the Amazon basin and it is a close relative of cacao. Cotyledons of seeds contain 0.25% liberine (Baumann and Wanner 1980). Extracts of the pulp of the fruit, which tastes like pear with a hint of banana, is used to flavour ice cream, juices and jelly. The seeds are used to produce a good-quality chocolate, and the shell is the basis of animal feed with a high protein content (Miranda and Lessi 1995).

Purine alkaloids also occur in the leaves of yerba maté (*Ilex paraguariensis*). The use of a diacritic on the final letter ('*maté*') is a hypercorrection intended to indicate that the word is distinct from the common English word 'mate', meaning a partner. Young maté leaves are infused with boiling water to produce a beverage that was originally restricted to rural areas of South America, such as the Brazilian Panthanal and the Pampas in Argentina, but which is now increasingly consumed elsewhere, especially in the United States. The young leaves contain $\sim$1% caffeine, 0.1–0.2% theobromine and <0.02% theophylline (Mazzafera 1994).

2.3 Biosynthesis of purine alkaloids

Caffeine is synthesised in young leaves of *C. sinensis* from xanthosine, which is produced from IMP, AMP, GMP and *S*-adenosyl-L-methionine (SAM; Figure 2.2). The AMP → IMP → XMP → xanthosine and the GMP → guanosine → xanthosine pathways appear to predominate. The principal route for caffeine biosynthesis in *tea* is the four-step xanthosine → 7-methylxanthosine → 7-methylxanthine → theobromine → caffeine pathway in which SAM serves as the methyl donor for the three methylation steps (Figure 2.3). This pathway also operates in coffee and young leaves of maté. However, in cacao, the conversion of theobromine to caffeine is limited and, as a consequence, theobromine accumulates in much higher concentrations than caffeine. Several genes encoding the *N*-methyltransferases operating in the caffeine biosynthetic pathway have been cloned including those sequencing (i) the bifunctional caffeine synthase that catalyses the two-step conversion of 7-methylxanthine to caffeine, (ii) a theobromine synthase from coffee that converts theobromine to caffeine and (iii) the methylxanthine synthase responsible for the conversion of xanthosine to 7-methylxanthosine (Ashihara *et al.* 2008, 2011).

2.4 Degradation of purine alkaloids

Purine alkaloids are produced in young leaves and immature fruits, and continue to accumulate gradually during the maturation of these organs. Therefore, leaves of tea and maté, and beans of coffee and cacao, harvested for beverage production contain sizable amounts of caffeine and/or theobromine, which are degraded very slowly via removal of the methyl groups, resulting in the formation of xanthine. The major caffeine catabolic pathway in tea, coffee and maté is a caffeine → theophylline → 3-methyxanthine → xanthine pathway. In cacao, a theobromine → 3-methylxanthine → xanthine pathway is operative (Mazzafera 2004; Ashihara *et al.* 2008). Xanthine is further degraded, by the conventional purine catabolism pathway, to CO_2 and NH_3 via

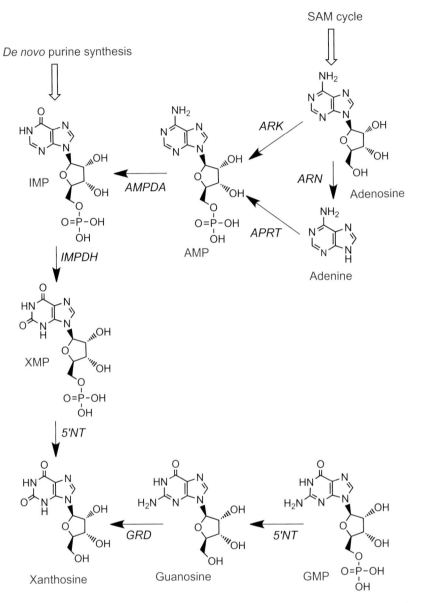

Figure 2.2 Caffeine is produced from xanthosine derived from four routes: (i) IMP originating from *de novo* purine synthesis (*de novo* route), (ii) adenosine released from the SAM cycle (SAM route), (iii) the cellular adenine nucleotide pool (AMP route) and (iv) the guanine nucleotide pool (GMP route). Enzymes: AMPDA, AMP deaminase; APRT, adenine phosphoribosyltransferase; ARK, adenosine kinase; ARN, adenosine nucleosidase; GRD, guanosine deaminase; IMPDH, IMP dehydrogenase 5'NT, 5'-nucleotidase.

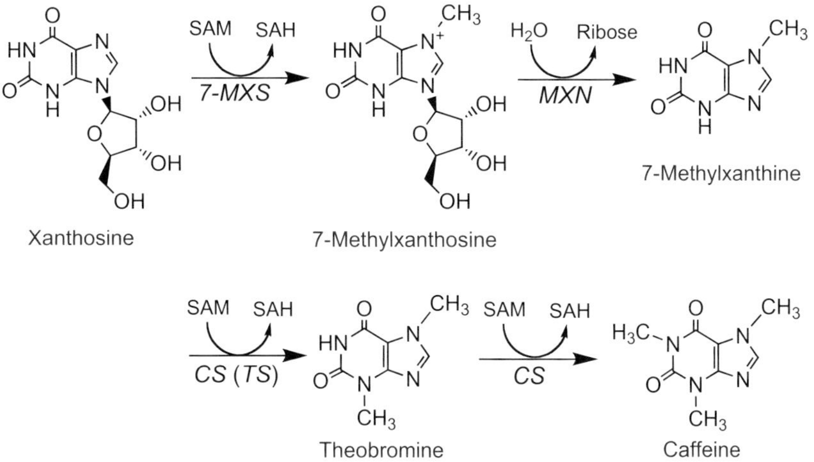

Figure 2.3 The caffeine biosynthesis pathway. SAM, S-adensoyl-L-methionine; SAH, S-adenosyl-L-homocysteine. Enzymes: 7-MXS, methylxanthine synthase; MXN, N-methylxanthine nucleosidase; CS, caffeine synthase; TS, theobromine synthase.

uric acid, allantoin and allantoic acid (Figure 2.4). *C. sinensis* and *C. arabica*, and most probably maté, accumulate caffeine because the conversion of caffeine to theophylline is the rate-limiting catabolic step (see Zrenner and Ashihara 2011).

2.5 Decaffeinated tea and coffee

Because some of the side effects of caffeine are undesirable, at least for some people (see Section 2.7), there has been a growing demand for a decaffeinated coffee beverage with 'decafe' now accounting for around 10% of coffee consumption worldwide. There is a somewhat smaller market for decaffeinated teas. These days 'decafe' is manufactured mainly by supercritical fluid extraction with liquid CO_2, but while reducing the caffeine content other ingredients relating to fragrance and taste are also removed, resulting in a taste that does not suit the palate of many coffee connoisseurs. One approach to this problem, which has received much media attention because of the interest of potential consumers, is research to produce genetically modified decaffeinated coffee. An alternative route that has been more low profile is the search for naturally caffeine-deficient coffee strains that yield a quality beverage and produce berries in quantities that are commercially viable.

The cloning of genes encoding enzymes for caffeine biosynthesis facilitated attempts to use genetic engineering to produce decaffeinated coffee, and transgenic *C. canephora* seedlings have been obtained with a 70% reduced caffeine content (Ogita *et al.* 2003). As yet, there are no reports on the caffeine content of beans produced by such plants. When transgenic beans are produced, a more substantial suppression of caffeine production will be required to qualify for commercial sale under the 'decaffeinated' label as a

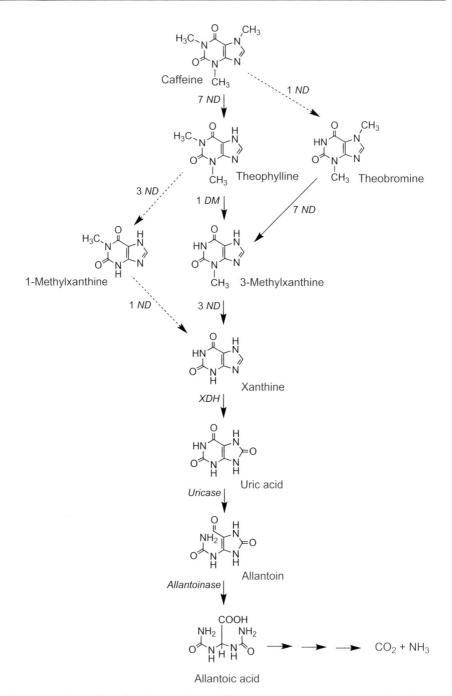

Figure 2.4 Catabolism of caffeine in plants. Caffeine is catabolised mainly to xanthine via theophylline and 3-methylxanthine. Xanthine is further degraded to CO_2 and NH_3 by the conventional oxidative purine catabolic pathway. The conversion of caffeine to theophylline is the rate-limiting step in caffeine-accumulating species such as *Coffea arabica* and *Camellia sinensis*. Solid arrows indicate major routes and dotted arrow minor conversions. Enzymes: 1ND, 1N-demethylase; 3ND, 3N-demethylase; 7ND, 7N-demethylase; XDH, xanthine dehydrogenase.

>90% reduction is a legal requirement. It will also be more appropriate to produce transgenic decaffeinated *C. arabica*, which yields a more superior beverage than Robusta coffee from *C. canephora*.

Silvarolla *et al.* (2004) reported the discovery of naturally decaffeinated mutant plants in the progeny of *C. arabica* accessions originating from Ethiopia in a coffee gene bank at the Agronomic Research Institute in Campinas, Sao Paulo, Brazil. Seeds from three of these plants contained only 0.08% caffeine. However, they also contained ∼0.6% theobromine, which when ingested is capable of causing similar physiological effects to caffeine (Eteng *et al.* 1997) (see Section 2.7). Recently, a new low-caffeine hybrid coffee was established in Madagascar as the result of an 11-year breeding programme with tetraploid inter-specific hybrids from *C. eugenioides*, *C. canephora* and *C. arabica*. Green beans were obtained that contained 0.4% caffeine and no detectable theobromine. The beans have a high cupping score, so beverage quality is similar to Arabica coffee (Nagai *et al.* 2008).

2.6   Metabolism of caffeine by humans

After drinking tea or coffee, caffeine is rapidly and totally absorbed in the upper gastrointestinal tract by humans, being detected in the bloodstream about 5 minutes after consumption and reaching peak values after some 20–30 minutes. It readily crosses the blood–brain barrier and enters all body fluids including serum, saliva, milk and semen. Caffeine is extensively metabolised in the liver by microsomal enzymes and xanthine oxidase, which together mediate demethylations and oxidations yielding products including three dimethylxanthines (theobromine, theophylline and paraxanthine), three monomethyl xanthines, the corresponding trimethyl, dimethyl and monomethyl uric acids, and three uracil derivatives formed by opening of the five-membered ring. The available evidence suggests that these metabolites are produced in humans by the pathways illustrated in Figure 2.5. More detailed information on the absorption and metabolism of caffeine in humans and model animal systems can be found in reviews by Bättig (1985), Clarke and Macrae (1988), Marks (1992) and Arnaud (1993, 2011).

The half-life of ingested caffeine in humans is typically 2.5–4.5 hours, hence the common adage that coffee should only be drunk before 4.00 pm if you want to sleep well later. However, normal sleep patterns can be restored through habituation to regular consumption. Exercise may increase the rate of elimination, whereas alcohol may retard it. Excretion is also slower in the grossly obese and those suffering from liver disease. Oral contraceptives delay clearance by about one-third. Smoking enhances clearance by inducing the caffeine-metabolising enzymes in the liver, and the more rapid reduction in serum caffeine levels with consequent need for more frequent replenishment may explain the strong statistical association between smoking and consumption of caffeinated beverages. The newborn and especially the premature infant may have a greatly extended half-life for caffeine (>100 hours) because of incomplete development of the metabolising enzymes in the liver. This phenomenon is not unique to caffeine and applies generally to xenobiotics. Caffeine metabolism is slower, and the half-life is extended up to 10 hours, in pregnancy, and much longer in the developing foetus. Thus, the occurrence of unwanted, unpleasant effects, or of

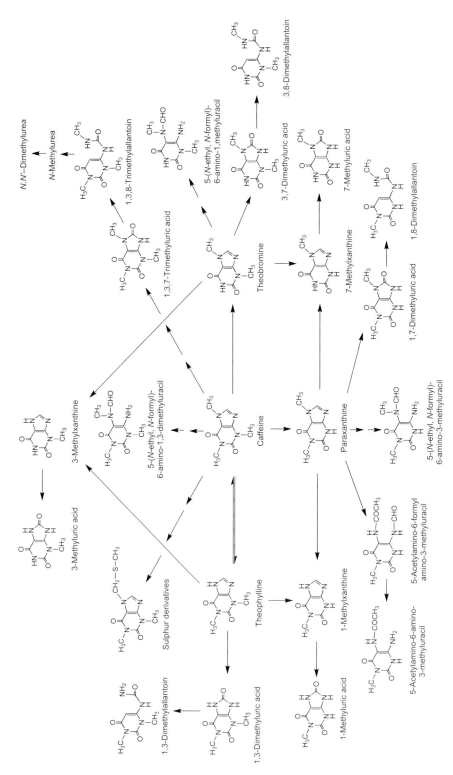

Figure 2.5 Pathways for human metabolism of caffeine to di- and monomethylxanthines, the corresponding tri-, di- and monomethyl uric acids and three uracil derivatives formed by opening of the five-membered ring. For further details and routes for caffeine metabolism in humans and animals, see Arnaud (2011).

frank toxicity, is more likely in pregnancy, potentially in the foetus, in young children and in those with liver disease.

2.7 Effects of caffeine consumption on human health

Coffee is still the largest cash commodity in the United States, besides oil. In the United Kingdom, around 50 million cups of tea are drunk daily. Drinking an estimated 1.7 cups per capita daily, US consumption is around 400 million cups of coffee each day: 90% of adults consume caffeine-containing drinks daily, with an average intake of about 200–300 mg. Whatever the impact of caffeine on individual health, the overall effect of tea and coffee on public health is integral to the social function and economic success of both Western and Eastern civilisation. Caffeine has become a major commercial commodity as a food additive in many carbonated, sweetened drinks, especially those popular with young people. Chocolate is responsible for less dietary caffeine but is the main source of theobromine. Green tea is a major source of caffeine, but it is consumed by few people in most Western countries. Black tea has a wide range of caffeine contents, depending on method of brewing, but can be a major contributor. The UK Food Standards Agency (2004) reviewed the caffeine contents of common beverages as usually consumed and found a wide variation. Tea ranged from less than 1 to 90 mg of caffeine per serving (mean 40 mg), instant coffee from 21 to 120 mg of caffeine per serving (mean 54 mg) and ground coffee from 15 to 254 mg of caffeine per serving (mean 105 mg). These are very much 'ballpark' figures, as tea and coffee preparation habits vary considerably. Tea, in particular, is often drunk very weak but in very large amounts, as is coffee in North America.

Arabica coffee beans have almost half the caffeine content of Robusta and yield a higher quality beverage. Roasting coffee beans results in some degradation of caffeine and can generate a powerful, burned flavour (see Chapter 5). Thus, more heavily roasted beans can go further, with commercial advantages, if the burned flavour is accepted. This, and the use of Robusta beans, have become the accepted norm in much of North America, particularly in 'instant' freeze-dried 'coffee'. More sophisticated coffee lovers, and those from North Africa, the Middle East and the French and Spanish diaspora, tend to prefer less severely roasted beans and accept only Arabica beans with their more complex flavour.

The extraordinary demand for tea, coffee and chocolate attracted the suggestion that they might actually contain something with an addictive property, which could be exploited by the food industry. That addictive principle turned out to be caffeine. The word 'addiction' carries a host of negative connotations, but addiction is a perfectly normal biological mechanism that underpins a number of survival mechanisms, where repeated behaviour becomes self-rewarding, and continues without conscious decision-making or effort. Addictions to oxygen, or breathing, to food/eating, water/drinking, and even sex, are all useful for a species. Appetite is a manifestation of addiction, but the term is usually reserved for an appetite that has gone wrong, when the mechanisms of addiction become exaggerated and attached to behaviours that become harmful to health, i.e. when people like, or want, or need, something too much.

The addictive power of caffeine has obvious commercial attraction and is the main reason for the addition of synthetic caffeine in a huge range of carbonated beverages. When the highly addictive cocaine had to be removed from the original formulation of Coca-Cola, caffeine replaced it. Coffee is safe, or produces only minor and acceptable adverse effects, for most people under normal conditions of consumption. The US Food and Drug Administration (2003) lists caffeine as a 'Multiple Purpose Generally Recognized as Safe Food Substance'. However, its role in causing hyperactivity in children still attracts concern and there have been isolated deaths apparently from excessive caffeine consumption by young people. In Europe, caffeine as a food additive is still classed as a flavouring. It does have some flavour, but unpleasant and bitter, such that it is very unattractive, especially to young children. It has to be heavily sweetened, and this is the main reason why carbonated soft drinks contain an enormous 10% sugar, or equivalent sweetness from artificial sweeteners. It has been established by an Australian group, using a careful randomised controlled trial design, that the caffeine content of soft drinks increases consumption in young adults (Riddell *et al.* 2010). Although added supposedly as a flavouring, careful taste-panel studies have shown that no flavour is detectable from the amount of caffeine added to soft drinks (Keast and Riddell 2007; Hattersley *et al.* 2009). The assumption has to be that the reason caffeine is added to 60% of all soft drinks in the United States is for its addictive properties, not as a flavouring (Keast and Riddell 2007).

Recognising its addictive property, commercial companies have produced a bewildering range of new caffeine-containing drinks, commonly and misleadingly called 'Energy Drinks'. When added to calorific foods, caffeine contributes to greater energy intakes (i.e. calorie intakes). It has been added to a surprising range of sweetened foods, such as ice creams, and cookies, as well as to some alcoholic drinks such as Buckfast tonic wine (281 mg per bottle). Thus, caffeine has been blamed, with some justification, for rising obesity in children and for delinquent alcoholic behaviour (Scotsman 2010). The evidence linking consumption of sweetened soft drinks with childhood obesity is particularly well established (Bray 2010). The progressive increase in bottle and can sizes over the last 50 years is clearly a major factor, but caffeine is also likely to have contributed to the escalation of intakes (Gibson 2008).

2.7.1 Biochemical and biological actions of caffeine

The effects and mechanisms of caffeine and its metabolites, paraxanthine, theobromine, and theophylline are similar, and they generally occur together. The main biological effects of caffeine in humans are listed in Table 2.1. They are numerous and affect many body systems. Saliva concentration of caffeine and theobromine can be used as a biomarker of consumption (Ptolemy *et al.* 2010). As with many bioactive compounds, or drugs, the actions of caffeine are often a double-edged sword, such that there may be a balance to be drawn between effects whose experience is positive or beneficial and others that have negative impacts on well-being. As discussed in Section 2.7, the extended retention of caffeine by pregnant women, the developing foetus, new

Table 2.1 Biological properties of caffeine; most are mediated by augmentation of catecholamine effects

Biological properties of caffeine

CNS and sympathetic nervous system stimulant
    Alertness, heightened awareness
    Agitation, anxiety
    Tremor
    Sleep disturbances
    Addiction
    Lowered seizure threshold

Diuretic
    Polyuria, nocturia
    Relative dehydration

Cardiac stimulant
    Sinus tachycardia, (palpitations)
    Increased cardiac muscle contractility (treatment of heart failure)
    Arrhythmias: ventricular extrasystole ('missed beats', palpitations)

Smooth muscle relaxant
    Gastro-oesophageal (reflux, heartburn)
    Bronchodilatation (asthma treatment, illegal sports performance enhancement)
    Uterine muscle relaxation (?? miscarriage)

Vasodilator
    Headaches on caffeine withdrawal
    Synergism with nitrites
    Synergism with analgesics

born and young children, as well as those with liver disease, makes these groups more susceptible to the effects of caffeine toxicity.

Legend has it that the main effect of caffeine was initially recognised from the invigorating effect of coffee berries on goats (see Chapter 1). Centuries later, this has led to the extraordinary attraction of tea and coffee to many consumers, who exhibit increased alertness and a capacity to remain awake for longer periods without sleep. The biochemical mechanism underlying this effect, and indeed many other caffeine actions, is a synergistic interaction with catecholamines (adrenaline and noradrenaline – the major signals for the sympathetic nervous system), and antagonisation of the actions of adenosine, which elevates another neurotransmitter in the brain, i.e. dopamine. Elevation of dopamine concentration in the brain is responsible for many of the central nervous system stimulant and addictive properties of caffeine in common with various psychoactive drugs. Dopamine also enhances the action of serotonin on raising mood. Caffeine increases sympathetic activity and nor/adrenaline levels, and if catecholamines are elevated for any reason, e.g. stress, then the biological actions of caffeine are exaggerated. Its interaction with the synthetic catecholamine drug, amphetamine, and related drugs is well recognised. An important catecholamine action augmented by caffeine is inhibition of the enzyme cyclic AMP-phosphodiesterase, which results in cyclic AMP accumulation in cells, and promotes glucose synthesis. It also leads to excessive gastric acid secretion, which can cause symptoms of heartburn and indigestion (Benowitz 1990).

Consumed in its natural context, as coffee or tea, caffeine produces an immediate and usually pleasant feeling of alertness. However, excessive consumption results in an unpleasant state of excitement and anxiety. This change can occur at different doses. For some people, even a single cup (~50–60 mg) may be acutely unpleasant and cause sleeplessness with a racing mind. For others, and through tolerance to increasing exposure, ten times this amount may still be positive, and not interfere with sleep. Sleep may still be disturbed by the diuretic effect of caffeine on the renal tubules. For some individuals, unpleasant palpitations are the limiting effect, either a fast racing pulse (sinus tachycardia) or 'missed beats' (ventricular extrasystoles). For most, these go unnoticed and are not dangerous. For some people, caffeine can contribute, through anxiety and hyperventilation, to a sense of respiratory distress. Others find that caffeine steadies their breathing. This may relate to the well-documented pharmacological actions of caffeine on dilating the respiratory smooth muscle, or on stimulating increased cardiac contractility to help overcome incipient heart failure. There is substantial variability in the effects of caffeine on individuals, at least partly the result of genetic variation of susceptibility (Rogers et al. 2010).

These effects of caffeine and its derivatives on respiratory and cardiac function have been exploited to improve function in patients with asthma and heart failure. Synthetic theophylline is very widely used as the medically prescribed drug 'aminophilline' in divided doses of up to 1100 mg per day for asthma and up to 500 mg per day for heart failure. Studies with caffeine itself in patients with asthma have shown that moderate (6 mg/kg) to high (9 mg/kg) doses produce a significant reduction in post-exercise bronchospasm, which could potentially overcome the tolerance that commonly develops to adrenergic beta-2 agonist inhalers (van Haitsma et al. 2010).

Inevitably, researchers have been interested in whether consumptions of coffee, and to a lesser extent tea, so by proxy caffeine, might affect risks of major diseases such as heart disease, stroke and cancer. Observational studies suggest that caffeine has no major net effect on heart disease, which is reassuring. Caffeine, or perhaps the oils present in coffee, tends to increase blood pressure (Hartley et al. 2004). In a high dose of 5 mg/kg body weight, caffeine prolongs QRS complex, thus QT interval on EGG in healthy volunteers, and it commonly generates ventricle ectopies. Caffeine does not generally appear to aggravate risk of hazardous ventricular arrhythmia (Myers 1991), including in patients after myocardial infarction (Sutherland et al. 1985). However, there do appear to be some individuals who are prone to caffeine-induced arrhythmias, and for whom stopping caffeine ingestion relieves the problem (Myers and Harris 1990). There are some short-term studies that suggest that coffee might impair endothelial function – another component of vascular disease (Buscemi et al. 2010). On the other hand, caffeine improves insulin sensitivity and glucose tolerance in healthy subjects (Moisey et al. 2010) and consumption is associated with reduced risk of developing type 2 diabetes (Rosengren et al. 2004; Salazar Martinez et al. 2004). A caffeine dose of 5 mg/kg body weight has been shown to improve insulin sensitivity and glucose tolerance in healthy volunteers (Moisey et al. 2010).

The literature on caffeine and cancer is also mainly observational in humans and, in fact, relates to caffeine-containing foods and drinks, mostly coffee. If a large number of cancers are included in outcomes of long-term cohort studies, or of experiments in animals, then by random chance alone some (about 1 in 20) will appear to be associated with caffeine. This has led to a variety of scares over the years. The net effect on all cancers is probably neutral (van Dam 2008; Tang *et al.* 2009). Isolated studies claiming that caffeine helps in cancer treatment are also likely to be biased.

The major metabolite of caffeine, paraxanthine (see Figure 2.5), causes an increase in circulating free fatty acids, which may be responsible for cardiac arrhythmias, as well as releasing fuel for muscle action. Theophylline is often said to be mainly responsible for the relaxation of bronchial smooth muscle; however, one randomised controlled trial found caffeine to be equally effective in treating apnea in premature infants, for which theophylline is the standard treatment (Skouroliakou *et al.* 2009).

2.7.2 Mental performance enhancement

Observational evidence indicates coffee drinking is associated with a better health outlook in old age. It has been associated with better cognitive function (Johnson-Kozlow *et al.* 2002) and with a 30% lower likelihood of Parkinson's disease (Martyn and Gale 2003). These studies do not prove causality.

There is limited evidence that caffeine, when accompanied by a source of glucose, may improve attention, learning and verbal memory consolidation (Adan and Serra-Grabulosa 2010). This literature is bedevilled by small studies with low power, and by publication bias, whereby studies with negative results are much less likely to be published than positive ones. There are clear commercial interests that can accentuate this potential source of bias. It has been argued that since most subjects in such studies are already regular caffeine consumers, any observed effect from caffeine consumption in fact represents correction of a withdrawal state. The effects of caffeine on caffeine-naïve subjects are not so widely available.

2.7.3 Physical performance enhancement

The potentially potent and overtly drug-like pharmacological actions of caffeine present regulatory bodies with difficulties in distinguishing between 'drugs' and bioactive food constituents, and this is particularly relevant for sports performances. As described in Section 2.7.1, caffeine and its metabolites relax smooth bronchial muscle, increase cardiac output and blood flow, release fatty acids into the circulation and help glucose synthesis. These effects would be expected to improve performance. In trained athletes, a dose of 100 mg (~2 cups of coffee) given to elite sprint swimmers in a robust randomised controlled trial showed a small but significant 1.3% improvement in performance time (van den Bogaerde and Hopkins 2010). A modest dose of caffeine, 5 mg/kg, improved cognitive performance, passing accuracy and jump performance of footballers (Foskett et al. 2009). Caffeine may have less effect in longer events, i.e. rowing or road cycling events (Desbrow *et al.* 2009; Skinner *et al.* 2010).

As with research in other fields, the published literature may be biased in favour of studies with positive results, and it is possible that effects of caffeine interpreted as beneficial in fact represent the correction of withdrawal from a dependency state:

some studies in caffeine-naïve subjects have failed to show ergometric benefits (Woolf *et al.* 2009).

Caffeine was in the past listed as a banned substance by International Olympic Committee, and athletes were tested for caffeine as an illegal performance-enhancing drug. The urine concentration considered illegal was 12 μg/mL. This level can be easily reached with five strong coffees (Nutritional Reviews 2007). In 2004, after it became clear that the ban was unworkable, the World Anti-Doping Agency removed the ban, which, of course, allowed some to experiment with very high doses of caffeine. If there are indeed benefits for sports performance, it is likely that they would only be detectable or useful for trained athletes at peak performance, not by untrained or casual participants.

2.7.4 Caffeine toxicity

Chronic caffeine consumption in daily amounts above ∼400 mg not infrequently causes a low-grade state of malaise described as 'caffeinism', exaggerated physiological effects causing a state of chronic anxiety, nervousness and insomnia, with headaches, palpitations and muscular tension or tremor. Indigestion is common and peptic ulcer may develop. As many as one person in ten may exhibit signs of caffeinism (Gilliland and Andress 1981; Bradley and Petree 1990). Toxicity is more likely when the plasma half-life is prolonged, such as in children, in pregnancy and in those with liver disease. A variety of symptoms can develop (Table 2.1). The World Health Organisation (2007) 'International Classification of Diseases' lists a number of conditions secondary to caffeine on mental and behavioural disorders due to psychoactive substance use (Table 2.2). Caffeine has been blamed for hyperactivity in children and there is some evidence for this (Linnet *et al.* 2009; Oddy and O'Sullivan 2009; Bekkhus *et al.* 2010).

Caffeine poisoning, or acute intoxication, with large acute doses, is characterised by extreme anxiety, and delirium, hallucinations flushing, dehydration, tachycardia. The amount required may be relatively low; for example, a 17-year-old student had to be hospitalised after seven double espressos (BBC 2007). Epileptic seizures and muscle damage (rhabdomyolysis, leading to kidney failure), cardiac arrhythmias and ultimately ventricular fibrillation are further potentially fatal consequences. The fatal dose for humans varies, but may be as little as 2 g for some people. In a sad

Table 2.2 World Health Organisation (2007) classification of disease diagnostic codes (ICD-9-CM) for caffeine toxicity

Condition	ICD-9-CM code
Caffeine intoxication	305.90
Caffeine-induced anxiety disorder	292.89
Caffeine-induced sleep disorder	292.85
Caffeine-related disorder not otherwise specified	292.90

well-documented recent case, a 23-year-old man died after taking about 5.6 g of pure caffeine, achieving a supra-lethal blood concentration of 251 mg/L (1.29 mM) (BBC 2010). The average blood concentration observed from consuming caffeine in normal dietary amounts is ~5 mg/L (25 µM) (MacDonald et al. 1991). Isolated deaths in young people have been attributed to coffee, and to sudden adult death syndrome from the so-called energy drink, Red Bull, which contains ~80 mg caffeine per can (BBC 2001; Metro 2008; Daily Telegraph 2011). However, the doses consumed from normal foods, even in very abnormal amounts, cannot reach the fatal levels achieved from pure caffeine ingestion (Donnerstein et al. 1998). Some individuals may be highly susceptible to toxicity through pre-existing cardiac condition and condition abnormalities or through having a low seizure threshold. Evidence on QT interval is mixed. A caffeine intake of 400 mg increases QT interval during sleep (Bonnet et al. 2005) but not during waking time (Ammar et al. 2001). It is not known whether caffeine presents special hazards to people with prolonged QT intervals.

2.7.5 Tolerance, withdrawal and dependence

A factor in the commercial success of caffeine is that tolerance to its adverse effects, especially on adenosine antagonism and sleep disturbance, develops very rapidly – in as little as 7 days with some people, which permits continued high consumption. However, the development of tolerance always comes at a cost, and that is the predictable and unpleasant withdrawal symptoms if regular caffeine consumption is interrupted for a day or two. When adenosine action is released from caffeine antagonism, there is reflex intracranial vasodilation, causing headache (Guieu et al. (1998). This is coupled with reduced catecholamine activity, causing fatigue and reduced serotonin, causing mild depression and impaired cognitive function, especially with sleep deprivation (Griffiths et al. 2003; Killgore et al. 2007; Childs and de Wit 2008). As with any drug withdrawal, the response is a cry for more caffeine in whatever form it is usually supplied, and so the dependence continues. A relatively high proportion of people recognise that their caffeine use satisfies DSM-IV criteria for substance dependence (http://www.apas.org.uk/docs/dsm.pdf). If the caffeine vehicle-food contains extra calories, it can be seen how caffeine tolerance could fuel obesity.

2.7.6 Caffeine in pregnancy

Evidence is mixed, with some studies positive but others negative that caffeine may contribute to risk of low birth weight and to miscarriage. This is biologically plausible, as a smooth muscle relaxant that has a prolonged half-life in pregnancy. A recent meta-analysis has reviewed this literature and found no significant adverse associations between caffeine consumption in pregnancy and foetal growth or pre-term birth, but it may still contribute to miscarriage (Bech et al. 2007; Maslova et al. 2010). Current advice in the United Kingdom from the Food Standards Agency is for pregnant women to restrict caffeine to below 200 mg per day, equivalent to about 4 cups of strong coffee (Wadge 2009).

2.7.7 Toxicity in other species

A number of species, including dogs, foxes, badgers, horses and the New Zealand kea (a somewhat malevolent alpine parrot) have very limited capacity to metabolise theobromine, which tends to accumulate and becomes toxic with potentially fatal consequences (Jannsson *et al.* 2001; Gartrell and Reid 2007). Efforts to teach keas not to steal chocolate or mountaineers boots laid in the sun to dry out have never been successful. Depending on the size of dog, as little as 100–200 mg of theobromine, the content of 200 g of milk chocolate or 20 g of dark chocolate, can be fatal. Caffeine consumption by horses leads to theophylline accumulation and hyperactivity (Vickroy *et al.* 2008).

2.8 Summary

Purine alkaloids are produced in high concentration by a limited number of plant species including *Camellia sinensis*, *Coffea arabica* and *Theobroma cacao*, which are used on worldwide basis to produce the well-known beverages tea, coffee and cocoa. How this came about and our fondness for purine alkaloid-containing beverages is described in Chapter 1, while purine alkaloid metabolism in plants and humans is outlined in this chapter. Arguments will continue to attack, and defend, the record of caffeine and its derivatives on human health. While there are clearly specific actions of caffeine that can be troublesome to some individuals, the health records of tea and coffee are good – probably rather better than most other beverages. The role of beverages such as tea, coffee and cocoa as a social lubricant for adults clearly outweighs any adverse effects of caffeine. The addition of caffeine into children's diets, heavily disguised by sugar or other sweeteners, has become accepted worldwide, but requires better independent assessment for health outcomes particularly for hyperactivity and/or for obesity. Because of an extended elimination half-life, caffeine intake should be restricted in pregnancy, in people with liver disease, and in young children.

References

Adan, A. and Serra-Grabulosa, J.M. (2010) Effects of caffeine and glucose, alone and combined, on cognitive performance. *Human Psychopharmacol.*, **25**, 310–317.

Ammar, R., Song, J.C., Kluger, J. *et al.* (2001) White CM. Evaluation of electrocardiographic and hemodynamic effects of caffeine with acute dosing in healthy volunteers. *Pharmacotherapy*, **21**, 437–442.

Arnaud, M.J. (1993) Metabolism of caffeine and other components of coffee. In S. Garattini (ed.), *Caffeine, Coffee and Health*. Raven Press, New York, pp. 43–95.

Arnaud, M.J. (2011) Pharmacokinetic and metabolism of natural methylxanthines in animal and man. *Handb. Exp. Pharmacol.*, **200**, 33–91.

Ashihara, H. (2006) Metabolism of alkaloids in coffee plants. *Braz. J. Plant Physiol.*, **18**, 1–8.

Ashihara, H., Sano, H. and Crozier, A. (2008) Caffeine and related purine alkaloids: biosynthesis, catabolism, function and genetic engineering. *Phytochemistry*, **69**, 841–856.

Ashihara, H., Ogita, S. and Crozier, A. (2011) Purine alkaloid biosynthesis. In H. Ashihara, A. Crozier and A. Komamine (eds), *Plant Metabolism and Biotechnology*. John Wiley & Sons, Ltd., Oxford, pp. 155–181.

Bättig, K. (1985) The physiological effects of coffee consumption. In M.N. Clifford and K.C. Willson (eds), *Coffee: Botany, Biochemistry and Production of Beans and Beverage*. Croom Helm, London, pp. 394–439.

BBC (2001) Red Bull in suspected link to deaths. http://news.bbc.co.uk/1/hi/1435409.stm.

BBC (2007) Girl overdoses on expresso coffee. http://news.bbc.co.uk/1/hi/england/wear/6944026.stm.

BBC (2010) Caffeine death sparks alert by Nottinghamshire coroner. http://www.bbc.co.uk/news/uk-england-nottinghamshire-11645363.

Baumann, T.W. and Wanner, H. (1980) The 1,3,7,9-tetramethyluric acid content of cupu (*Theobroma grandiflorum*). *Acta Amaz.*, **10**, 425.

Baumann, T.W., Oechslin, M. and Wanner, H. (1976) Caffeine and methylated uric acids: chemical patterns during vegetative development of *Coffea liberica*. *Biochem. Physiol. Pflanzen*, **170**, 217–225.

Bech, B.H., Obel, C., Henriksen, T.B. *et al.* (2007) Effect of reducing caffeine intake on birth weight and length of gestation: randomised controlled trial. *BMJ*, **334**, 409.

Bekkhus, M., Skjoshaug, T., Nordhagen, R. *et al.* (2010) Intrauterine exposure to caffeine and inattention/overactive children. *Acta Paediatr.*, **99**, 925–928.

Benowitz, N. (1990) Clinical pharmacology of caffeine. *Annu. Rev. Med.*, **41**, 277–288.

Bonnet, M., Tancer, M., Uhde, T. *et al.* (2005) Effects of caffeine on heart rate and QT variability during sleep. *Depress Anxiety*, **22**, 150–155.

Bradley, J.R. and Petree, A. (1990) Caffeine consumption, expectancies of caffeine-enhanced performance, and caffeinism symptoms among university students. *J. Drug Educ.*, **20**, 319–328.

Bray, GA. (2010) Soft drink consumption and obesity: it is all about fructose. *Curr. Opin. Lipidol.*, **21**, 51–57.

Buscemi, S., Verga, S., Batsis, J.A. *et al.* (2010) Acute effects of coffee on endothelial function in healthy subjects. *Eur. J. Clin. Nutr.*, **64**, 483–489.

Childs, E. and de Wit, H. (2008) Enhanced mood and psychomotor performance by a caffeine-containing energy capsule in fatigued individuals. *Exp. Clin. Psychopharmacol.*, **16**, 13–21.

Clarke, R.J. and Macrae, R. (1988) *Coffee-3 Physiology*. Elsevier Applied Science, Amsterdam.

Clifford, M.N., Williams, T. and Bridson, D. (1989) Chlorogenic acids and caffeine as possible taxonomic criteria in *Coffea* and *Psilanthus*. *Phytochemistry*, **28**, 829–838.

Daily Telegraph (2011) Student died after sharing three cans of Red Bull. http://www.telegraph.co.uk/health/1374291/Student-died-after-sharing-three-cans-of-Red-Bull.html.

Desbrow, B., Barrett, C.M., Minahan, C.L. *et al.* (2009) Caffeine, cycling performance, and exogenous CHO oxidation: a dose-response study. *Med. Sci. Sports Exerc.*, **41**, 1744–1751.

Donnerstein, R.L., Zhu, D., Samson, R. *et al.* (1998) Acute effects of caffeine ingestion on signal-averaged electrocardiograms. *Am. Heart J.*, **136**, 643–646.

Eteng, M.U., Eyong, E.U., Akpanyung, E.O. *et al.* (1997) Recent advances in caffeine and theobromine toxicities: a review. *Plant Food Hum. Nutr.*, **51**, 231–243.

Foskett, A., Ali, A. and Gant N. (2009) Caffeine enhances cognitive function and skill performance during simulated soccer activity. *Int. J. Sport Nutr. Exerc. Metab.*, **19**, 410–423.

Gartrell, B.D. and Reid, C. (2007) Death by chocolate: a fatal problem for an inquisitive wild parrot. *N.Z. Vet. J.*, **55**, 149–151.

Gibson, S. (2008) Sugar-sweetened soft drinks and obesity: a systematic review of the evidence from observational studies and interventions. *Nutr. Res. Rev.*, **21**, 134–147.

Gilliland, K. and Andress, D. (1981) *Ad lib* caffeine consumption, symptoms of caffeinism, and academic performance. *Am. J. Psychiatry*, **138**, 512–514.

Griffiths, R.R., Juliano, L.M. and Chausmer, A.L. (2003) Caffeine pharmacology and clinical effects. In A.W. Graham, T.K. Schultz, M.F. Mayo-Smith, R.K. Ries and B.B. Wilford (eds). *Principles of Addiction Medicine*, 3rd edition. American Society of Addiction, Chevy Chase, MD, pp. 193–224.

Guieu, R., Devaux, C. and Henry, H. (1998) Adenosine and migraine. *Can J. Neuro. l Sci.*, **25**, 55–58.

Hartley, T.R., Lovallo, W.R. and Whitsett, T.L. (2004) Cardiovascular effects of caffeine in men and women. *Am. J. Cardiol.*, **93**, 1022–1026.

Hattersley, L., Irwin M., King L. *et al.* (2009) Determinants and patterns of soft drink consumption in young adults: a qualitative analysis. *Public Health Nutr.*, **12**, 1816–1822.

Jannsson, D.S., Galgan, V., Schubert, B. *et al.* (2001) Theobromine intoxication in a red fox and a European badger in Sweden. *J. Wildl. Dis.*, **37**, 362–365.

Johnson-Kozlow, M., Kritz-Silverstein, D., Barrett-Connor, E. *et al.* (2002) Coffee consumption and cognitive function among older adults. *Am. J. Epidemiol.*, **156**, 842–850.

Keast, R.S. and Riddell, L.J. (2007) Caffeine as a flavour additive in soft-drinks. *Appetite*, **49**, 255–259.

Killgore, W.D., Kahn-Greene, E.T., Killgore, D.B. *et al.* (2007) Effects of acute caffeine withdrawal on Short Category Test performance in sleep-deprived individuals. *Percept. Mot. Skills*, **105**, 1265–1274.

Linnet, K.M., Wisborg, K., Secher, N.J. *et al.* (2009) Coffee consumption during pregnancy and the risk of hyperkinetic disorder and ADHD: a prospective cohort study. *Acta Paediatr.*, **98**, 173–179.

MacDonald, T.M., Sharpe, K., Fowler, G. *et al.* (1991) Caffeine restriction: effect on mild hypertension. *BMJ*, **303**, 1235–1238.

Marks, V. (1992) Physiological and clinical effects of tea. In K.C. Willson and M.N. Clifford (eds), *Tea: Cultivation to Consumption*. Chapman and Hall, London, pp. 707–739.

Martyn, C. and Gale, G. (2003) Tobacco, coffee and Parkinson's disease. *BMJ*, **326**, 561–562.

Maslova, E., Bhattacharya, S., Lin, S.-W. *et al.* (2010) Caffeine consumption during pregnancy and risk of preterm birth: a meta-analysis. *Am. J. Clin. Nutr.*, **92**, 1120–1123.

Mazzafera, P. (1994) Caffeine, theobromine and theophylline distribution in *Ilex paraguariensis*. *Rev. Brasil Fisiol. Veg.*, **6**, 149–151.

Mazzafera, P. (2004) Catabolism of caffeine in plants and microorganisms. *Front. Biosci.*, **1**, 1348–1359

Mazzafera, P. and Carvalho, A. (1992) Breeding for low seed caffeine content of coffee (C*offea* L.) by interspecific hybridization. *Euphytica*, **59**, 55–60.

Metro (2008) Man killed by 4 cans of Red Bull a day. http://www.metro.co.uk/news/144543-man-killed-by-4-cans-of-red-bull-a-day.

Miranda, R. de. M. and Lessi, E. (1995) Conservation of the cupuassu (*Theobroma grandiflorum* Schum) with cold storage. *Acta Hort.*, **370**, 231–223.

Moisey, L.L., Robinson, L.E. and Graham, T.E. (2010) Consumption of caffeinated coffee and a high carbohydrate meal affects postprandial metabolism of a subsequent oral glucose tolerance test in young, healthy males. *Br. J. Nutr.*, **103**, 833–841.

Myers, M.G. (1991) Caffeine and cardiac arrhythmias. *Ann. Intern. Med.*, **114**, 147–150.

Myers, M.G. and Harris, L. (1990) High dose caffeine and ventricular arrhythmias. *Can. J. Cardiol.*, **6**, 95–98.

Nagai, C., Rakotomalala, J.J., Katahira, R., Li, Y. *et al.* (2008) Production of a new low-caffeine hybrid coffee and the biochemical mechanism of low caffeine accumulation. *Euphytica*, **164**, 133–142.

Nagata, T. and Sakai, S. (1984) Differences in caffeine, flavonols, and amino acids contents in leaves of cultivated species of Camellia. *Jpn. J. Breed.*, **34**, 459–467.

Nutritional Reviews (2007) Caffeine effects on weight loss and sport. http://www.nutritionalreviews.org/caffeine.htm.

Oddy, W.H. and O'Sullivan, T.A. (2009) Energy drinks for children and adolescents. *BMJ*, **339**, b5268.

Ogita, S., Uefuji, H., Yamaguchi, Y. et al. (2003) Producing decaffeinated coffee plants. *Nature*, **423**, 823.

Petermann, J. and Baumann, T.W. (1983) Metabolic relations between methylxanthines and methyluric acids in Coffea. *Plant Physiol.*, **73**, 961–964.

Ptolemy, A.S., Tzioumis, E., Thomke, A. et al. (2010) Quantification of theobromine and caffeine in saliva, plasma and urine via liquid chromatography–mass spectrometry: a single analytical protocol applicable to cocoa intervention studies. *J. Chromatogr. B.*, **878**, 409–416.

Riddell, L.J., Keast, R.S.J. and Swinburn, B. (2010) Caffeine increases consumption of sugar sweetened beverages. *Obesity Rev.*, **11**, 368–369.

Rogers, P.J., Hohoff, C., Heatherley, S.V. et al. (2010) Association of the anxiogenic and alerting effects of caffeine with ADORA2A and ADORA1 polymorphisms and habitual level of caffeine consumption. *Neuropsychopharmacology*, **35**, 1973–1983.

Rosengren, A., Dotevall, A., Wilhelmsen, L. et al. (2004) Coffee and incidence of diabetes in Swedish women: a prospective 18-year follow-up study. *J. Intern. Med.*, **255**, 89–95.

Salazar Martinez, E., Willett, W.C., Ascherio, A. et al. (2004) Coffee consumption and risk for Type 2 diabetes mellitus. *Ann. Intern. Med.*, **140**, 1–8.

Scotsman (2010) Crime link as Buckfast revealed to have as much caffeine as eight colas. http://thescotsman.scotsman.com/news/Crime-link-as-Buckfast-revealed.5989472.jp.

Senanayake, U.M. and Wijesekera, R.O.B. (1971) Theobromine and caffeine content of the cocoa bean during its growth. *J. Sci. Food Agric.*, **22**, 262–263.

Silvarolla, M.B., Mazzafera, P. and Fazuoli, L.C. (2004) Plant biochemistry: a naturally decaffeinated arabica coffee. *Nature*, **429**, 826.

Skinner, T.L., Jenkins, D.G., Coombes, J.S. et al. (2010) Dose response of caffeine on 2000-m rowing performance *Med. Sci. Sports Exerc.*, **42**, 571–576.

Skouroliakou, M., Bacopoulou, F. and Markantonis, S.L. (2009) Caffeine versus theophylline for apnea of prematurity: a randomised controlled trial. *J. Paediatr. Child Health*, **45**, 587–592.

Sutherland, D.J., McPherson, D.D., Renton, K.W. et al. (1985) The effect of caffeine on cardiac rate, rhythm and ventricular repolarization. Analysis of 18 normal subjects and 18 patients with primary ventricular dysrhythmia. *Chest*, **87**, 319–324

Tang N, Zhou B, Wang B. et al. (2009) Coffee consumption and risk of breast cancer: a meta-analysis. *Am. J. Obstet. Gynecol.*, **200**, 290.e1-9.

US Food and Drug Administration (2003) US Code of Federal Regulations. U.S. Office of the Federal Register. http://edocket.access.gpo.gov/cfr_2003/aprqtr/21cfr182.1180.htm.

UK Food Standards Agency (2004) Survey of caffeine levels in hot beverages. http://www.food.gov.uk/science/surveillance/fsis2004branch/fsis5304.

van Dam, R.M. (2008) Coffee consumption and risk of type 2 diabetes, cardiovascular diseases, and cancer. *Appl. Physiol. Nutr. Metab.*, **33**, 1269–1283.

van den Bogaerde, T.J. and Hopkins, W.G. (2010) Monitoring acute effects on athletic performance with mixed linear modelling. *Med. Sci. Sports Exer.*, **42**, 1339–1344.

van Haitsma, T.A., Mickleborough, T., Stager, J.M. et al. (2010) Comparative effects of caffeine and albuterol on the bronchoconstrictor response to exercise in asthmatic athletes. *Int. J. Sports. Med.*, **32**, 231–236.

Vickroy, T.W., Chang, S.K. and Chou, C.C. (2008) Caffeine-induced hyperactivity in the horse: comparisons of drug and metabolite concentrations in blood and cerebrospinal fluid. *J. Vet. Pharmacol. Ther.*, **31**, 156–166.

Wadge, A. (2009) Food Standard Agency's advice on caffeine. *BMJ*, **338**, b299.

Woolf, K., Bidwell, W.K. and Carlson, A.G. (2009) Effect of caffeine as an ergogenic aid during anaerobic exercise performance in caffeine naïve collegiate football players. *J. Strength Cond. Res.*, **23**, 1363–1369.

World Health Organisation (2007) *International Classification of Diseases*, Chapter 5. http://apps.who.int/classifications/apps/icd/icd10online/gF10.htm#s05f10.

Zrenner, R. and Ashihara, H. (2011) Nucleotide metabolism. In H. Ashihara, A. Crozier and A. Komamine (eds), *Plant Metabolism and Biotechnology*. John Wiley & Sons, Ltd., Oxford, pp. 155–181.

Chapter 3
Phytochemicals in Teas and Tisanes and Their Bioavailability

Michael N. Clifford[1] and Alan Crozier[2]

[1]Food Safety Research Group, Centre for Nutrition and Food Safety, Faculty of Health and Medical Sciences, University of Surrey, Guildford, Surrey GU2 7XH, UK
[2]School of Medicine, College of Medical, Veterinary and Life Sciences, University of Glasgow, Glasgow G12 8QQ, UK

3.1 Introduction

This chapter provides information on the diversity of phytochemicals in teas and tisanes. It also examines the bioavailability of selected compounds, most notably flavonoids and related phenolics, which have been linked with the potentially protective effects of these beverages on health, many of which are discussed in detail in Chapter 4.

3.2 Phytochemical content of teas and tisanes

3.2.1 Camellia *teas*

Camellia-based teas, containing flavan-3-ols and derived compounds (Figure 3.1), are one of the most widely consumed beverages in the world. Grown in about 30 countries, the botanical classification of *Camellia* spp. is complex and confused, with many forms of commercial tea that may or may not be distinct species (Kaundun and Matsumoto 2002). The main forms recognised are *Camellia sinensis* var. *sinensis*, which originated on the northern slopes of the Himalayas and has small leaves a few centimetres in length, and *C. sinensis* var. *assamica*, with leaves 10–15 cm or more in length, which developed on the southern slopes of the Himalayas (Willson 1999). As examples of the variability, there is a large-leafed var. *sinensis* found in Yunnan

---

Teas, Cocoa and Coffee: Plant Secondary Metabolites and Health, First Edition.
Edited by Professor Alan Crozier, Professor Hiroshi Ashihara and Professor F. Tomás Barberán.
© 2012 Blackwell Publishing Ltd. Published 2012 by Blackwell Publishing Ltd.

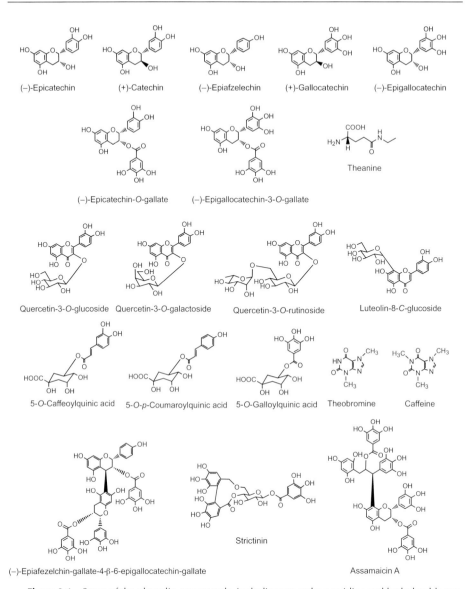

Figure 3.1 Some of the phenolic compounds, including proanthocyanidins and hydrolysable tannins, present in green tea (*Camellia sinensis*) infusions. The predominant purine alkaloid is caffeine with trace levels of theobromine.

that is rich in (−)-epicatechin-3-O-gallate (Shao *et al.* 1995, Obuchowicz *et al.* 2011), a form of var. *sinensis* comparatively rich in methylated flavan-3-ols (Chiu and Lin 2005) and the so-called var. *assamica* × var. *sinensis* hybrids used for Japanese green tea production with a comparatively low flavan-3-ol content but rich in theanine (*N*-ethylglutamine) (Figure 3.1) and other amino acids (Table 3.1; Takeo 1992, Obuchowicz *et al.* 2011).

Table 3.1 Approximate composition of green tea shoots (% dry weight)

	Var. assamica	Small-leafed var. sinensis	So-called hybrid of small-leafed var. sinensis and var. assamica	Large-leafed var. sinensis
Substances soluble in hot water				
Total polyphenols	25–30	14–23		32–33
Flavan-3-ols				
(−)-Epigallocatechin gallate	9–13	7–13	11–15	7–8
(−)-Epicatechin gallate	3–6	3–4	3–6	13–14
(−)-Epigallocatechin	3–6	2–4	4–6	1–2
(−)-Epicatechin	1–3	1–2	2–3	2–3
(+)-Catechin				4
Other flavan-3-ols	1–2			2
Flavonols and flavonol glycosides	1.5	1.5–1.7		1
Flavandiols	2–3			1
Phenolic acids and esters (despides)	5			
Caffeine	3–4	3		
Amino acids	2	4–5	2–3	
Theanine	2	2–5	2	
Simple carbohydrates (e.g. sugars)	4			
Organic acids	0.5			
Substances partially soluble in hot water				
Polysaccharides	1–2			
Starch, pectic substances	12			
Pentosans, etc.	15			
Proteins	5			
Ash				
Substances insoluble in water				
Cellulose	7			
Lignin	6			
Lipids	3			
Pigments (chlorophyll, carotenoids, etc.)	0.5			
Volatile substances	0.01–0.02			

Note: Blanks in the table indicate that data are not available.

Tea is generally consumed in one of three forms, green, oolong or black, but there are many more variations that arise, through differences in the nature of the leaf used and the method of processing (Hampton 1992; Takeo 1992), including some, such as pu-erh tea, that involve a microbial transformation stage (Shao *et al.* 1995). In 2010, ~3.2 million metric tonnes of dried tea were produced, 61% of which was black tea and 31% green tea, with the balance made up of minor teas, such as oolong and pu-erh. In all cases, the raw material is young leaves, the tea flush, which are preferred as they have a higher flavan-3-ol content and elevated levels of active enzymes. The highest quality teas utilise 'two leaves and a bud', with progressively lower quality taking four or even five leaves (Willson 1999).

There are basically two types of green tea (Takeo 1992). The Japanese type utilises shade-grown hybrid leaf that after harvesting is steamed rapidly to inhibit polyphenol oxidase and other enzymes. Chinese green tea traditionally uses selected forms of var. *sinensis* and dry heat (firing) rather than steaming, giving a less efficient inhibition of the polyphenol oxidase activity and allowing some transformation of the flavan-3-ols.

In the production of black tea, there are again two major processes (Hampton 1992): the so-called orthodox and the more recently introduced, but now well-established, cut–tear–curl process. In both processes, the objective is to achieve efficient disruption of cellular compartmentation, thus bringing phenolic compounds into contact with polyphenol oxidases and at the same time activating many other enzymes. The detailed preparation of the leaf, known as withering, time and temperature of the fermentation stage, and the method of arresting the fermentation to give a relatively stable product, all vary geographically across the black tea-producing areas. However, oxidation for 60–120 minutes at about 40°C before drying gives some idea of the conditions employed.

When harvested, the fresh tea leaf is unusually rich in polyphenols (∼30% dry weight) (Table 3.1), and this changes with processing even during the manufacture of commercial green tea, and progressively through semi-fermented teas to black teas and those with a microbial processing stage. Flavan-3-ols are the dominant polyphenols of fresh leaf. Usually (−)-epigallocatechin-3-*O*-gallate dominates, occasionally taking second place to (−)-epicatechin-3-*O*-gallate, together with smaller but still substantial amounts of (−)-epigallocatechin, (+)-gallocatechin, (+)-catechin, (−)-epicatechin and (−)-epiafzelechin (Figure 3.1). The minor flavan-3-ols also occur as gallates, and (−)-epigallocatechin may occur as a digallate, esterified with *p*-coumaric acid or caffeic acid and with various levels of methylation (Hashimoto *et al.* 1992). There are at least 15 flavonol glycosides comprising mono-, di- and tri-*O*-glycosides based upon kaempferol, quercetin and myricetin, and various permutations of glucose, galactose, rhamnose, arabinose and rutinose (Finger *et al.* 1991; Engelhardt *et al.* 1992; Price *et al.* 1998; Lakenbrink *et al.* 2000a), three apigenin-*C*-glycosides (Sakamoto 1967, 1969, 1970), several caffeoyl- and *p*-coumaroylquinic acids (chlorogenic acids) and galloylquinic acids (theogallins), and at least 27 proanthocyanidins, including some with epiafzelechin units (Nonaka *et al.* 1983; Lakenbrink *et al.* 1999). In addition, some forms have a significant content of hydrolysable tannins, such as strictinin (Nonaka *et al.* 1984), perhaps indicating an affinity with *Camellia japonica*, *Camellia sasanqua* and *Camellia oleifera* (Hatano *et al.* 1991; Han *et al.* 1994; Yoshida *et al.* 1994), whereas others contain chalcan–flavan dimers known as assamaicins (Hashimoto *et al.* 1989a). Relevant structures are illustrated in Figure 3.1, and data on the levels of some of these compounds in green tea shoots are presented in Table 3.1.

In green teas, especially Japanese production, most of these various polyphenols survive and can be found in the marketed product. In Chinese green teas and the semi-fermented teas such as oolong, some transformations occur, leading, for example, to the production of theasinensins (flavan-3-ol dimers linked 2→2′), oolong homo-bis-flavans (linked 8→8′ or 8→6′), oolongtheanin and 8*C*-ascorbyl-epigallocatechin-3-*O*-gallate (Figure 3.2; Hashimoto *et al.* 1987, 1988, 1989a, 1989b). In black tea production, the transformations are much more extensive,

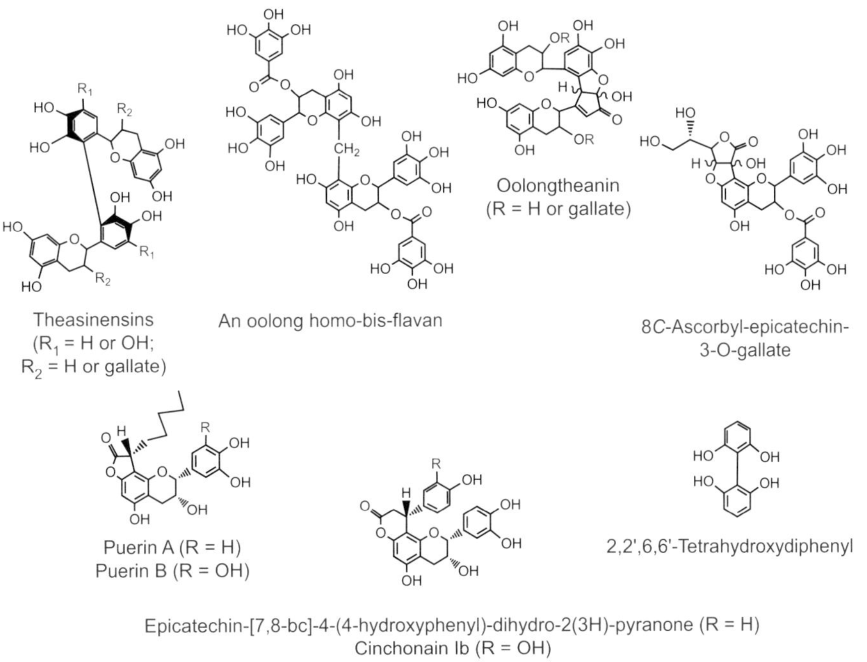

Figure 3.2 Transformation products found in Chinese green teas and semi-fermented teas, such as oolong.

with some 90% destruction of the flavan-3-ols in orthodox processing and even greater transformation in cut–tear–curl processing. Some losses of galloylquinic acids, quercetin glycosides and especially myricetin glycosides have been noted, and recent studies on thearubigins suggest that theasinensins and possibly proanthocyanidins may also be transformed. Pu-erh tea is produced by a microbial fermentation of black tea, and it contains some novel compounds which, it has been suggested, are formed during fermentation (Zhou *et al.* 2005). These include two 8-*C*-substituted flavan-3-ols, puerins A and B, and two known cinchonain-type phenols, epicatechin-[7,8-bc]-4-(4-hydroxyphenyl)-dihydro-2(3H)-pyranone and cinchonain Ib, and 2,2′,6,6′-tetrahydroxydiphenyl (Figure 3.2). However, various cinchonains have previously been reported in unfermented plant material (Nonaka *et al.* 1982; Nonaka and Nishioka 1982; Chen *et al.* 1993).

It is generally considered that polyphenol oxidase, which has at least three isoforms, is the key enzyme in the fermentation processes that produce black teas, but there is also evidence for an important contribution from peroxidases, with the essential hydrogen peroxide being generated by polyphenol oxidase (Subramanian *et al.* 1999). The primary substrates for polyphenol oxidase are the flavan-3-ols that are converted to quinones. These quinones react further, and may be reduced back to phenols by oxidising other phenols, such as gallic acid, flavonol glycosides and theaflavins, which are not direct substrates for polyphenol oxidase (Opie *et al.* 1993, 1995).

The nature of these transformations has been the subject of a detailed review by Drynan et al. (2010).

Many of the transformation products are still uncharacterised. The best known are the various theaflavins and theaflavin gallates (Figure 3.3), characterised by their bicyclic undecane benztropolone nucleus, reddish colour and solubility in ethyl acetate.

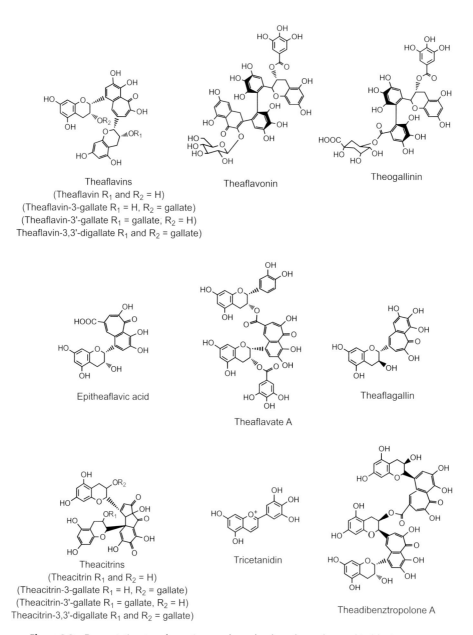

Figure 3.3 Fermentation transformation products that have been detected in black tea.

These form through the Michael addition of a B-ring trihydroxy (epi)gallocatechin quinone to a B-ring dihydroxy (epi)catechin quinone prior to carbonyl addition across the ring and subsequent decarboxylation (Goodsall *et al.* 1996), but it is now accepted that the theasinensins (Figure 3.2) form more rapidly and may actually be theaflavin precursors (Hashimoto *et al.* 1992; Tanaka *et al.* 2002a). Theaflavonins and theogallinin, (2→2'-linked theasinensin analogues) (Figure 3.3) formed from (−)-epigallocatechin/(−)-epigallocatechin-3-O-gallate and isomyricetin-3-glucoside or 5-galloylquinic acid, respectively, have also been found in black tea (Hashimoto *et al.* 1992).

Coupled oxidation of free gallic acid or ester gallate produces quinones that can replace (epi)gallocatechin quinone, leading to (epi)theaflavic acids and various theaflavates (Figure 3.3; Wan *et al.* 1997). Interaction between two quinones derived from trihydroxy precursors can produce benztropolone-containing theaflagallins (Hashimoto *et al.* 1986) or yellowish theacitrins that have a tricyclic dodecane nucleus (Davis *et al.* 1997). Mono- or digallated analogues are similarly formed from the appropriate gallated precursors, and in the case of theaflavins-coupled oxidation of benztropolone gallates can lead to theadibenztropolones and higher homologues, at least in model systems. Oxidative degallation of (−)-epigallocatechin gallate produces the pinkish-red desoxyanthocyanidin, tricetanidin (Figure 3.3; Coggon *et al.* 1973).

The brownish water-soluble thearubigins are the major phenolic fraction of black tea. Masses certainly extend to ∼2000 daltons. Early reports that these were polymeric proanthocyanidins (Brown *et al.* 1969) probably arose through detection of proanthocyanidins that had passed through unchanged from the fresh leaf. Only slow progress was made over the next 40 years. The few structures that were identified included dibenztropolones (Figure 3.3) where the 'chain extension' has involved coupled oxidation of ester gallate (Sang *et al.* 2002, 2004), theanaphthoquinones formed when a bicylco-undecane benztropolone nucleus collapses back to a bicyclo-decane nucleus (Tanaka *et al.* 2000, 2001), and dehydrotheasinensins (Figure 3.4; Tanaka *et al.* 2005a).

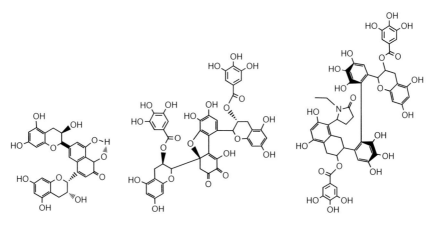

Theanaphthoquinone   Dehydrotheasinensin AQ   8'-Ethylpyrrolidinonyl-theasinensin A

Figure 3.4 Dimeric phenolics from black tea that have been associated with the production of thearubigins.

It was suggested that production of higher mass thearubigins could involve coupled oxidation of gallate esters, yielding tribenztropolones, etc., coupled oxidation of large mass precursors such as proanthocyanidin gallates or theasinensin gallates rather than flavan-3-ol gallates (Menet et al. 2004), or interaction of quinones with peptides and proteins. Although long anticipated, 8′-ethylpyrrolidinonyl-theasinensin A (Figure 3.4), the first such product containing an N-ethyl-2-pyrrolidinone moiety, was not isolated from black tea until 2005 (Tanaka et al. 2005b). It is probably formed from a theasinensin and the quinone-driven Strecker aldehyde produced by decarboxylation of theanine. Model system studies have led to the characterisation of some additional structures (Tanaka et al. 2001, 2002b, 2002c, 2003), but since their relevance to commercial black tea is currently unclear, they are not discussed further in this chapter.

A significant breakthrough in the characterisation of thearubigins was reported in 2010 as the result of collaborative studies involving researchers at the University of Surrey in the United Kingdom and Jacobs University Bremen in Germany that made use of modern mass spectrometric procedures, in particular ion trap–MS^n and Fourier transform ion cyclotron resonance mass spectrometry (FTICR-MS) (Kuhnert et al. 2010a, 2010b). These studies of the thearubigins from 15 commercial black teas, selected so as to represent the major variations found in production worldwide, revealed on average 5000 thearubigin components in the mass range between 1000 and 2100 daltons. FTICR-MS data revealed the presence of a maximum of 9428 peaks in the mass range of 300–1000 daltons. The higher mass components are still under investigation. This vast number of products easily explains why the thearubigins do not resolve during chromatography, appearing as a streak during paper or thin-layer chromatography or as a broad Gaussian-like hump when analysed by reverse-phase high performance liquid chromatography (HPLC). A typical cup of black tea contains approximately 100 mg of thearubigins, estimated by subtracting flavan-3-ols, flavonol glycosides and theaflavins, determined by HPLC, from the total phenolics measure with the Folin–Ciocalteau assay Graham 1992; Astill et al. 2001; Bond et al. 2003) suggesting that very few if any individual thearubigins will exceed 100 μg per cup.

The accurate high-resolution mass data obtained by FTICR-MS allowed molecular formulas to be assigned to 1517 thearubigins and this facilitated the development of a novel hypothesis to explain the immense number of compounds formed by the action of primarily a single enzyme, polyphenol oxidase, and a handful of flavan-3-ol substrates. The hypothesis assumes that the thearubigins form from the four classes of well-defined flavan-3-ol dimers, i.e. theaflavins, theacitrins, theasinensins and theanaphthoquinones (Figures 3.2, 3.3 and 3.4). These dimers could be enlarged by the creation of new B-ring C–C bonds in the case of the theasinensins or by the repeat of the dimerisation where ester gallate replaces an (epi)gallocatechin unit. Oligomers containing up to seven (epi)catechin units or up to four (epi)catechin gallate units can be accommodated within the mass range observed. Any one of these oligomers can be converted to an *ortho*-quinone, either directly by polyphenol oxidase or, if not, then indirectly where a non-substrate is oxidised by a polyphenol oxidase-derived quinone. The resultant quinone then reacts with water, the most abundant nucleophile in the fermenting leaf. Formally, this nucleophilic addition inserts an oxygen, raising the molecular mass of the oligomers by 16 amu. This new product is amenable to a further oxidation step to produce a new quinone, which then reacts with water as described previously. This sequence of reactions can continue until all available C–H sites, including those in ester gallate, have acquired an oxygen and become C–OH. Collectively, these products are described as polyhydroxylated oligomers (polyhydroxylated theasinensins, polyhydroxylated theaflavins, polyhydroxylated theacitrins, polhydroxylated theanaphthoquinones and their associated mono- and digallates). With the exception where all available sites have been oxygenated, regioisomers are feasible and many have been observed, as discussed later in this section.

It is also clear that the analogous products formed by addition of hydrogen peroxide are also present. Hydrogen peroxide is a more powerful nucleophile than water and has been observed in fermenting black tea at concentrations in the range 30–60 μM (Subramanian *et al.* 1999), but because of its reactivity, this might be an underestimate of the amount generated during fermentation and consumed in such transformations.

Any one of the polyhydroxylated oligomers is likely to exist in a redox equilibrium with its corresponding *ortho*- and *para*-quinoid structures, and evidence for such mono- and diquinones was obtained by mass spectrometry (MS) and circular dichroism. Nitrogen-containing compounds were not detected, but there were indications of oligomers containing sugar units that might indicate involvement of flavonol glycosides in theaflavonin-type products.

To better appreciate how such transformations might generate thousands of products it is instructive to consider the theoretical yields from theaflavin and its isomers. For theaflavin (Figure 3.3), there is one fully hydroxylated product plus 30 partially hydroxylated products if all the possible regioisomers are included. Black tea contains at least three of the four possible theaflavins, namely theaflavin (produced from (–)-epicatechin and (–)-epigallocatechin), isotheaflavin (produced from (–)-epicatechin and (+)-gallocatechin) and neotheaflavin (produced from (+)-catechin and (–)-epigallocatechin) (Drynan *et al.* 2010). Accordingly, there is the possibility that black tea contains 96 theaflavins (three parent compounds plus 93 hydroxylated derivatives).

Adding to this theoretical total, the corresponding number of products that might form from at least four theaflavin monogallates and at least one theaflavin digallate to say nothing of the corresponding theasinensins, theacitrins and theanaphthoquinones, plus as yet uncharacterised derivatives, it is not difficult to envisage how several thousand products can arise. That said, evidence has so far been obtained for only 29 hydroxylated theaflavins (with between one and six oxygen insertions), 12 theaflavin monogallates (with between one and six oxygen insertions), nine theaflavin digallates (with between one and four oxygen insertions) and ten theacitrin monogallates (with between one and four oxygen insertions). Evidence has also been obtained for at least ten mono- or diquinone forms of the parent compounds and hydroxylated derivatives in each of these homologous series (Kuhnert *et al.* 2010a). The oxidative cascade hypothesis is under critical scrutiny and much remains to be elucidated, but the intransigent thearubigins are gradually yielding their secrets.

Aqueous infusions of tea leaves contain the purine alkaloid caffeine and traces of theobromine. Green and semi-fermented teas retain substantial amounts of the flavan-3-ols, but they decline progressively with increased fermentation and are lowest in cut–tear–curl black teas. Beverages from green, semi-fermented and black teas also have significant contents of flavonol glycosides and smaller amounts of chlorogenic acids, flavone-C-glycosides and 5-O-galloquinic acid (theogallin) (Figure 3.1), which are less affected by processing but may vary more markedly with the origin of the fresh leaf (Engelhardt *et al.* 1992; Shao *et al.* 1995; Lin *et al.* 1998; Price *et al.* 1998; Lakenbrink *et al.* 2000b; Luximon-Ramma *et al.* 2005). Black tea beverage uniquely contains theaflavins and to a greater extent the high-molecular-weight thearubigins, which are responsible for the astringent taste of black tea and the characteristic red-brown colour. As noted in the preceding text, thearubigins are difficult to analyse, but indirect estimates indicate that they comprise around 80% of the phenolic components in black tea infusions (Stewart *et al.* 2005). Details of how some of the phenolic compounds in green tea are modified by fermentation to produce black tea are presented in Table 3.2.

The polyphenol and purine composition of green tea, pu-erh tea and white tea (tea prepared from shade-grown leaf buds with low chlorophyll content) have recently been determined by ultra-high performance liquid chromatography (UHPLC)-MS (Zhao *et al.* 2011). Seventy-seven components were at least partially characterised and fifty-nine were quantified. Among the notable minor components were four caffeoylquinic acids (one of which was uncharacterised, but plausibly *cis*-5-O-caffeoylquinic acid characteristic of plant tissues exposed to comparatively intense UV), caffeoyl-malic acid, (+)-catechin-3-O-gallate, and eight kaempferol glycosides bearing one or two *p*-coumaroyl residues or one acetyl residue.

Further changes may occur during the domestic brewing process and production of instant tea beverages. The flavan-3-ols may epimerise, producing, for example, (+)-epicatechin and (−)-catechin, (−)-gallocatechin gallate, etc. (Figure 3.5; Wang and Helliwell 2000; Ito *et al.* 2003). Black tea brew can form either scum or cream as it cools. Scum formation requires hard water containing calcium bicarbonate that facilitates oxidation of brew phenols at the air–water interface (Spiro and Jaganyi 1993). Tea cream is a precipitate formed as black tea cools, being more obvious in strong infusions, and involves an association between theaflavins, some thearubigins and caffeine, exacerbated by the calcium in hard water (Jöbstl *et al.* 2005).

3.2.2 Yerba maté tea

Yerba maté is a herbal tea prepared from the dried leaves of *Ilex paraguariensis*, which contain both caffeine and theobromine (Clifford and Ramìrez-Martìnez 1990). Originally, the drink was consumed by Guarani Indians in the forests of Paraguay and the habit was adopted by settlers in rural areas of South America, such as the Brazilian Panthanal and the Pampas in Argentina, where there was a belief that it ensured health, vitality and longevity. Its consumption is now becoming more widespread, with the annual maté trade in the United States in 2010 being around $1 billion and an estimated 5% of all tea products sold worldwide containing leaves of *I. paraguariensis* (Folch 2010). Maté, unlike *C. sinensis*, is a rich source of chlorogenic acids, containing substantial amounts of 3-, 4- and 5-O-caffeoylquinic acid (Figure 3.6; Clifford and Ramìrez-Martìnez 1990). A recent study identified 42 chlorogenic acids, including eight caffeoylquinic acids, five dicaffeoylquinic acids, six feruloylquinic acids, two

Table 3.2 Concentration of the major phenolics in infusions of green and black tea manufactured from the same batch of *Camellia sinensis* leaves

Compound	Green tea	Black tea	Black tea content as a percentage of green tea content
Gallic acid	6.0 ± 0.1	125 ± 7.5	2083
5-*O*-Galloylquinic acid	122 ± 1.4	148 ± 0.8	121
Total gallic acid derivatives	**128**	**273**	**213**
(+)-Gallocatechin	383 ± 3.1	n.d.	0
(−)-Epigallocatechin	1565 ± 18	33 ± 0.8	2.1
(+)-Catechin	270 ± 9.5	12 ± 0.1	4.4
(−)-Epicatechin	738 ± 17	11 ± 0.2	1.5
(−)-Epigallocatechin-3-*O*-gallate	1255 ± 63	19 ± 0.0	1.5
(−)-Epicatechin-3-*O*-gallate	361 ± 12	26 ± 0.1	7.2
Total flavan-3-ols	**4572**	**101**	**2.2**
3-*O*-Caffeoylquinic acid	60 ± 0.2	10 ± 0.2	17
5-*O*-Caffeoylquinic acid	231 ± 1.0	62 ± 0.2	27
4-*O*-*p*-Coumaroylquinic acid	160 ± 3.4	143 ± 0.2	89
Total hydroxycinammate quinic esters	**451**	**215**	**48**
Quercetin-*O*-rhamnosylgalactoside	15 ± 0.6	12 ± 0.2	80
Quercetin-3-*O*-rutinoside	131 ± 1.9	98 ± 1.4	75
Quercetin-3-*O*-galactoside	119 ± 0.9	75 ± 1.1	63
Quercetin-*O*-rhamnose-*O*-hexose-*O*-rhamnose	30 ± 0.4	25 ± 0.1	83
Quercetin-3-*O*-glucoside	185 ± 1.6	119 ± 0.1	64
Kaempferol-*O*-rhamnose-*O*-hexose-*O*-rhamnose	32 ± 0.2	30 ± 0.3	94
Kaempferol-*O*-galactoside	42 ± 0.6	29 ± 0.1	69
Kaempferol-*O*-rutinoside	69 ± 1.4	60 ± 0.4	87
Kaempferol-3-*O*-glucoside	102 ± 0.4	69 ± 0.9	68
Kaempferol-*O*-arabinoside	4.4 ± 0.3	n.d.	0
Unknown quercetin-*O*-conjugate	4 ± 0.1	4.3 ± 0.5	108
Unknown quercetin-*O*-conjugate	33 ± 0.1	24 ± 0.9	73
Unknown kaempferol-*O*-conjugate	9.5 ± 0.2	n.d.	0
Unknown kaempferol-*O*-conjugate	1.9 ± 0.0	1.4 ± 0.0	74
Total flavonols	**778**	**570**	**73**
Theaflavin	n.d.	64 ± 0.2	∞
Theaflavin-3-*O*-gallate	n.d.	63 ± 0.6	∞
Theaflavin-3′-*O*-gallate	n.d.	35 ± 0.8	∞
Theaflavin-3,3′-*O*-digallate	n.d.	62 ± 0.1	∞
Total theaflavins	**n.d.**	**224**	**∞**

After Del Rio *et al.* 2004.
Data expressed as mg/L ± standard error ($n = 3$); n.d., not detected. Green and black teas prepared by infusing 3 g of leaves with 300 mL of boiling water for 3 minutes.

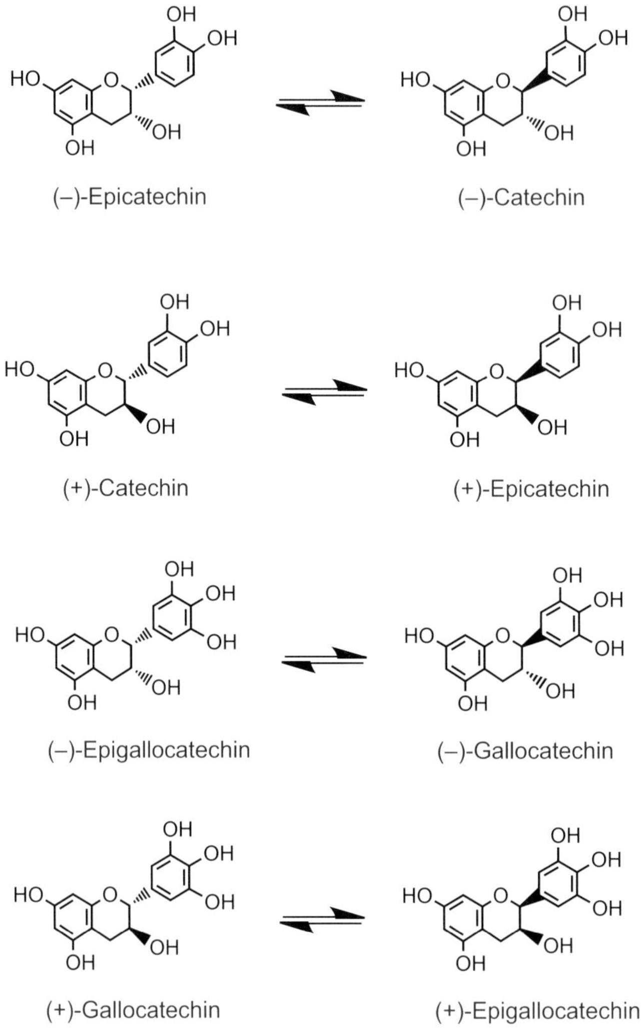

Figure 3.5 During brewing and production of instant tea beverages, flavan-3-ols such as (+)-catechin, (−)-epicatechin and (+)-gallocatechin may epimerise.

diferuloylquinic acids, five p-coumaroylquinic acids, four caffeoyl-p-coumaroylquinic acids, seven caffeoyl-feruloylquinic acids, three caffeoyl-sinapoylquinic acids and one tricaffeoylquinic acid, in a methanolic extract of dry maté leaves. Also detected were four caffeoylshikimic acids, three dicaffeoylshikimic acids and one tricaffeoylshikimic acid and one feruloylshikimic acid (Jaiswal et al. 2010).

Maté leaves have yielded three new saponins named matesaponins 2, 3 and 4, which have been characterised as ursolic acid 3-O-[β-D-glucopyranosyl-(1→3)-[α-L-rhamnopyranosyl-(1→2)]]-α-L-arabinopyranosyl]-(28 → 1)-β-D-glucopyranosyl ester (Figure 3.6), ursolic acid 3-O-[β-D-glucopyranosyl-(1→3)-α-L-arabinopyranosyl]-(28→1)-[β-D-glucopyranosyl-(1→6)-β-D-glucopyranosyl] ester,

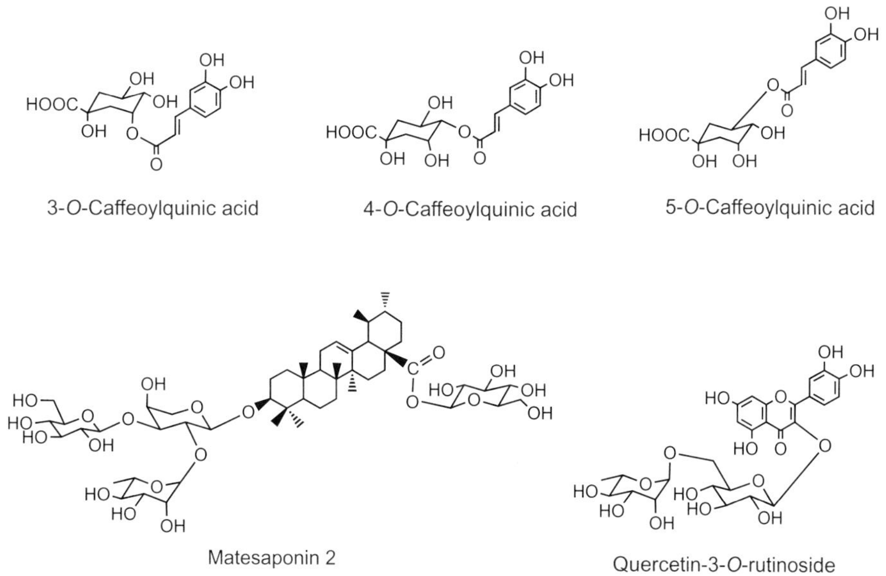

Figure 3.6 Maté infusions contain chlorogenic acids, saponins and trace amounts of quercetin-3-O-rutinoside.

and ursolic acid 3-O-[β-D-glucopyranosyl-(1→3)-[α-L-rhamnopyranosyl-(1→2)]]-α-L-arabinopyranosyl]-(28 → 1)-[β-D-glucopyranosyl-(1→6)-β-D-glucopyranosyl] ester, respectively (Gosmann *et al.* 1995). Maté infusions also contain quercetin-3-O-rutinoside (Figure 3.6) and glycosylated derivatives of luteolin and caffeic acid (Carini *et al.* 1998). Despite claims to the contrary (Morton 1989), neither condensed tannins nor hydrolysable tannins are present (Clifford and Ramìrez-Martìnez 1990).

The traditional method of brewing and consumption is to add the dry, sometimes roasted, leaf to water boiling vigorously in a gourd or metal vessel. To avoid the boiling liquid burning the lips, it is drawn into the mouth using a straw and deposited at the back of the throat where there are comparatively few pain receptors. Although this practice avoids much of the discomfort otherwise associated with the consumption of a hot beverage, it still damages the oesophagus. Exposure of the damaged tissue during the error-prone stage of tissue repair makes such consumers unusually susceptible to oesophageal cancer (IARC 1991). While there seems little doubt that the primary causative agent is boiling hot water, the involvement of other substances is less certain. As a consequence of the repeated and ultimately severe tissue damage, many substances will gain access to the tissue in an unmetabolised form, something that would not happen in healthy tissue. Redox cycling of dihydroxyphenols such as the caffeoylquinic acids may have a role to play, as may carcinogens from tobacco or alcohol (Castelletto *et al.* 1994). It is reassuring that consumption of maté beverage in a manner more closely resembling tea drinking is not associated with an increased risk of oesophageal cancer, even in South American populations where this phenomenon was first observed (Rolon *et al.* 1995; Sewram *et al.* 2003).

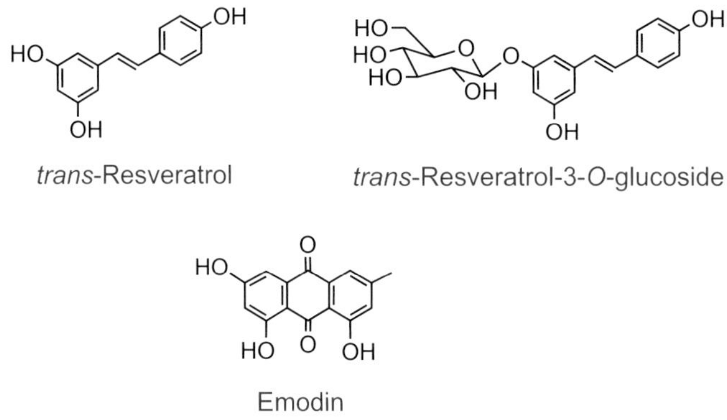

Figure 3.7 The dried woody root of *Polygonum cuspidatum* is used to make Itadori tea, which contains stilbenes, principally in the form of *trans*-resveratrol and its 3-O-glucoside, and the anthraquinone emodin.

3.2.3 Itadori tea

Polygonum cuspidatum Sieb et Zucc is more commonly known as Japanese knotweed. It is well known to gardeners as it is an extremely noxious weed that has invaded many areas of Europe and North America. In its native Asia, the root is dried and infused with boiling water to produce caffeine-free Itadori tea. Itadori means 'well-being' in Japanese, and the tea has been used for centuries in Japan and China as a traditional herbal remedy for many diseases including heart disease and stroke (see Chapter 4). The active agents are believed to be *trans*-resveratrol and its 3-O-glucoside (Figure 3.7) that are present in high concentrations in the woody root and the aerial parts of the plant (Burns *et al.* 2002). *P. cuspidatum* is also a source of the anthraquinone emodin (Figure 3.7), which is a mild laxative. The data in Table 3.3 show that a methanolic extract of dried *P. cuspidatum* root chippings, purchased in Japan, had a stilbene content of 6.5 µmol/g principally in the form of *trans*-resveratrol-3-O-glucoside. Itadori tea prepared by infusing 1 g of the commercial root preparation with 100 mL of boiling water for 5 minutes contained 0.3 and 2.3 µmol/100 mL, respectively, of the stilbene aglycone and its 3-O-glycoside (Burns *et al.* 2002).

Table 3.3 Stilbene content of *Polygonum cuspidatum* root and Itadori tea

Stilbene	Root (µmol/g)	Itadori tea (µmol/100 mL)
trans-Resveratrol	2.3 ± 0.03	0.30 ± 0.01
trans-Resveratrol-3-O-glucoside	4.2 ± 0.04	2.3 ± 0.01
Total	**6.5 ± 0.04**	**2.6 ± 0.01**

After Burns *et al.* 2002.
Data expressed as mean value ± standard error ($n = 3$).

3.2.4 Rooibos tea

Rooibos tea is a caffeine-free beverage made with infusions prepared from the leaves of South African shrub *Aspalathus linearis*, which belongs to the Fabaceae family. Traditionally, the tea is made from either the 'red' oxidised leaves (also referred to as fermented) or unfermented 'green' leaves. Rooibos or red bush tea, popularised by the 'The No. 1 Ladies' Detective Agency' novels of the Edinburgh University Emeritus Professor of Medical Law, Alexander McCall Smith (1999, 2000, 2009), is consumed in small but increasing quantities in the United Kingdom. Rooibos contains 2′,3,4,4′,6′-pentahydroxydihydrochalcone-3′-*C*-glucoside (aspalathin) and 2′,4,4′,6′-tetrahydroxydihydrochalcone-3′-*C*-glucoside (nothofagin), (*R*)/(*S*)-eriodictyol *C*-6 and *C*-8 glucosides (flavanones), and small amounts of a range of compounds including the flavones luteolin, chrysoeriol, luteolin-6-*C*-glucoside (isoorientin *aka* homoorientin), luteolin-8-*C*-glucoside (orientin), apigenin-6-*C*-glucoside (isovitexin) and apigenin-8-*C*-glucoside (vitexin), the flavan-3-ol (+)-catechin, and the flavonols quercetin, quercetin-3-*O*-glucoside, quercetin-3-*O*-galactoside and quercetin-3-*O*-rutinoside (Figure 3.8), as well as numerous hydroxybenzoic acids and hydroxycinnamates (Joubert *et al.* 2008; Krafczyk and Glomb 2008).

Qualitative and quantitative information on the flavonoid content of unfermented and fermented rooibos teas based on the data of Stalmach *et al.* (2009b) are presented in Table 3.4. Fermentation results in an approximately tenfold decline in the levels of the dihydrochalcones, aspalathin and nothofagin, which is accompanied by a fourfold increase in the concentration of the four eriodictyol-*C*-glucosides. There were only minor changes in the levels of other compounds in the fermented tea, namely the *C*-6 and *C*-8 glucosides of luteolin and apigenin, four quercetin-*O*-glycosides, and the aglycones luteolin and quercetin (Table 3.4). This is in keeping with the reported conversion of aspalathin to eriodictyol-*C*-glucosides during fermentation but not their further conversion to orientin and isoorientin (Krafczyk and Glomb 2008). Presumably, this is a reflection of differences in the oxidation conditions to which the *A. linearis* leaves were subjected during fermentation. Although not detected by Stalmach *et al.* (2009b), Krafczyk *et al.* (2009) report that during the early browning stages of rooibos fermentation, aspalathin forms two dimers in a reaction analogous to the formation of theaflavins during the fermentation of green tea. These dimers are, however, B-ring–A-ring linked, whereas the well-known theaflavin dimers in black tea produced from *C. sinensis* are B-ring–B-ring linked. Accordingly, these rooibos dimers more closely resemble assamaicins the chalcan–flavan dimers of green tea (Hashimoto *et al.* 1989a).

3.2.5 Honeybush tea

Caffeine-free honeybush tea is prepared from several *Cyclopia* species, principally *Cyclopia genistoides* and *Cyclopia subternata*, short woody shrubs endemic to the Cape region of South Africa. Unfermented tea is made by infusing leaves, stems and flowers, while fermentation in either a curing heap or a preheated oven changes the colour of the plant material from green to dark brown as phenolic compounds are oxidised (McKay and Blumberg 2007).

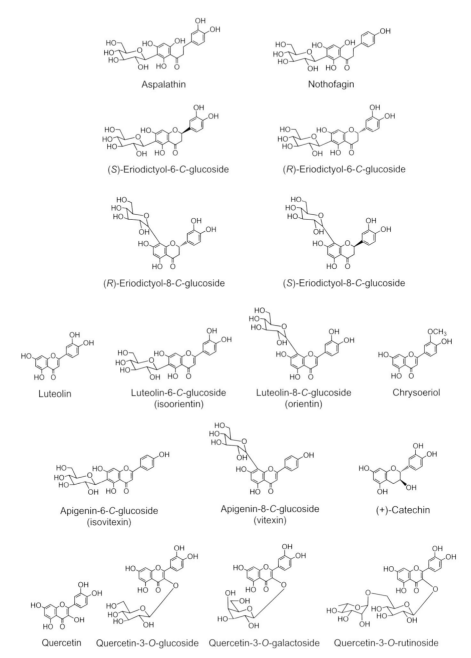

Figure 3.8 Rooibos tea prepared from leaves of *Aspalathus linearis* contains a wide spectrum of flavonoids.

Table 3.4 Quantification of flavonoids in 500 mL of unfermented and fermented rooibos teas

Flavonoids	Unfermented	Fermented
Aspalathin	90 ± 4.4	8.0 ± 0.6
Nothofagin	16 ± 0.9	1.9 ± 0.1
Eriodictyol-C-glucosides *(4 isomers)*	5.5 ± 0.0	23.1 ± 0.1
Luteolin-8-C-glucoside	15 ± 0.5	16 ± 0.6
Luteolin-6-C-glucoside	13 ± 0.5	12 ± 0.5
Apigenin-8-C-glucoside	2.2 ± 0.1	2.7 ± 0.1
Apigenin-6-C-glucoside	2.3 ± 0.1	2.2 ± 0.1
Quercetin-3-O-galactoside	1.5 ± 0.0	2.4 ± 0.1
Quercetin-3-O-glucoside	2.5 ± 0.1	2.2 ± 0.1
Quercetin-3-O-rutinoside	4.0 ± 0.1	2.6 ± 0.1
Quercetin-O-rutinoside isomer	7.1 ± 0.2	10 ± 0.4
Luteolin	0.5 ± 0.0	0.4 ± 0.1
Quercetin	0.1 ± 0.0	1.0 ± 0.1
Total	**159 ± 6.5**	**84 ± 2.9**

After Stalmach *et al.* 2009b.
Data expressed in μmol/500 mL as mean value ± standard error ($n = 3$).

The main secondary metabolites in green shoots of *C. subternata* are the xanthone C-glucosides mangiferin and isomangiferin together with the flavanone eriodictyol-7-O-rutinoside (eriocitrin) and the flavone luteolin-7-O-rutinoside (scolymoside) (Figure 3.9), and relatively minor changes occurred during a variety of steaming, heating and storage treatments (Joubert *et al.* 2010). A more detailed analysis of

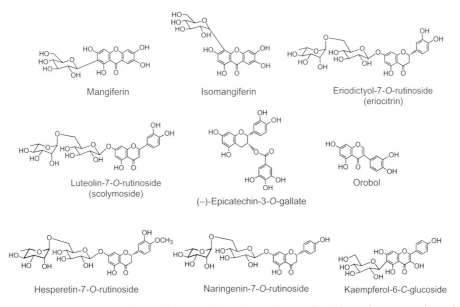

Figure 3.9 Structures of secondary metabolites in green leaves of *Cyclopia subternata* used to make honeybush tea. The main components are mangiferin and isomangiferin together with eriodictyl-7-O-rutinoside and luteolin-7-O-rutinoside.

unfermented *C. subternata* leaves revealed the presence of further compounds, including (–)-epicatechin-3-O-gallate, the isoflavone orobol, hesperetin-7-O-rutinoside, naringenin-7-O-rutinoside and kaempferol-6-C-glucoside (Figure 3.9; Kamara *et al.* 2004).

3.2.6 Chamomile tea

Dried flower heads of German chamomile (*Matricaria reticulata*) are brewed to produce a popular tisane. Although widely perceived as having relaxing effects, like many herbal preparations, the health benefits of chamomile are based on folklore rather than scientific evidence (see Chapter 4). Volatile oil from the flowers contains the sesquiterpenes α-bisabolol, bisabolol oxides A and B, and chamazulene, which is a thermal decomposition product from matricin. Matricin has anti-inflammatory activities and appears to be metabolised by the body to chamazulene carboxylic acid, a natural analogue of the synthetic analgesic ibuprofen (Figure 3.10; Dewick 2009). Recent analysis of chamomile tea extracts by HPLC-MS^2 revealed that the most abundant phenolic compounds were 5-O-caffeoylquinic acid, umbelliferone, apigenin and its 7-O-glucoside, quercetin-3-O-rutinoside and quercetin-3-O-rhamnoside (Figure 3.11; Nováková *et al.* 2010).

3.2.7 Hibiscus tea

The calyces of flowers of hibiscus (*Hibiscus sabdariffa*) are used to produce a herbal tea that has been attributed with a number of health-promoting effects (see Chapter 4). Analysis of an aqueous extract of the calyces of *H. sabdariffa* by electrophoresis-time-of-flight MS identified delphinidin-3-O-sambubioside and cyanidin-3-O-sambubioside

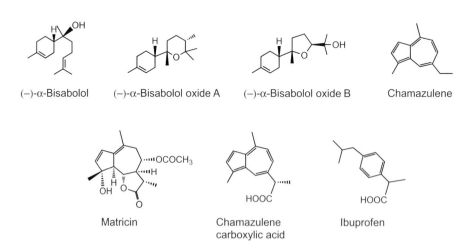

Figure 3.10 Chamomile flowers, which are used to make a herbal tea, contain several sesquiterpenes. Chamazulene is a breakdown product of matricin which in the body is converted to chamazulene carboxylic acid a natural analogue of the synthetic analgesic ibuprofen.

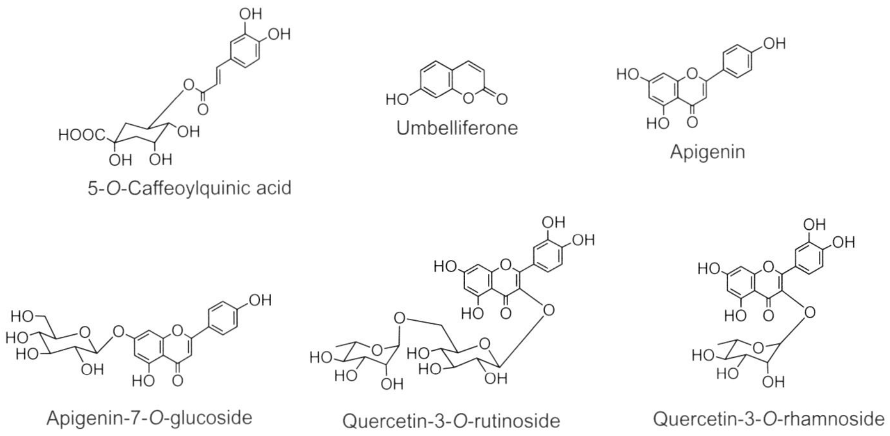

Figure 3.11 The principal phenolic compounds in chamomile (*Matricaria reticulata*) tea.

as major components along with small amounts of cyanidin-3-*O*-rutinoside, cyanidin-3,5-*O*-diglucoside, delphinidin-3-*O*-glucoside and 5-*O*-caffeolyquinic acid (Segura-Carretero *et al.* 2008). Subsequent analyses by HPLC-MS, by the same group, identified additional compounds including hydroxycitric acid, hibiscus acid, 7-hydroxycoumarin (umbelliferone), 5-*O*-caffeolyshikimic acid, *N*-feruloyltyramine (Figure 3.12) and several flavonol-*O*-glycosides (Rodríguez-Medina *et al.* 2009).

3.2.8 Fennel tea

Fennel (*Foeniculum vulgare*) contains a broad range of phytochemicals that vary with the tissue. These include benzoic acid-4-*O*-glucoside, polyacetylenes, various chlorogenic acids, rosmarinic acid, a range of glycosides and glucuronides of flavones, flavanones and flavonols, the phenylpropanoids, *trans*-anethole and estragole, and a number of terpenes including fenchone (Figure 3.13; Dirks and Herrmann 1984; Parejo *et al.* 2004; Zidorn *et al.* 2005; Krizman *et al.* 2007; Cosge *et al.* 2008; Miguel *et al.* 2010; Tschiggerl and Bucar 2010). Fennel tea is prepared as an infusion of crushed seeds and reputedly alleviates symptoms of flatulence and spasms in nursing babies. Zeller and Rychlik (2006) showed that the aroma of the tea is due principally to fenchone and *trans*-anethole, which occur in higher concentrations than estragole. Analysis of several commercial fennel tea infusions also revealed the presence of 5-*O*-caffoylquinic acid, quercetin-3-*O*-glucuronide and *p*-anisaldehyde (Figure 3.13), a breakdown produce of *trans*-anethole (Bilia *et al.* 2000).

3.2.9 Anastatica tea

A herbal tea is made from seeds of *Anastatica hirerochuntica*, a plant found in the Sahara-Arabian deserts and used to treat a variety of ailments. The beverage prepared

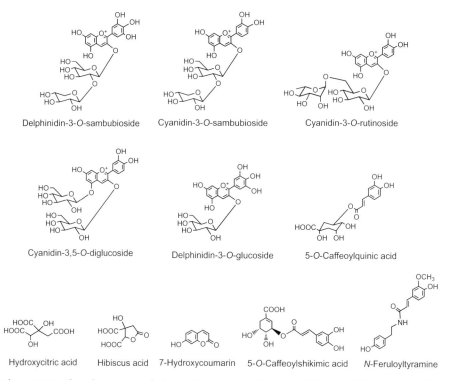

Figure 3.12 Phenolic compounds detected in calyces of flowers of hibiscus (*Hibiscus sabdariffa*) used to produce a herbal tea.

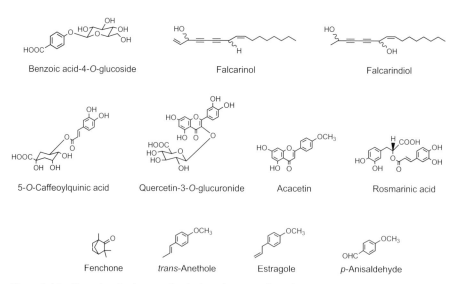

Figure 3.13 Phytochemicals occurring in fennel (*Foeniculum vulgare*).

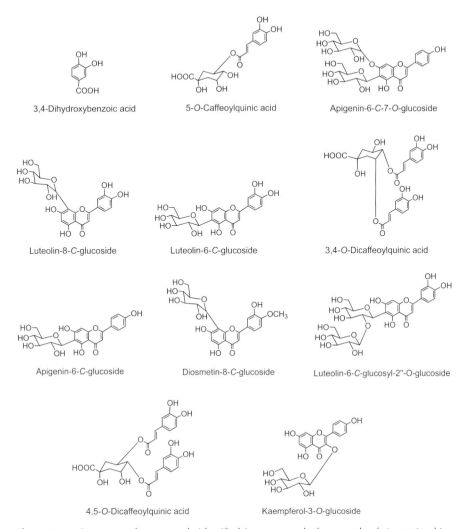

Figure 3.14 Structures of compounds identified in a tea made from seeds of *Anastatica hierochuntica*.

from an aqueous infusion of powdered seeds contained a range of flavonoids, benzoic acids and chlorogenic acids, with diosmetin-8-*C*-glucoside and 3,4-dihydroxybenzoic acid dominant (Figure 3.14; AlGamdi *et al.* 2011). There have been no studies on the absorption and metabolism of the (poly)phenolic compounds in this tea. Methanolic extracts of the whole plant also contain some unusual flavonoids, anastatin A and B, containing a dihydroxybenzofuran element fused to the A-ring (Yoshikawa *et al.* 2003), as well as silybin A and isosilybins A and B where a C_6–C_3 element is fused to the B-ring through a 1,4-dioxane element (Figure 3.15; Nakashima *et al.* 2010).

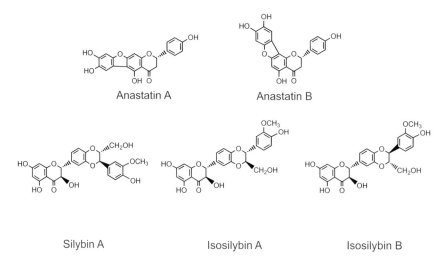

Figure 3.15 A methanolic extract of whole plants of *Anastatica hirerochuntica* was found to contain anastatin A and B, as well as silybin A and isosilybins A and B.

3.2.10 Ficus *tea*

In Malaysia, aqueous infusion of leaves of *Ficus deltoidea* are used to produce a popular herbal tea that is attributed with various medicinal effects including aphrodisiac activity and anti-hypertensive, anti-diabetic and anti-cancer properties (Hakiman and Maziah 2009). The tea contains flavan-3-ol monomers and proanthocyanidin dimers and trimers comprised of (epi)catechin and (epi)afzelechin units, which occur along with several flavone-C-glycosides (Figure 3.16; Omar *et al.* 2011). There are no reports on the fate of these compounds in the body following ingestion of the tea.

3.3 Bioavailability – absorption, distribution, metabolism and excretion

Following the ingestion of dietary flavonoids that, with the notable exception of flavan-3-ols, exist *in planta* predominantly as glycoside conjugates, absorption of some, but not all, components into the circulatory system occurs in the small intestine (Crozier *et al.* 2009). Typically, this is associated with hydrolysis, releasing the aglycone, as a result of the action of lactase phlorizin hydrolase (LPH) in the brush border of the small-intestine epithelial cells. LPH exhibits broad substrate specificity for flavonoid-O-β-D-glucosides and the released aglycone may then enter the epithelial cells by passive diffusion as a result of its increased lipophilicity and its proximity to the cellular membrane (Day *et al.* 2000). An alternative site of hydrolysis is a cytosolic β-glucosidase (CBG) within the epithelial cells. In order for CBG-mediated hydrolysis to occur, the polar glucosides must be transported into the epithelial cells, possibly with the involvement of the active sodium-dependent glucose transporter SGLT1

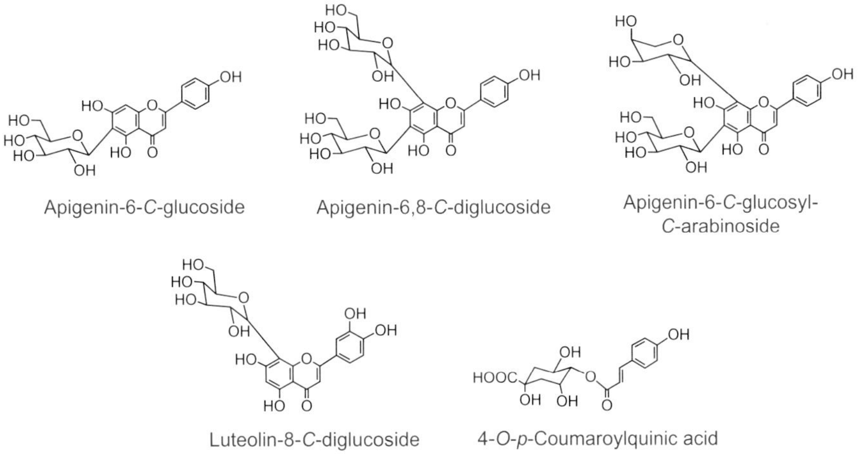

Figure 3.16 Aqueous infusions of leaves of *Ficus deltoidea*, a Malaysian herbal tea, contain a diversity of compounds including several flavone C-glycosides and 4-O-p-coumaroylquinic acid.

(Gee *et al.* 2000). Thus, it has been accepted that there are two possible routes by which the glucoside conjugates are hydrolysed and the resultant aglycones appear in the epithelial cells, namely 'LPH/diffusion' and 'transport/CBG'. However, a recent investigation, in which SGLT1 was expressed in *Xenopus laevis* oocytes, indicated that SGLT1 does not transport flavonoids and that glycosylated flavonoids, and some aglycones, have the capability to inhibit the glucose transporter (Kottra and Daniel 2007), as previously reported by Johnston *et al.* 2005.

Prior to passage into the bloodstream, the aglycones undergo metabolism, forming sulphate, glucuronide and/or methylated metabolites through the respective action of sulphotransferases (SULT), uridine-5′-diphosphate glucuronosyltransferases (UGT) and catechol-O-methyltransferases (COMT). There is also efflux of at least some of the metabolites back into the lumen of the small intestine and this is thought to involve members of the adenosine triphosphate (ATP)-binding cassette (ABC) family of transporters including multidrug resistance protein (MRP) and P-glycoprotein (P-gp). Once in the bloodstream, metabolites can be subjected to phase II metabolism, with further conversions occurring in the liver, where enterohepatic transport in the bile may result in some recycling back to the small intestine (Donovan *et al.* 2006b). Flavonoids and their metabolites not absorbed in the small intestine can be absorbed in the large intestine where they will also be subjected to the action of colonic microflora that will cleave conjugating moieties and the resultant aglycones will undergo ring fission, leading to the production of phenolic acids and hydroxycinnamates. These can be absorbed and may be subjected to phase II metabolism in the enterocyte and/or the liver before being excreted in urine in substantial quantities that, in most instances, are well in excess of the flavonoid metabolites that enter the circulatory system via the small intestine (Jaganath *et al.* 2006; Roowi *et al.* 2009; Williamson and Clifford 2010).

A detailed review on the bioavailability of polyphenols in humans was published in 2005 (Manach *et al.* 2005). Much of the research covered involved feeding volunteers a single supplement and monitoring the levels of flavonoids in plasma and urine over a 24-hour period. As flavonoid metabolites were, and indeed are still, rarely available commercially, analysis almost invariably involved treatment of samples with mollusc glucuronidase/sulphatase preparations and subsequent quantification of the released aglycones by HPLC using either absorbance, fluorescence or electrochemical detection. Some subsequent bioavailability studies have analysed samples directly by HPLC with tandem MS detection without recourse to enzyme hydrolysis. The availability of reference compounds enables specific metabolites to be identified by HPLC-MS^2 and MS^3 (Mullen *et al.* 2006). In the absence of standards, it is not always possible to distinguish between isomers and ascertain the position of conjugating groups on the flavonoid skeleton. Nonetheless, a metabolite that in reality is, say, pelargonidin-3-*O*-glucuronide can be partially identified as a pelargonidin-*O*-glucuronide on the basis of its MS fragmentation pattern (Mullen *et al.* 2008a). The use of MS in this way represents a powerful HPLC detection system, as with low ng quantities of sample, it provides structural information on analytes of interest that is not obtained with other detectors.

Quantification of partially identified metabolites by MS using consecutive reaction monitoring (CRM) or selected ion monitoring (SIM) is, of necessity, based on a calibration curve of a related compound, which in the instance cited in the preceding text could be pelargonidin-3-*O*-glucoside as it is readily available from commercial sources. In such circumstances, as the slopes of the glucoside and glucuronide SIM dose–response curves are unlikely to be identical, this approach introduces a potential source of error in the quantitative estimates and there is a view that quantitative estimates based on enzyme hydrolysis are, therefore, much more accurate. However, enzyme hydrolysis makes use of glucuronidase/sulphatase preparations that contain a mixture of enzyme activities and there can be substantial batch-to-batch variation in their specificity (Donovan *et al.* 2006b). There are no reports of flavonoid bioavailability studies using glucuronidase/sulphatase preparations where information on the identity, number and quantity of the individual sulphate and glucuronide conjugates in the samples of interest has been obtained. As a consequence, there are no direct data on the efficiency with which the enzymes hydrolyse the individual metabolites and release the aglycone. This introduces a varying, unmeasured error factor. The accuracy of quantitative estimates based on the use of glucuronidase/sulphatase preparations are, therefore, probably no better, and possibly much worse, than those based on HPLC-CRM/SIM. The fact that enzyme hydrolysis results in very reproducible data is not relevant as reproducibility is a measure of precision, although it is frequently mistaken for accuracy (Reeve and Crozier 1980). These shortcomings of analyses based on enzyme hydrolysis apply to bioavailability studies with all dietary flavonoids, and it is interesting to note that the one publication on the subject to date reports that the use of enzyme hydrolysis results in an underestimation of isoflavone metabolites (Gu *et al.* 2005).

3.3.1 Green tea

3.3.1.1 Studies with healthy subjects

Green tea as mentioned in Section 3.2 is an extremely rich source of flavan-3-ol monomers. In a recent study, 500 mL of a bottled green tea was given as an acute supplement to ten volunteers after which plasma and urine were collected over a 24-hour period. The tea contained a total of 648 μmol of flavan-3-ols principally in the form of 257 μmol of (−)-epigallocatechin, 230 μmol of (−)-epigallocatechin-3-O-gallate, 58 μmol of (−)-epicatechin, 49 μmol of (−)-epicatechin-3-O-gallate and 36 μmol of (+)-gallocatechin (Stalmach et al. 2009c).

Two of the native green tea flavan-3-ols, (−)-epicatechin-3-O-gallate and (−)-epigallocatechin-3-O-gallate, were identified in plasma along with glucuronide, methyl-glucuronide and methyl-sulphate metabolites of (epi)gallocatechin and glucuronide, sulphate and methyl-sulphate metabolites of (epi)catechin[1]. The pharmacokinetic profiles of the eight groups of flavan-3-ols and their metabolites are illustrated in Figure 3.17. The C_{max} values ranged from 25 to 126 nM and T_{max} values from 1.6 to 2.3 hours (Table 3.5). These T_{max} values and the pharmacokinetic profiles are indicative of absorption in the small intestine. The appearance of unmetabolised flavonoids in plasma is unusual. The passage of (−)-epicatechin-3-O-gallate and (−)-epigallocatechin-3-O-gallate through the wall of the small intestine into the circulatory system without metabolism could be a consequence of the presence of the 3-O-galloyl moiety, as gallic acid *per se* is readily absorbed with a reported urinary excretion of 37% of intake (Shahrzad and Bitsch 1998; Shahrzad et al. 2001).

Urine collected 0–24 hours after green tea ingestion contained an array of flavan-3-ol metabolites similar to that detected in plasma, except for the presence of minor amounts of three additional (epi)gallocatechin-O-sulphates and the absence of (−)-epicatechin-3-O-gallate and (−)-epigallocatechin-3-O-gallate (Table 3.6). This indicates that the flavan-3-ols do not undergo extensive phase II metabolism. In total, 52.4 μmol of metabolites were excreted, which is equivalent to 8.1% of the ingested green tea flavan-3-ols. When the urinary (epi)gallocatechin and (epi)catchin metabolites are considered separately, a somewhat different picture emerges. The 33.3 μmol excretion of (epi)gallocatechin metabolites is 11.4% of the ingested (−)-epigallocatechin and (+)-gallocatechin, while the 19.1 ± 2.2 μmol recovery of (epi)catechin represents 28.5% of intake (Table 3.6). These are figures in keeping with high recoveries obtained in earlier studies with green tea and cocoa products (Manach et al. 2005; Auger et al. 2008), confirming that (−)-epicatechin and (+)-catechin, in particular, are highly bioavailable being absorbed and excreted (also see Chapter 7 Donovan et al. 2006b).

The absence of detectable amounts of (−)-epigallocatechin-3-O-gallate in urine, despite its presence in plasma, an event observed by several investigators (Unno et al. 1996; Chow et al. 2001; Henning et al. 2005), is difficult to explain. It is possible that the kidneys are unable to remove (−)-epigallocatechin-3-O-gallate from the bloodstream, but if this is the case, there must be other mechanisms that result in its rapid decline after reaching C_{max}. Studies with rats have lead to speculation that (−)-epigallocatechin-3-gallate may be removed from the bloodstream in the liver and returned to the small intestine in the bile (Kida et al. 2000; Kohri et al. 2001). To what extent enterohepatic recirculation of (−)-epigallocatechin-3-O-gallate, and also (−)-epicatechin-3-O-gallate, occurs in humans remains to be established. However, it is of note that feeding studies

---

[1] Note that without reference compounds and chiral chromatography (Donovan et al. 2006a) reversed-phase HPLC, even with MS^3 detection, is unable to distinguish between the four possible enantiomers of any (epi)catechin, or any (epi)gallocatechin derivative (see Figure 3.5).

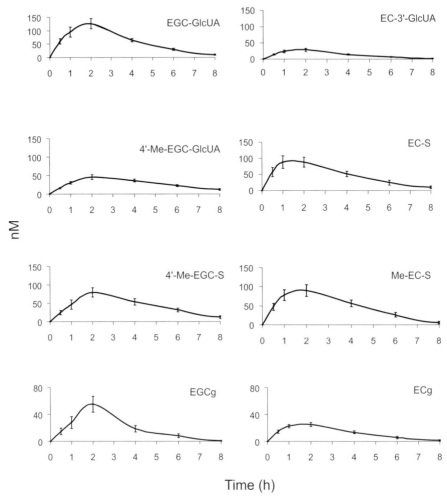

Figure 3.17 Concentrations of (epi)gallocatechin-O-glucuronide (EGC-GlcUA), 4′-O-methyl-(epi)gallocatechin-O-glucuronide (4′-Me-EGC-GlcUA), 4′-O-methyl-(epi)gallocatechin-O-sulphates (4′-Me-EGC-S), (−)-epicatechin-3′-O-glucuronide (EC-3′-GlcUA), 3′- and 4′-O-methyl-(epi)catechin-O-sulphates (Me-EC-S), (−)-epigallocatechin-3-gallate (EGCg) and (−)-epicatechin-3-gallate (ECg) in the plasma of human subjects 0–8 hours after the ingestion of 500 mL of green tea. Data expressed as mean values with their standard errors ($n = 10$) depicted by vertical bars. Note that no flavan-3-ols or their metabolites were detected in plasma collected 24 hours after ingestion of the green tea.

with [2-^{14}C]resveratrol, discussed in Section 3.3.3, indicate that stilbene metabolites do undergo enterohepatic recycling in humans (Walle et al. 2004).

Summing the C_{max} values for the individual plasma flavan-3-ols and metabolites in Table 3.5 results in an overall maximum plasma concentration of 538 nM being attained after the ingestion of green tea (Stalmach et al. 2009c). This is lower than the 1313 nM C_{max} of quercetin metabolites obtained following the ingestion of onions containing 250 μmol of quercetin-4′-O-glucoside and quercetin-3,4′-O-diglucoside (Mullen et al.

Table 3.5 Pharmacokinetic analysis of flavan-3-ols and their metabolites detected in plasma of ten volunteers following the ingestion of 500 mL of green tea

Flavan-3-ols	C_{max}(nM)	T_{max}(h)	$T_{1/2}$(h)
(Epi)gallocatechin-O-glucuronide	126 ± 19	2.2 ± 0.2	1.6
4'-O-Methyl-(epi)gallocatechin-O-glucuronide	46 ± 6.3	2.3 ± 0.3	3.1
(Epi)catechin-O-glucuronide	29 ± 4.7	1.7 ± 0.2	1.6
(Epi)catechin-O-sulphates	89 ± 15	1.6 ± 0.2	1.9
4'-O-Methyl-(epi)gallocatechin-O-sulphates	79 ± 12	2.2 ± 0.2	2.2
O-Methyl-(epi)catechin-O-sulphates	90 ± 15	1.7 ± 0.2	1.5
(−)-Epigallocatechin-3-O-gallate	55 ± 12	1.9 ± 0.1	1.0
(−)-Epicatechin-3-O-gallate	25 ± 3.0	1.6 ± 0.2	1.5

After Stalmach et al. 2009c.
Data expressed as mean value ± standard error ($n = 10$).

2006) and is also less than the 922 nM C_{max} of the hesperetin-O-glucuronides that appear in plasma after the ingestion of orange juice containing 168 μmol of hesperetin-7-O-rutinoside (Mullen et al. 2008b). Despite the relatively low concentration of the green tea, flavan-3-ol metabolites in plasma (Table 3.5; Figure 3.17), the data on urinary excretion (Table 3.6) demonstrate that they are absorbed in substantial quantities, especially (−)-epicatechin. Their failure to accumulate in comparable concentrations in plasma suggests that they are in a state of flux and are being more rapidly turned over in the circulatory system and, rather than accumulating, are excreted via the kidneys. In the circumstances, urinary excretion provides a more realistic assessment of absorption, but as this does not include the possibility of metabolites being sequestered in body tissues, this too is theoretically an underestimate of absorption, but to what degree remains to be determined. However, the fact that tissue sequestration has yet to be convincingly demonstrated suggests that it can only be at low levels, if at all.

Table 3.6 Quantification of the major groups of flavan-3-ol metabolites excreted in urine 0–24 hours after the ingestion of 500 mL of green tea by ten volunteers

Flavan-3-ol metabolites *(number of isomers)*	0–24 h excretion (μmol)
(Epi)gallocatechin-O-glucuronide *(1)*	6.5 ± 1.2
4'-O-Methyl-(epi)gallocatechin-O-glucuronide *(1)*	4.4 ± 1.5
4'-O-Methyl-(epi)gallocatechin-O-sulphates *(2)*	19.8 ± 0.3
(Epi)gallocatechin-O-sulphates *(3)*	2.6 ± 3.0
Total (epi)gallocatechin metabolites	*33.3 (11.4%)*
(Epi)catechin-O-glucuronide *(1)*	1.5 ± 0.3
(Epi)catechin-O-sulphates *(2)*	6.7 ± 0.7
O-Methyl-(epi)catechin-O-sulphates *(5)*	10.9 ± 1.2
Total (epi)catechin metabolites	*19.1 (28.5%)*
Total flavan-3-ol metabolites	**52.4 *(8.1%)***

After Stalmach et al. 2009c.
Data expressed as mean value ± standard error ($n = 10$). Italicised figures in parentheses indicate amount excreted as a percentage of intake.

A further point of note is that the plasma T_{max} times of the (epi)catechin metabolites following absorption in the small intestine are all in excess of 1.6 hours (Table 3.5), while with one exception, the T_{max} of the flavonol metabolites absorbed in the small intestine and derived from onion quercetin glucosides was 0.6–0.8 hour. This is unexpected, as in contrast to the onion flavonols, the green tea flavan-3-ols were already in solution and did not have to be solubilised post-ingestion. Furthermore, they were aglycones, not conjugates, and therefore, they did not have to be hydrolysed prior to absorption and metabolism. The delayed T_{max} of the green tea flavonols is unlikely to be due to a slower rate of absorption as their excretion is well in excess of the quantity of flavonol metabolites that appear in urine. Although further investigation is required, this does raise the possibility that (i) the flavan-3-ols may be absorbed distal and flavonols in the proximal part of the small intestine or (ii) some component(s) in the green tea may be slowing transport through the gastrointestinal tract.

3.3.1.2 Studies with ileostomists and in vitro faecal incubations

Stalmach *et al.* (2010a) carried out studies with ileostomists who drank green tea containing 634 μmol of flavan-3-ols – a very similar intake to that used in their studies with healthy subjects (Stalmach *et al.* 2009c) that were discussed in Section 3.3.1.1. The plasma profile of (epi)catechins and their metabolites was similar to that illustrated in Figure 3.17, which was obtained with healthy subjects with an intact-functioning colon. Urinary excretion by the ileostomists was 8.0% of intake for (epi)gallocatechin metabolites and 27.4% of (epi)catechin metabolites, values that are similar to that observed with healthy subjects (Table 3.6). This demonstrates unequivocally that the flavan-3-ol monomers are absorbed in the upper part of the gastrointestinal tract.

Despite the substantial absorption of green tea flavan-3-ols in the upper gastrointestinal tract, Stalmach *et al.* (2010a) found that 69% of intake was recovered in 0–24 hours ileal fluid as a mixture of native compounds and metabolites (Table 3.7). Thus, in volunteers with a functioning colon, most of the ingested flavan-3-ols will pass from the small to large intestine where their fate is a key part of the overall bioavailability equation. To mimic these events, two sets of experiments were carried out (Roowi *et al.* 2010). Firstly, 50 μmol of (−)-epicatechin, (−)-epigallocatechin and (−)-epigallocatechin-3-O-gallate were incubated under anaerobic conditions *in vitro* with faecal slurries and their degradation to phenolic acid by the microbiota monitored. A limitation of *in vitro* fermentation models is that they may not fully depict the *in vivo* conditions. The use of faecal material may alter the bacterial composition and, thus, may not fully represent the microbiota present in the colonic lumen and mucosa, where catabolism occurs *in vivo*. Obviously, the accumulation and retention of the degradation products in the fermentation vessel makes collection, identification and quantification of the metabolites easier, but it is not necessarily representative of the events that occur *in vivo* where the actual concentration of a metabolite at any time interval is dependent on the combined rates of catabolism and absorption. However, the use of an *in vitro* model provides information on the types of breakdown products formed, helps elucidate the pathways involved, and the rate of catabolism can be determined.

Table 3.7 Quantities of flavan-3-ols and metabolites recovered in ileal fluid 0–24 hours after the ingestion of 300 mL of green tea

Flavan-3-ols and metabolites *(number of isomers)*	Amount ingested (µmol)	Recovered in ileal fluid	
		(µmol)	% of amount ingested
(+)-Catechin	18	1.2 ± 0.2	6.8 ± 1.2
(−)-Epicatechin	69	7.9 ± 1.7	11 ± 2.5
(+)-Gallocatechin	50	13 ± 2.1	27 ± 4.2
(−)-Epigallocatechin	190	35 ± 5.8	18 ± 3.1
(−)-Epigallocatechin-3-O-gallate	238	116 ± 5.1	49 ± 2.1
(+)-Gallocatechin-3-O-gallate	5.2	4.6 ± 0.3	89 ± 5.2
(−)-Epicatechin-O-gallate	64	29 ± 3.3	45 ± 5.2
Total parent flavan-3-ols	*634*	*206 ± 11*	*33 ± 1.8*
(Epi)catechin-O-sulphates *(3)*		72 ± 4.1	
O-Methyl-(epi)catechin-O-sulphates *(5)*		5.4 ± 0.5	
(Epi)catechin-O-glucuronide *(1)*		0.5 ± 0.2	
Total (epi)catechin metabolites		*78 ± 4.3*	*90 ± 5.0*
(Epi)gallocatechin-O-sulphates *(3)*		108 ± 5.2	
O-Methyl-(epi)gallocatechin-O-sulphates *(3)*		40 ± 4.3	
O-Methyl-(epi)gallocatechin-O-glucuronide *(1)*		1.5 ± 0.5	
(Epi)gallocatechin-O-glucuronide *(1)*		1.5 ± 0.4	
Total (epi)gallocatechin metabolites		*151 ± 8.9*	*63 ± 3.7*
(Epi)gallocatechin-3-O-gallate-O-sulphate *(1)*		1.8 ± 0.4	
O-Methyl-(epi)gallocatechin-3-O-gallate-O-sulphates *(2)*		0.9 ± 0.2	
Total (−)-epigallocatechin-3-O-gallate metabolites		*2.8 ± 0.5*	*1.1 ± 0.2*
(Epi)catechin-3-O-gallate-O-sulphate *(1)*		0.4 ± 0.1	
Methyl-(epi)catechin-3-O-gallate-O-sulphates *(2)*		0.4 ± 0.1	
Total (−)-epicatechin-3-O-gallate metabolites		*0.9 ± 0.2*	*1.4 ± 0.3*
Total metabolites		232 ± 13	37 ± 2.1
Total parent flavan-3-ols and metabolites		439 ± 13	69 ± 2.0

After Stalmach et al. 2010a.
Data expressed as mean value ± standard error ($n = 5$).

To complement the *in vitro* incubations, phenolic acids excreted in urine 0–24 hours after (i) the ingestion of green tea and water by healthy subjects in a crossover study and (ii) the consumption of green tea by ileostomists were also investigated (Roowi et al. 2010). The data obtained in these studies provided the basis for the operation of the catabolic pathways that are illustrated in Figure 3.18. Some of these catabolites, such as 4′-hydroxyphenylacetic acid and hippuric acid, were detected in urine from subjects, with an ileostomy indicating that they are produced in the body by additional routes unrelated to colonic degradation of flavan-3-ols. It is, for instance, well known

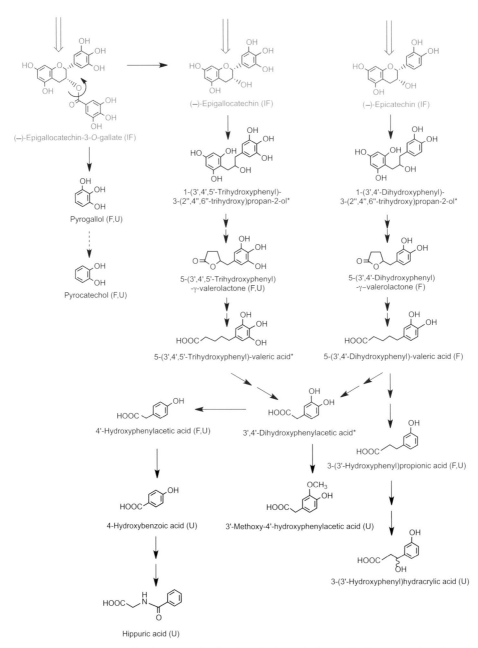

Figure 3.18 Proposed pathways involved in the colonic catabolism and urinary excretion of green tea flavan-3-ols. Following consumption of green tea, more than 50% of the ingested (−)-epicatechin, (−)-epigallocatechin and (−)-epigallocatechin-3-O-gallate (grey structures) pass into the large intestine. When incubated with faecal slurries, these compounds are catabolised by the colonic microflora probably via the pathways illustrated with red structures. Analysis of urine after green tea consumption indicates that some of the colonic catabolites enter the circulation and undergo further metabolism before being excreted in urine. Black structures indicate catabolites detected in urine but not produced by faecal fermentation of (−)-epicatechin, (−)-epigallocatechin or (−)-epigallocatechin-3-O-gallate. The dotted arrow between pyrogallol and pyrocatechol indicate this

that there are pathways to hippuric acid (*N*-benzoyl-glycine) from compounds such as benzoic acid, quinic acid (Clifford *et al.* 2000), tryptophan, tyrosine and phenylalanine (Self *et al.* 1960; Grumer 1961; Bridges *et al.* 1970). Nonetheless, the elevated urinary excretion of hippuric acid and 4'-hydroxyphenylacetic acid, occurring after green tea consumption, is likely to be partially derived from flavan-3-ol degradation. Earlier research showing statistically significant increases in urinary excretion of hippuric acid after consumption of both green and black tea by human subjects (Clifford *et al.* 2000; Mulder *et al.* 2005) supports this hypothesis.

Quantitative estimates of the extent of ring fission of the flavan-3-ol skeleton are difficult to assess because, as discussed in the preceding text, the production of some of the urinary phenolic acids was not exclusive to colonic degradation of flavan-3-ols. If these compounds, along with pyrogallol and pyrocatechol that are derived from cleavage of the gallate moiety from (−)-epigallocatechin-3-*O*-gallate rather than ring fission, are excluded, the excretion of the remaining urinary phenolic acids, namely 4-hydroxybenzoic acid, 3'-methoxy-4'-hydroxyphenylacetic acid, 3-(3'-hydroxyphenyl)hydracrylic acid and 5-(3',4',5'-trihydroxyphenyl)-γ-valerolactone, after ingestion of green tea was 210 μmol compared to 38 μmol after drinking water (Roowi *et al.* 2010). The 172 μmol difference between these figures corresponds to a 27% of the 634 μmol of flavan-3-ols present in the ingested green tea (Table 3.7; Stalmach *et al.* 2010a). Added to this is the ∼8% excretion of glucuronide, sulphate and methylated flavan-3-ols originating from absorption in the small intestine. This estimate of a 35% recovery is nonetheless a minimum value because with the analytical methodology used, some urinary catabolites will have escaped detection (Roowi *et al.* 2010). This will include glucuronide and sulphate metabolites of 5-(3',4',5'-trihydroxyphenyl)-γ-valerolactone, 5-(3',4'-dihydroxyphenyl)-γ-valerolactone and 5-(3',5'-dihydroxyphenyl)-γ-valerolactone, which were detected after green tea consumption with a cumulative 0- to 24-hour excretion corresponding to 16% of flavan-3-ol intake (Li *et al.* 2000; Meng *et al.* 2002; Sang *et al.* 2008). More recently, in a similar study in which urine was collected for 48 hours after green tea intake, valerolactone metabolites were excreted in quantities equivalent to 36% of intake (Del Rio *et al.* 2010a). When added to the 35% recovery of Roowi *et al.* (2010), this gives a total of 71% of intake. While this figure is obviously an approximation because of factors such as different volunteers, flavan-3-ol intakes and analytical methodology, it does demonstrate that, despite substantial modification as they pass through the body, there is a very high urinary recovery of flavan-3-ols, principally in the form of colon-derived catabolites.

Figure 3.18 (*continued*) is a minor conversion. Double arrows indicate conversions where the intermediate(s) did not accumulate and are unknown. Compounds detected in ileal fluid after green tea consumption (IF); catabolites detected in faecal slurries (F) and in urine (U); potential intermediates that did not accumulate in detectable quantities in faecal slurries (*). An alternative route to 3-(3'-hydroxyphenyl)hydracrylic acid is from cinnamic acid by the addition of water across the double bond in a hydratase catalysed reaction. (After Roowi *et al.* 2010.)

Table 3.8 Urinary excretion of flavan-3-ol metabolites after the ingestion of increasing doses of (–)-epicatechin and (–)-epigallocatechin in Polyphenon E by humans with an ileostomy. Metabolites excreted over a 24-hour period after ingestion expressed as μmol and italicised figures in parentheses represent the percentage of glucuronide, sulphate and methylated metabolites

	Dose		
(Epi)gallocatechin metabolites	22 μmol	55 μmol	165 μmol
Glucuronides	1.8 *(17%)*	1.0 *(18%)*	1.7 *(19%)*
Sulphates	3.9 *(37%)*	2.0 *(36%)*	3.6 *(40%)*
Methylated	4.8 *(46%)*	2.6 *(46%)*	4.7 *(51%)*
Total (epi)gallocatechin metabolites	**5.7 ± 1.9**[a]	**3.0 ± 0.8**[a]	**5.3 ± 1.2**[a]
	Dose		
(Epi)catechin metabolites	77 μmol	192 μmol	577 μmol
Glucuronides	3.4 *(7%)*	14 *(9%)*	38 *(10%)*
Sulphates	33 *(64%)*	93 *(58%)*	224 *(59%)*
Methylated	15 *(29%)*	53 *(33%)*	120 *(31%)*
Total (epi)catechin metabolites	**36 ± 9**[a]	**107 ± 27**[b]	**262 ± 26**[c]

Based on data of Auger *et al.* 2008.
Values for total (epi)gallocatechin and (epi)catechin metabolites with different superscripts are significantly different ($P < 0.05$).

3.3.1.3 Dose effects

Auger *et al.* (2008) fed ileostomists increasing doses of Polyphenon E, a green tea extract containing a characteristic array of flavan-3-ols, and urinary excretion of metabolites was used as a measure of absorption in the small intestine. The data obtained with (epi)gallocatechin and (epi)catechin metabolites are summarised in Table 3.8. At a dose of 22 μmol, the 0- to 24-hour excretion of (epi)gallocatechin metabolites was 5.7 ± 1.9 μmol and this figure did not increase significantly with intakes of 55 and 165 μmol. There is therefore a strict limit on the extent to which (epi)gallocatechins can be absorbed. Following the ingestion of 77 μmol of (epi)catechins, 36 ± 9 μmol were excreted, and with doses of 192 and 577 μmol, urinary excretion increased significantly to 107 ± 27 and 262 ± 26 μmol. Thus, even at the highest dose, (epi)catechins, unlike (epi)gallocatechins, are still readily absorbed. The addition of a 5'-hydroxyl group to (epi)catechin, therefore, markedly reduces the extent to which the molecule can enter into the circulatory system from the small intestine. It is also of note that at the three doses that were administered, the ratio of the urinary glucuronide, sulphate and methylated (epi)catechin metabolites changes little (Table 3.8). This indicates that even at the highest intake, the UGT, SULT and COMT enzymes involved in the formation of the (epi)catechin metabolites do not become saturated and limit conversions.

3.3.1.4 In vivo *and* in vitro *stability of green tea flavan-3-ols*

The necessity of volunteer studies is also indicated by claims in the literature that green tea flavan-3-ols are poorly bioavailable because of instability under digestive conditions, with >80% losses being observed with *in vitro* digestion models simulating gastric and small intestine conditions (Zhu *et al.* 1997; Yoshino *et al.* 1999; Record and Lane 2001; Green *et al.* 2007). It is clear that the data obtained in these investigations do not accurately reflect the *in vivo* fate of flavan-3-ols following ingestion as they are at variance with the high urinary excretion observed in green tea feeding studies (Manach *et al.* 2005; Stalmach *et al.* 2009c, 2010a) and the substantial recovery of flavan-3-ols in ileal fluid after the consumption of Polyphenon E, a green tea extract (Auger *et al.* 2008).

3.3.1.5 Inhibition of adrenaline metabolism by (−)-epigallocatechin-3-O-gallate

In vitro investigation of the behaviour of rat liver cytosolic COMT has demonstrated that it is inhibited by S-adenosyl-L-homocysteine, a demethylated product of the methyl donor S-adenosyl-L-methionine. The IC_{50} was approximately 10 μM. This observation has led to the suggestion that some of the biological effects of tea flavan-3-ols may be exerted by their O-methylated products or may result from their potential inhibition of the COMT-catalysed O-methylation of endogenous catecholamines and catechol oestrogens (Zhu *et al.* 2001). It was subsequently demonstrated *in vitro* that (−)-epigallocatechin-3-O-gallate and its 4″-monomethyl and 4′,4″-dimethyl metabolites inhibit COMT with IC_{50} concentrations of 0.07, 0.10 and 0.15 μM, respectively, with (−)-epigallocatechin-3-O-gallate achieving ∼30% inhibition even at 50 nM (Chen *et al.* 2005).

Unmetabolised (−)-epigallocatechin-3-O-gallate can transiently reach 55 ± 12 nM in plasma after the consumption of 500 mL of green tea (see Table 3.5), and although there are no data for its concentration in liver cytosol, it has been suggested that such inhibition could occur *in vivo* and impact on thermogenesis and body mass loss, especially when the flavan-3-ols are consumed with caffeine. A recent meta-analysis suggested that the effect is only modest and possibly less pronounced in caffeine-habituated subjects (Hursel *et al.* 2009). Furthermore, a caffeine-free (−)-epigallocatechin-3-O-gallate supplement (405 mg per day for 2 days) had no effect on weight loss (Lonac *et al.* 2011).

A study in which overweight breast cancer survivors consumed 960 mL of decaffeinated green or placebo tea daily for 6 months produced a slight reduction in body weight that did not reach statistical significance (Stendell-Hollis *et al.* 2010). Although this conclusion might be disappointing for those who seek an easy way to reduce weight, it should be reassuring to people taking medication that relies upon COMT for clearance.

3.3.2 Black tea

Although consumed far more extensively in Europe than is green tea, the absorption from black tea beverage of polyphenols, their gut flora catabolism and human metabolism have been less extensively and less thoroughly studied. In part, this can be explained by the lack of knowledge of the thearubigins and the non-availability

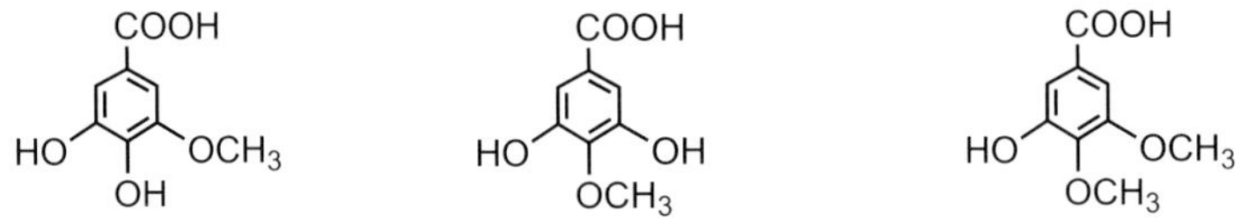

3-O-Methylgallic acid   **4-O-Methylgallic acid**   **3,4-O-Dimethylgallic acid**

Figure 3.19 Metabolites of gallic acid excreted in urine after the consumption of black tea.

of commercial standards for thearubigins or theaflavins. Such studies, as have been reported, have focused on the absorption of flavan-3-ols (Henning et al. 2004) or flavonol glycosides (de Vries et al. 1998; Hollman et al. 2001). The appearance in urine of the gallic acid metabolites 3-O-methylgallic acid, 4-O-methylgallic acid and 3,4-di-O-methylgallic acid (Figure 3.19) has also been reported and used as an index of black tea consumption (Hodgson et al. 2000, 2004), although these metabolites are also to be expected following not only green tea consumption but also black tea consumption and the ingestion of certain fruits, such as grapes, and the associated wines. The absorption and metabolism of flavan-3-ols and flavonol glycosides from black tea is not obviously different from that observed after green tea consumption, although *pro rata* the dose of flavan-3-ols is much reduced. The dose of flavonol glycosides is similar because these are not so greatly transformed during fermentation (see Table 3.2).

To date, only one study has investigated the absorption of mixed theaflavins (theaflavin 17.7%, theaflavin-3-gallate 31.8%, theaflavin-3′-gallate 16.7% and theaflavin-3,3′-digallate 31.4%) (Figure 3.3; Mulder et al. 2001). An extremely high dose (700 mg mixed theaflavins, equivalent to about 30 cups of black tea) was given to two healthy volunteers, one male and one female. Plasma and urine were analysed by HPLC-MS^2 after enzymatic deconjugation with β-glucuronidase and sulphatase, and extraction of the unconjugated/deconjugated theaflavins into ethyl acetate. Only theaflavin was detected because the enzyme treatment also removed ester gallate. Maximum theaflavin concentrations detected in the plasma of the female and the male volunteer were 1.0 and 0.5 µg/L, respectively (1.8 and 0.9 fM), and maximum urine concentrations were 0.6 and 4.2 µg/L, respectively (1.1 and 7.4 fM), all at 2 hours. These values should be doubled to correct for the relatively poor recovery observed with standard theaflavin, but even so, the total amount of theaflavin excreted was considerably less than 0.001% of the very large dose consumed (Mulder et al. 2001).

Attempts to investigate gut flora catabolism of mixed thearubigins *in vitro*, using conditions that were suitable for flavonols, flavan-3-ols and proanthocyanidin B2 (Stoupi et al. 2010a, 2010b), failed to observe the production of phenolic acids, even when using a low-protein medium to minimise effects from protein binding (Knight 2004; Stoupi 2010). In contrast, volunteer studies have clearly demonstrated that the consumption of black tea beverage results in a substantially increased excretion of hippuric acid relative to baseline, suggesting that a combination of gut flora catabolism and post-absorption metabolism results in a significant production of benzoic acid (Clifford et al. 2000; Daykin et al. 2005; Mulder et al. 2005). The yield of benzoic acid excreted as hippuric acid is such that it points to thearubigins and theaflavins, serving as substrates *in vivo* and being degraded to phenolic acids. The yield of hippuric acid was not significantly affected by the presence of caffeine in the black tea or by the addition of milk, but varied quite markedly, approximately fourfold, between individuals for a given intake of black tea (Clifford *et al.* 2000).

The influence of added milk on polyphenol absorption has been investigated by several groups. In one study, 18 healthy volunteers consumed eight cups of beverage per day at 2-hour intervals. It was observed that when 15 mL of milk was added to 135 mL of black tea beverage, containing ~100 μmol quercetin glycosides and some 60–70 μmol kaempferol glycosides, there were no significant differences in the plasma concentrations of flavonol metabolites (after enzymatic deconjugation) as judged by the areas under the concentration–time curves. The plasma C_{max} values were ~50 nM for quercetin and 30–45 nM for kaempferol (Hollman *et al.* 2001).

In a similar study, nine healthy participants drank 400 mL of black tea (the average daily UK consumption) prepared from 3 g of black tea (the average content of a UK tea bag). The beverage was diluted before consumption by the addition of 100 mL of either water (control) or semi-skimmed cows' milk. Plasma was sampled for up to 3 hours and samples were analysed colorimetrically for flavan-3-ols (using dimethylaminocinnamaldehyde) and by HPLC with fluorescence detection, after enzymatic deconjugation, for flavonols. The addition of milk produced no significant effects on plasma concentrations of total flavan-3-ols (~0.3–0.4 μM), quercetin (~40–60 nM) or kaempferol (~80–180 nM) (Kyle *et al.* 2007).

A third study in which 12 healthy volunteers consumed 3 g of freeze-dried tea solids reconstituted in either 500 mL water or 500 mL water plus 100 mL semi-skimmed milk produced essentially identical results for plasma total flavan-3-ols as measured with dimethylaminocinnamaldehyde (C_{max} 0.17–0.18 μM), but interestingly suggested that the rate of elimination of total flavan-3-ols after consumption of black tea with milk was somewhat slower than for black tea and both were significantly slower than elimination after consumption of green tea ($t_{e\frac{1}{2}} = 8.6$, 6.9 and 4.8 hours, respectively) (van het Hof *et al.* 1998). In the absence of data for the elimination flavan-3-ols given intravenously, such differences in the elimination rate constants must be interpreted cautiously because they might reflect a situation where the rate of absorption is the limiting factor on the rate of excretion.

In contrast to this, a fourth study using five healthy volunteers who consumed either 300 mL of black tea beverage or 300 mL of black tea beverage to which 100 mL of milk was added concluded that an increase in plasma antioxidant potential was completely abolished by the addition of milk (Serafini *et al.* 1996). In this study, not only was the rate of milk addition appreciably higher, but efficacy was judged by measuring effects on human plasma antioxidant capacity (TRAP). When these data are taken in combination with the results from the other studies it suggests that (i) polyphenol metabolites in human plasma have little effect on plasma antioxidant capacity, as discussed more fully elsewhere (Clifford 2004; Clifford and Brown 2006) and (ii) the TRAP assay is not specific for polyphenols. A similar study in which volunteers consumed cocoa with and without added milk has clearly demonstrated that added milk significantly reduces urinary excretion of (epi)catechin metabolites (Mullen *et al.* 2009), as discussed in more detail in Chapter 7.

3.3.3 Itadori tea

While there are few, if any, investigations in which the fate of *trans*-resveratrol and 3-*O*-glucoside has been monitored following ingestion of Itadori tea, there are numerous reports of bioavailability studies with animals and humans that have involved ingestion of *trans*-resveratrol, and these have been summarised in a recent review by Andres-Lacueva *et al.* (2010).

In a rare radiolabelled study with humans, reported by Walle *et al.* (2004), volunteers ingested [2-^{14}C]*trans*-resveratrol (110 μmol, 50 μCi) after which plasma, urine and faeces were collected over a 72-hour period. Total radioactivity in plasma reached a C_{max} equivalent to ∼2 μM after 1 hour with a second peak of ∼1.3 μM after 6 hours that was followed by an exponential decline. The plasma $T_{1/2}$ was 9.2 ± 0.6 hours, which is much longer than the flavan-3-ol $T_{1/2}$ values obtained after drinking green tea (Table 3.5). Indirect evidence suggested that the main components in plasma were glucuronide and sulphate metabolites of resveratrol. Most radioactivity over the 72-hour period was recovered in urine with smaller amounts in faeces, and when urine and faecal figures were combined, there was a mean overall recovery of 82.3% of the administered dose (Table 3.9). The main metabolites in urine were tentatively identified as monoglucuronides and sulphates of resveratrol and dihydroresveratrol. In a parallel study, 0.8 μmol of [2-^{14}C]resveratrol was administered to volunteers intravenously and a $T_{1/2}$ of 11.4 ± 1.1 hours was calculated based on levels of radioactivity in the plasma, suggesting that the comparatively long $T_{1/2}$, relative to that of many flavonoids, is not a consequence of a slow rate of absorption. Radioactivity recoveries of 64.1% and 10.4% were obtained in 0- to 72-hour urine and faeces, respectively. The appearance of radioactivity in faeces after intravenous administration of [^{14}C]resveratrol is compelling evidence of enterohepatic recycling from the circulatory system to the gastrointestinal tract via the bile duct.

Boocock *et al.* (2007) fed 0.5–5.0 g (2.2–22 mmol) doses of *trans*-resveratrol to volunteers and monitored the appearance of stilbenes in plasma and urine over a 24-hour period. Even the lowest dose is in excess of the resveratrol content of ∼30 bottles of a red wine (Burns *et al.* 2002). With the 2.2 mmol intake, traces

Table 3.9 Total radioactivity recovered in urine and faeces 0–72 hours after the ingestion of [2-^{14}C]resveratrol (110 μmol, 50 μCi) by six volunteers

Volunteer	Urine	Faeces	Total
1	74.3	23.3	97.6
2	66.5	4.0	70.5
3	73.9	2.8	76.7
4	70.1	7.5	77.5
5	53.4	38.1	91.5
6	84.9	0.3	85.2
Total ± standard error	70.5 ± 4.2	12.7 ± 5.9	83.2 ± 4.1

After Walle *et al.* 2004.
Data expressed as percentage of the ingested dose.

Table 3.10 Pharmacokinetic analysis of resveratrol and its metabolites detected in plasma of ten volunteers following the ingestion of 500 mg (2.2 mmol) and 5 g (22 mmol) of resveratrol

	2.2 mmol intake			22 mmol intake		
	$C_{max}(\mu M)$	$T_{max}(h)$	$T_{1/2}(h)$	$C_{max}(\mu M)$	$T_{max}(h)$	$T_{1/2}(h)$
Resveratrol	0.3	0.8	2.8	2.4	1.5	8.5
Resveratrol glucuronide-1	1.0	2.0	2.9	3.2	2.0	7.9
Resveratrol glucuronide-2	0.9	1.5	3.1	4.3	2.5	5.8
Resveratrol-3'-O-sulphate	3.7	1.5	3.2	13.9	2.1	7.7

After Boocock et al. 2007.
Data presented as mean values.

of resveratrol along with two resveratrol-O-glucuronides and larger amounts of a resveratrol-O-sulphate were detected in both plasma and urine. Pharmacokinetic analyses of the stilbene plasma profiles obtained with the 2.2 and 22 mmol intakes are presented in Table 3.10. The tenfold increase in dose resulted in higher C_{max} values, an eightfold with resveratrol, but only approximately fourfold with the sulphate and glucuronide metabolites. The higher stilbene intake had relatively little impact on T_{max} times, which are indicative of absorption in the small intestine. The higher intake did, however, increase $T_{1/2}$ times by 3.7–5.7 hours, suggesting some potential for accumulation of resveratrol metabolites in the circulatory system on repeat dosing during the waking day. The slow elimination of the stilbenes from the circulatory system compared to that of flavan-3-ols is also evident from the presence of metabolites in plasma 24 hours after ingestion (Boocock et al. 2007), which was not the case with the flavan-3-ols (Figure 3.17).

This group carried out a further study in which the impact of a daily intake of 0.5 and 1.0 g (2.2 and 4.4 mmol) of resveratrol for a period of 8 days on colon tumour cell proliferation was investigated. Resveratrol, resveratrol-3-O-glucuronide, resveratrol-4'-O-glucuronide, resveratrol-3-O-sulphate and resveratrol-4'-O-sulphate (Figure 3.20) were identified along with a resveratrol-O-sulphate-O-glucuronide and a resveratrol-O-disulphate in normal and malignant biopsy tissue samples (Patel et al. 2010). Resveratrol and resveratrol-3-O-glucuronde accumulated in tissues at maximal concentrations of 674 and 86 nmol/g, respectively, and tumour proliferation was reduced by 5%. Resveratrol and its metabolites were also detected in plasma after the 8-day intake. Doubling the resveratrol intake results in an approximately twofold increase in metabolite plasma levels with resveratrol-O-sulphate-O-glucuronide accumulating in especially high concentrations (Table 3.11).

3.3.4 Rooibos tea

As discussed in Section 3.2.3, the principal components in unfermented rooibos tea are the dihyrochalcone C-glucosides aspalathin and nothofagin with smaller amounts of flavanone and flavone C-glucosides and flavonol O-glycosides. Fermentation results in a decline in the dihydrochalcones and an increase in the flavanone C-glucosides,

Figure 3.20 *trans*-Resveratrol metabolites identified in normal and malignant colon tissue biopsies after the ingestion of *trans*-resveratrol by volunteers.

while the other flavonoids remained largely unchanged (see Table 3.4; Stalmach *et al.* 2009b).

Five hundred millilitre volumes of the unfermented and fermented teas that were analysed by Stalmach *et al.* (2009b) were, in the same study, also fed to volunteers after which plasma and urine collected at intervals over a 24-hour period were analysed by HPLC-MS^2. This was the first detailed study in which the bioavailability of flavonoid *C*-glycosides, as opposed to *O*-glycosides, was investigated in humans. Presumably because of either their low concentration in the teas and/or their low bioavailability, no flavonoid metabolites were detected in plasma and only aspalathin and eriodictyol metabolites were present in urine in detectable quantities.

Urine contained glucuronides, sulphates, methyl, methyl-sulphates and methyl-glucuronides metabolites of aspalathin. Most urinary excretion of the aspalathin metabolites occurred during the first 5 hours following intake (80–90% of the total amounts excreted), indicative of small intestine absorption. In total, 316 nmol

Table 3.11 Plasma concentrations of resveratrol metabolites in volunteers after receiving 2.2 and 4.4 mmol of resveratrol daily for 8 days

	Daily dose	
	2.2 mmol	4.4 mmol
Resveratrol-3-*O*-glucuronide	n.q.	0.24
Resveratrol-4'-*O*-glucuronide	0.04	0.24
Resveratrol-*O*-sulphate-*O*-glucuronide	13.4	22.3
Resveratrol-*O*-disulphate	0.31	0.60
Resveratrol-3-*O*-sulphate	0.13	0.59

After Patel *et al.* 2010.
Data expressed as mean values in mM; n.q., below limit of quantification.

Table 3.12 Summary of the urinary excretion of aspalathin and eriodictyol metabolites in urine collected from ten volunteers after drinking 500 mL of unfermented rooibos tea containing 90 μmol of aspalathin and 5.5 μmol of eriodictyol-C-glucosides (see Table 3.4)

Metabolites	Quantity excreted in urine (nmol)			
	0–5 h	5–12 h	12–24 h	0–24 h
Aspalathin-O-glucuronide	6.2 ± 1.1	5.0 ± 1.1	1.3 ± 0.6	12 ± 1.6
O-Methyl-aspalathin-O-glucuronide	180 ± 22	42 ± 9.1	7.2 ± 2.5	229 ± 26
Aspalathin-O-sulphate	40 ± 6.6	11 ± 2.3	1.3 ± 1.1	52 ± 6.8
O-Methyl-aspalathin-O-sulphate	20 ± 3.0	2.7 ± 0.9	n.d.	23 ± 3.2
Total aspalathin metabolites (% intake)				*0.35%*
Eriodictyol-O-sulphate	10 ± 4.4	22 ± 4.2	3.0 ± 1.3	35 ± 7.3
Eriodictyol-O-sulphate (% intake)				*0.64%*

After Stalmach et al. 2009b.
Data expressed as mean values in nmol ± standard error ($n = 10$). Bold, italicised figure indicate 0- to 24-hour excretion as a percentage of intake; n.d., not detected.

of aspalathin metabolites were excreted following an intake of 90 μmol with unfermented tea. This represents a recovery of 0.35% (Table 3.12). Following the ingestion of 8 μmol of aspalathin with the fermented tea, there was urinary recovery of 14.3 nmol of metabolites corresponding to 0.18% of intake (Table 3.13). Aspalathin, therefore, has very limited bioavailability.

Incubation of aspalathin with artificial gastric juice for up to 2 hours led to a recovery of ~100%, suggesting that aspalathin is not subject to degradation in the stomach. The very low urinary recoveries of aspalathin metabolites, and their failure to accumulate in detectable quantities in plasma, is probably a consequence of the C-glucoside moiety

Table 3.13 Summary of the urinary excretion of aspalathin and eriodictyol metabolites in urine collected from ten volunteers after drinking 500 mL of fermented rooibos tea containing 8.1 μmol of aspalathin and 23.1 μmol of eriodictyol-C-glucosides (see Table 3.4)

Metabolites	Quantity excreted in urine (nmol)			
	0–5 h	5–12 h	12–24 h	0–24 h
O-Methyl-aspalathin-O-glucuronide	8.8 ± 1.7	1.6 ± 0.6	n.d.	10 ± 2.0
Aspalathin-O-sulphate	2.6 ± 0.9	0.2 ± 0.2	n.d.	2.8 ± 0.9
O-Methyl-aspalathin-O-sulphate	1.5 ± 1.0	n.d.	n.d.	1.5 ± 1.0
Total aspalathin metabolites (% intake)				*0.18%*
Eriodictyol-O-sulphate	3.5 ± 2.4	52 ± 14	12 ± 4.2	68 ± 16
Eriodictyol-O-sulphate (% intake)				*0.29%*

After Stalmach et al. 2009b.
Data expressed as mean values in nmol ± standard error ($n = 10$). Italicised figure indicate 0- to 24-hour excretion as a percentage of intake; n.d., not detected.

not being readily cleaved by either LPH or CBG as the dihydrochalcone passes through the small intestine.

Urinary excretion of eriodictyol-O-sulphate was also relatively low with recoveries of 0.29% and 0.64%, following consumption of 23 and 5.5 μmol of eriodictyol-C-glucosides contained in the fermented and unfermented beverages, respectively (Tables 3.12 and 3.13). The majority of eriodictyol-O-sulphate excretion occurred during 5–12 hours following ingestion. This is indicative of low-level absorption occurring in the large rather than small intestine, in which case some of the eriodictyol may have been derived from colonic microflora-mediated biotransformation of aspalathin. It is of interest that in contrast to aspalathin, the eriodicyol-C-glucosides in the tea were subject to cleavage of the conjugating sugar prior to absorption and excretion as an eriodictyol-O-sulphate.

Kreuz *et al.* (2008) reported urinary excretion of aspalathin in the form of glucuronidated, methyl-glucuronidated, methylated and free aspalathin, as well as dimethyl eriodictyol glucuronide in pigs, following intakes of 15.3 g of aspalathin for 11 days – a daily intake equivalent to *ca.* 400 times that given as a single dose by Stalmach *et al.* (2009b). They reported urinary excretion levels ranging from 0.16% to 0.87% in three pigs after 7 days of treatment, which is in broad agreement with the observed level of human urinary excretion. This suggests that limited absorption occurs regardless of the dose ingested.

The data obtained by Stalmach *et al.* (2009b) indicate that the dihydrochalcone and flavanone C-glucosides in unfermented and fermented rooibos tea are poorly bioavailable, with only trace quantities of metabolites being excreted in urine up to 24 hours after consumption. The absence of metabolites in plasma in detectable quantities also reflects the low bioavailability of these compounds, coupled with a rapid rate of turnover and removal from the circulatory system. Although as yet there are no reports of human feeding studies with ileostomists, most of the rooibos flavonoids most probably pass from the small to the large intestine where, when subjected to the action of the colonic microflora, they undergo cleavage of the C-linked sugar moiety and ring fission producing low-molecular-weight phenolic acids. These catabolites are typically absorbed into the portal vein and, after passing through the body, are excreted in substantial amounts (Del Rio *et al.* 2010b).

3.3.5 Honeybush tea

There are two detailed reports dealing with the ingestion of a honeybush tea extract by pigs at a dose corresponding to 74 mg of mangiferin and 1 mg of hesperetin-7-O-rutinoside per kg body weight per day for a period of 11 days (Bock *et al.* 2008; Bock and Ternes 2010). Norathyriol, the aglycone of mangiferin, was detected in glucuronidase/sulphatase-treated plasma collected after 9 and 11 days of supplementation. Urine contained six metabolites of mangiferin and hesperetin-7-O-rutinoside, including methyl mangiferin, a norathyriol-O-glucuronide and O-glucuronides of both hesperetin and eriodictyol, with the recoveries on days 9 and 11 corresponding to ∼1.5% of the ingested mangiferin and 30% of hesperetin-7-O-rutinoside. The main metabolite detected in faeces was norathyriol in amounts equivalent to ∼8%

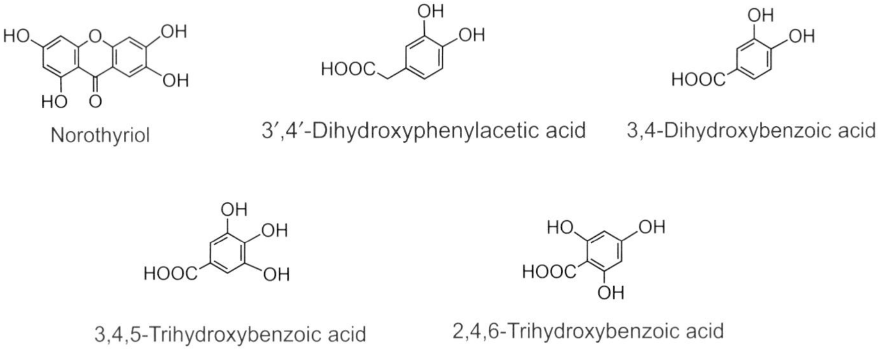

Figure 3.21 Among the metabolites detected in faeces after the ingestion of an extract of *Cyclopia genistoides* (honeybush tea) were norathyriol, 3,4-dihydroxybenzoic acid, 2,4,6-trihydroxybenezoic acid, 3,4,5-trihydroxybenzoic acid and 3′,4′-dihydroxyphenylacetic acid.

of the administered mangiferin (Bock *et al.* 2008). In this context, it is of interest to note that a human intestinal bacterium, *Bacteroides* sp. MANG produces an enzyme that cleaves the *C*-glucosyl bond of mangiferin under anaerobic conditions, yielding norathyriol (Sanugul *et al.* 2005). Trace levels of 3′,4′-dihydroxyphenylacetic acid, 3,4-dihydroxybenzoic acid, 2,4,6-trihydroxybenzoic acid and 3,4,5-trihydroxybenzoic acid (Figure 3.21) were detected in faeces after supplementation but were absent in blank samples (Bock and Ternes 2010).

3.3.6 Hibiscus tea

Frank *et al.* (2005) fed a *H. sabdariffa* extract containing 108 μmol of cyanidin-3-*O*-sambubioside, 137 μmol of delphinidin-3-*O*-sambubioside, and trace amounts of delphinidin-3-*O*-glucoside and cyanidin-3-*O*-glucoside to volunteers. The unmetabolised sambubiosides were detected in plasma with a combined C_{max} of ∼6 nM and a T_{max} of 1.5 hours indicative of absorption in the upper gastrointestinal tract. In keeping with the generally low bioavailability of anthocyanins, and anthocyanin disaccharides in particular (Crozier *et al.* 2009), 0- to 7-hour urinary excretion of the sambubiosides was equivalent to 0.018% of the administered dose.

3.3.7 Fennel tea

Fennel tea, which is appearing in increasing quantities on supermarket shelves, is prepared as an infusion of crushed seeds *F. vulgare* and is characterised by the presence of fenchone, estragole and *trans*-anethole (Zeller and Rychlik (2006). Although fenchone and *trans*-anethole occur in higher concentrations, bioavailability studies have focused on the fate of estragole because of concerns about its carcinogenicity, genotoxicity and hepatotoxicity.

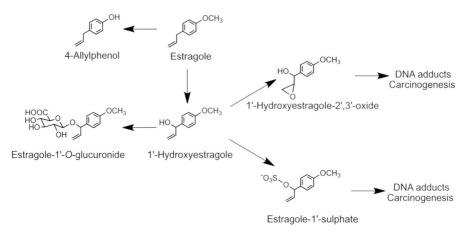

Figure 3.22 Proposed metabolism of estragole in rodents, a component of fennel tea, and the formation of potentially carcinogenic metabolites. (After Iyer et al. 2003.)

Particular attention has centred on the metabolites of estragole as potentially damaging species (Figure 3.22). In rodents, estragole is converted to 4-allylphenol and 1′-hydroxyestragole, and the latter is more carcinogenic than estragole (Drinkwater et al. 1976). The sulphate metabolite of 1′-hydroxyestragole, estragole-1′-O-sulphate forms DNA adducts and is considered to be responsible for the hepatocarcinogenic effects of estragole and 1′-hydroxyestragole (Wiseman et al. 1987). 1′-Hydroxyestragole also forms DNA adducts via the formation of 1′-hydroxyestragole-2′,3′-oxide, but this metabolite is probably not implicated in the carcinogenicity of estragole as it is rapidly degraded by epoxide hydrolases *in vivo* (Luo and Guenthner 1995; Guenthner and Luo 2001). 1′-Hydroxyestragole is also converted to a glucuronide conjugate that is excreted in urine (Figure 3.19; Drinkwater et al. 1976).

In a feed in which seven healthy volunteers consumed 500 mL fennel tea containing 1.3 mg of estragole, the excretion of 1′-hydroxyestragole in urine up to 10 hours varied from not detectable to 0.01% of intake, whereas excretion of its glucuronide conjugate accounted for between 0.17% and 0.41% (Zeller et al. 2006). 1′-Hydroxestragole is glucuronidated *in vitro* by the human liver glucuronyltransferases UGT2B7 and UGT2B15. Examination of 27 human liver samples indicated a considerable variation in activity (CV = 42%). UGT2B7 is polymorphic and some genotypes are likely to be more susceptible to estragole toxicity and this toxicity is likely to be potentiated by other substrates such as certain prescription drugs that may be taken chronically (Iyer et al. 2003).

The safety of herbal teas and other botanicals that in animal studies have shown carcinogenicity and genotoxicity pose particular difficulties with regard to risk assessment and control. The traditional view would be that no safe dose could be set, but the European Food Safety Authority (EFSA) Scientific Committee is of the opinion that where a botanical ingredient contains such substances the margin-of-exposure (MOE) approach can be applied, covering the botanicals under examination and any other dietary sources of exposure (European Food Safety Authority 2005). As explained by

Raffo *et al.* (2011), the MOE uses a toxicological reference point, often taken from an animal study and corresponding to a dose that causes a low, but measurable, response in animals and compares it with various dietary intake estimates in humans, obtained taking into account differences in consumption patterns. It is calculated as the ratio between a defined point on the dose–response curve for the adverse effect and the human intake. Regarding the interpretation of MOE values, the EFSA Scientific Committee is of the view that in general an MOE of 10,000 or higher would be of low concern from a public health point of view and might be considered as a low priority for risk management actions. However, Raffo *et al.* (2011) point out that 'a risk assessment related to the exposure to estragole and limited only to the consumption of fennel tea led to the calculation of MOE values of 34–1000, and to the conclusion that the use of fennel herbal tea is of high priority for risk management' (European Food Safety Authority 2009).

Estragole has been measured in infusions of fennel teas prepared as recommended by the producers. Five samples contained between 9 and 26 μg/L, while two samples yielded 898 and 912 μg/L. Infusions prepared from six samples of commercially available whole fennel seeds contained from 250 to 393 μg/L and two infusions prepared from cracked fennel seeds contained 1478 and 1718 μg/L. Infusions prepared from tea bags contained between 241 and 2058 μg/L and instant teas yield 10–912 μg/L (Raffo *et al.* 2011). On the basis of these data, the authors concluded that an adult consuming three cups of fennel tea per day would have a maximum exposure to estragole of 10 μg/kg body weight per day. The corresponding exposure of infants was up to 51 μg/kg body weight per day for teas from tea bags, and up to 23 μg/kg body weight per day for instant teas. Although 1′-hydroxyestragole and its glucuronide conjugate are rapidly eliminated from adults (Zeller *et al.* 2009), concern remains over the availability of estragole-rich fennel teas specifically designed for feeding to infants as young as 3 months, where the dose on a body weight basis is much higher and liver enzymes involved in detoxification might not be so active.

3.3.8 Other teas

There are no reports on compounds detected in plasma or urine after the consumption of chamomile, *Ficus* and *Anastatica* teas and likewise with maté. However, as maté is a rich source of chlorogenic acids, it is to be anticipated that they will be absorbed in the small and large intestine and metabolised in much the same way as chlorogenic acids in coffee, which have been investigated in some detail using volunteers with and without a functioning colon (Stalmach *et al.* 2009a, 2010b), as described in Chapter 5.

3.4 Summary

The teas and tisanes, some of which are popular beverages across the world while others have a more restricted consumption, contain a diversity of phytochemicals. Flavan-3-ol monomers predominate in green tea and are all but replaced in black tea by high-molecular-weight thearubigins whose complex structural diversity has only recently been appreciated (Kuhnert *et al.* 2010a, 2010b). Yerba maté tea is a rich source of chlorogenic acids, while Itadori tea, compared to other food, contains substantial amounts of the stilbenes *trans*-resveratrol and its 3-*O*-glucoside that are frequently and incorrectly attributed as being the compounds responsible for the protective effects of red wine (Crozier *et al.* 2010). Rooibos and honeybush teas from South Africa contain their own unusual spectrum of flavonoids and related compounds as do hibiscus, fennel, *Anastatica* and *Ficus* teas. Where data are available, the bioavailabilities of the principal components in these teas are discussed, with some compounds being absorbed in the small intestine and others entering the circulation system from the colon (Crozier *et al.* 2009). The parent compounds rarely enter the bloodstream without being converted to sulphate, glucuronide and/or methylated metabolites. Compounds passing from the small to the large intestine are also subject to degradation to phenolic acids as a result of the action of the colonic microflora. As a consequence it is clear that the potentially protective effects of dietary (poly)phenols are due to the action of their metabolites and catabolites rather than the original parent compounds that were ingested. A number of these compounds have already been shown to have anti-glycative and neuroprotective activity *in vitro* at low mM concentrations (Verzelloni *et al.* 2011).

References

AlGamdi, N., Mullen, W. and Crozier, A. (2011) Tea prepared from *Anastatica hirerochuntica* seeds contains a diversity of antioxidant flavonoids, chlorogenic acids and phenolic compounds. *Phytochemistry*, **72**, 248–254.

Andres-Lacueva, C., Urpi-Sarda, M., Zamora-Ros, R. *et al.* (2010) Bioavailability and metabolism of resveratrol. In C.G. Fraga (ed), *Plant Phenolics and Human Health*. John Wiley & Sons, Inc., Hoboken, NJ, pp. 265–297.

Astill, C., Birch, M.R., Dacombe, C. *et al.* (2001). Factors affecting the caffeine and polyphenol contents of black and green tea infusions. *J. Agric. Food Chem.*, **49**, 5340–5347.

Auger, C., Hara, Y. and Crozier, A. (2008) Bioavailability of Polyphenon E flavan-3-ols in humans with an ileostomy. *J. Nutr.*, **138**, 1535S–1542S.

Bilia, R.A., Fumarola, M., Gallori, S. *et al.* (2000) Identification by HPLC-DAD and HPLC-M analyses and quantification of constituents of fennel teas and decoctions. *J. Agric. Food Chem.*, **48**, 4734–4738.

Bock, C. and Ternes, W. (2010) The phenolic acids from bacterial degradation of mangiferin aglycone and quantified in feces of pigs after oral ingestion of an extract of *Cyclopia genistoides* (honeybush tea). *Nutr. Res.*, **30**, 348–357.

Bock, C., Waldmann, K.-H. and Ternes, W. (2008) Mangiferin and hesperidin metabolites are absorbed from the gastrointestinal tract of pigs after oral ingestion of a *Cyclopia genistoides* (honeybush tea) extract. *Nutr. Res.*, **28**, 879–891.

Bond, T.A., Lewis, J.R., Davis, A. *et al.* (2003). Analysis and purification of catechins and their transformation products. In C. Santos-Buelga and G. Williamson (eds), *Methods in Polyphenol Analysis*. Royal Society of Chemistry, Cambridge, pp. 229–266.

Boocock, D.J., Faust, G.E.S., Patel, K.R. *et al.* (2007) Phase I dose escalation pharmacokinetic study in health volunteers of resveratrol, a potential cancer chemopreventive agent. *Cancer Epidemiol. Biomarkers Prev.*, **16**, 1246–1252.

Bridges, J.W., French, M.R., Smith, R.L. *et al.* (1970) The fate of benzoic acid in various species. *Biochem. J.*, **118**, 47–51.

Brown, A.G., Eyton, W.B., Holmes, A. *et al.* (1969) Identification of the thearubigins as polymeric proanthocyanidins. *Nature*, **221**, 742–744.

Burns, J., Yokota, T., Ashihara, H. *et al.* (2002) Plant foods and herbal sources of resveratrol. *J. Agric. Food Chem.*, **50**, 3337–3340.

Carini, M., Facino, R.M., Aldini, G. *et al.* (1998) Characterization of phenolic antioxidants from maté (*Ilex paraguayensis*) by liquid chromatography/mass spectrometry and liquid chromatography/tandem mass spectrometry. *Rapid Commun. Mass Spectrom.*, **12**, 1813–1819.

Castelletto, R., Castellsague, X., Munoz, N. *et al.* (1994) Alcohol, tobacco, diet, maté drinking, and oesophageal cancer in Argentina. *Cancer Epidemiol. Biomarkers Prev.*, **3**, 557–564.

Chen, H.-F., Tanaka, T., Nonaka, G.-I. *et al.* (1993) Phenylpropanoid-substituted catechins from *Castanopsis hystrix* and structure revision of cinchonains. *Phytochemistry*, **33**, 183–187.

Chen, D., Wang, C.Y., Lambert, D.J. *et al.* (2005) Inhibition of human liver catechol-*O*-methyltransferase by tea catechins and their metabolites: structure–activity relationship and molecular-modeling studies. *Biochem. Pharmacol.*, **69**, 1523–1531.

Chiu, F.L. and Lin, J.K. (2005) HPLC analysis of naturally occurring methylated catechins, 3″- and 4″-methyl-epigallocatechin gallate, in various fresh tea leaves and commercial teas and their potent inhibitory effects on inducible nitric oxide synthase in macrophages. *J. Agric. Food Chem.*, **53**, 7035–7042.

Chow, H.H., Cai, Y., Alberts, D.S. *et al.* (2001) Phase I pharmacokinetic study of tea polyphenols following single-dose administration of epigallocatechin gallate and Polyphenon E. *Cancer Epidemiol. Biomarkers Prev.*, **10**, 53–58.

Clifford, M.N. (2004) Diet-derived phenols in plasma and tissues and their implications for health. *Planta Med.*, **12**, 1103–1114.

Clifford, M.N., Brown, J.E. (2006) Dietary flavonoids and health – broadening the perspective. In O. Andersen and K.R. Markham (eds), *Flavonoids: Chemistry, Biochemistry and Applications*. CRC Press, Boca Raton, pp. 320–370.

Clifford, M.N. and Ramìrez-Martìnez, J.R. (1990) Chlorogenic acids and purine alkaloid content of maté (*Ilex paraguariensis*) leaf and beverage. *Food Chem.*, **35**, 13–21.

Clifford, M.N., Copeland, E.L., Bloxsidge, J.P. *et al.* (2000) Hippuric acid as a major excretion product associated with black tea consumption. *Xenobiotica*, **30**, 317–326.

Coggon, P., Moss, G.A., Graham, H.N. *et al.* (1973) The biochemistry of tea fermentation: oxidative degallation and epimerisation of the tea flavanol gallates. *J. Agric. Food Chem.*, **21**, 727–733.

Cosge, B., Kiralan, M. and Gurbuz, B. (2008) Characteristics of fatty acids and essential oil from sweet fennel (*Foeniculum vulgare* Mill. var. dulce) and bitter fennel fruits (*F. vulgare* Mill. var. vulgare) growing in Turkey. *Natl. Prod. Res.*, **22**, 1011–1016.

Crozier, A., Jaganath, I.B. and Clifford, M.N. (2009) Dietary phenolics: chemistry, bioavailability and effects on health. *Natl. Prod. Rep.*, **26**, 1001–1043.

Crozier, A., Borges, G. and Ryan, D. (2010) The glass that cheers: phenolic and polyphenolic constituents and the beneficial effects of moderate red wine consumption. *The Biochemist*, **32**, 4–9.

Davis, A.L., Lewis, J.R., Cai, Y. *et al.* (1997) A polyphenolic pigment from black tea. *Phytochemistry*, **46**, 1397–1402.

Day, A.J., Canada, F.J., Diaz, J.C. *et al.* (2000) Dietary flavonoid and isoflavone glycosides are hydrolysed by the lactase site of lactase phlorizin hydrolase. *FEBS Lett.*, **468**, 166–170.

Daykin, C.A., Van Duynhoven, J.P., Groenewegen, A. *et al.* (2005) Nuclear magnetic resonance spectroscopic based studies of the metabolism of black tea polyphenols in humans. *J Agric. Food Chem.*, **53**, 1428–1434.

de Vries, J.H., Hollman, P.C., Meyboom, S. *et al.* (1998) Plasma concentrations and urinary excretion of the antioxidant flavonols quercetin and kaempferol as biomarkers for dietary intake. *Am. J. Clin. Nutr.*, **68**, 60–65.

Del Rio, D., Stewart, A.J., Mullen, W. *et al.* (2004) HPLC-MS^n analysis of phenolic compounds and purine alkaloids in green and black tea. *J. Agric. Food Chem.*, **52**, 2807–2815.

Del Rio, D., Calani, L., Cordero, C. *et al.* (2010a) Bioavailability and catabolism of green tea flavan-3-ols in humans. *Nutrition*, **11/12**, 1110–1116.

Del Rio, D., Costa, L.G., Lean, M.E.J. *et al.* (2010b) Polyphenols and health: what compounds are involved. *Nutr. Metabol. Cardiovasc. Disease*, **20**, 1–6.

Dewick, P.M. (2009) *Medicinal Natural Products. A Biosynthetic Approach*, 3rd edition. John Wiley & Sons, Ltd., Chichester.

Dirks, U. and Herrmann, K. (1984) 4-(β-D-lucopyranosyloxy)benzoic acid, a characteristic phenolic constituent of the Apiaceae. *Phytochemistry*, **23**, 1811–1812.

Donovan, J.L., Crespy, V., Oliveira, M. *et al.* (2006a) (+)-Catechin is more bioavailable then (−)-catechin: relevance to the bioavailability of catechin from cocoa. *Free Rad. Res.*, **40**, 1029–1034.

Donovan, J.L., Manach, C., Faulks, R.M. *et al.* (2006b) Absorption and metabolism of dietary secondary metabolites. In A. Crozier, M.N. Clifford and H. Ashihara (eds), *Plant Secondary Metabolites. Occurrence, Structure and Role in the Human Diet.* Blackwell Publishing Ltd., Oxford, pp. 303–351.

Drinkwater, N.R., Miller, E.C., Miller, J.A. *et al.* (1976) Hepatocarcinogenicity of estrgole (1-allyl-4mthoxybenzene) and 1′-hydroxyestragole in the mouse and mutagenicity of 1′-acetoxyestragole in bacteria. *J. Natl. Cancer Inst.*, **57**, 1323–1331.

Drynan, J.W., Clifford, M.N., Obuchowicz, J. *et al.* (2010) The chemistry of low molecular weight black tea polyphenols. *Natl. Prod. Rep.*, **27**, 417–462.

Engelhardt, U.H., Finger, A., Herzig, B. *et al.* (1992), Determination of flavonol glycosides in black tea. *Deusche Lebensmittel Rundschau*, **88**, 69–73.

European Food Safety Authority (2005) Opinion of the scientific committee on a request from EFSA related to a harmonised approach for risk assessment of substances which are both genotoxic and carcinogenic. *EFSA J.*, **282**, 1–31.

European Food Safety Authority (2009) ESCO working group on botanicals and botanical preparations. Advice on the EFSA guidance document for the safety assessment of botanicals and botanical preparations intended for use as food supplements, based on real case studies. *EFSA J.*, **7**, 280.

Finger, A., Engelhardt, U.H. and Wray, V. (1991) Flavonol triglycosides containing galactose in tea. *Phytochemistry*, **30**, 2057–2060.

Folch, C. (2010) Stimulating consumption: yerba maté myths, markets and meanings from conquest to present. *Comp. Studies Soc. Hist.*, **52**, 6–36.

Frank, T., Janßen, M., Netzel, M. *et al.* (2005) Pharmacokinetics of anthocyanin-3-glycoside following consumption of *Hibiscus sabdariffa* L. extract. *J. Clin. Pharm.*, **4**, 203–210.

Gee, J.M., DuPont, S.M., Day, A.J. *et al.* (2000) Intestinal transport of quercetin glycosides in rats involves both deglycosylation and interaction with the hexose transport pathway. *J. Nutr.*, **130**, 2765–2771.

Goodsall, C.W., Parry, A.D. and Safford, D.S. (1996) Investigations into the mechanism of theaflavin formation during black tea manufacture. In J. Vercauteren *et al.* (eds), *Polyphenols 96. XVIIIe Journeés Internationales Groupe Polyphénols*, Vol. 2, INRA editions. Paris, pp. 287–288.

Gosmann, G., Guillaume, D., Taketa, A.T. *et al.* (1995) Triterpenoid saponins from *Ilex paraguariensis*. *J. Natl. Prod.*, **58**, 438–441.

Graham, H.N. (1992) Green tea composition, consumption and polyphenol chemistry. *Prev. Med.*, **21**, 334–350.

Green, R.J., Murphy, A.S., Schulz, B. et al. (2007) Common tea formulations modulate *in vitro* digestive recovery of green tea catechins. *Mol. Nutr. Food Res.*, **51**, 1152–1162.

Grumer, H.D. (1961) Formation of hippuric acid from phenylalanine labelled with carbon-14 in phenylketonuric subjects. *Nature*, **189**, 63–64.

Gu, L., Laly, M., Chang, H.C. et al. (2005) Isoflavone conjugates are underestimated in tissues using enzymatic hydrolysis. *J. Agric. Food Chem.*, **53**, 6858–6863.

Guenthner, T.M. and Luo, G. (2001) Investigation of the role of the 2′,3′-epoxidation pathway in the bioactivation and genotoxicity of dietary allylbenzene analogs. *Toxicology*, **160**, 47–58.

Hakiman, M. and Maziah, M. (2009) Non-enzymatic and enzymatic antioxidant activities in aqueous extract of different *Ficus deltoidea accessions*. *J. Med. Plant Res.*, **3**, 120–131.

Hampton, M.G. (1992) Production of black tea. In K.C. Willson and M.N. Clifford (eds), *Tea: Cultivation to Consumption*. Chapman and Hall, London, pp. 459–512.

Han, L., Hatano, T., Yoshida, T. et al. (1994) Tannins of theaceous plants. V. Camelliatannins F, G and H, three new tannins from *Camellia japonica* L. *Chem. Pharm. Bull.*, **42**, 1399–1409.

Hashimoto, F., Nonaka, G.-I. and Nishioka, I. (1986) Tannins and related compounds XXXVI. Isolation and structures of theaflagallins, new red pigments from black tea. *Chem. Pharm. Bull.*, **34**, 61–65.

Hashimoto, F., Nonaka, G.-I. and Nishioka, I. (1987) Tannins and related compounds LVI. Isolation of four new acylated flavan-3-ols from oolong tea. *Chem. Pharm. Bull.*, **35**, 611–616.

Hashimoto, F., Nonaka, G.-I. and Nishioka, I. (1988) Tannins and related compounds LXIX. Isolation and structure elucidation of B,B′-linked bisflavanoids, theasinensins D–G and oolongtheanin from oolong tea. *Chem. Pharm. Bull.*, **36**, 1676–1684.

Hashimoto, F., Nonaka, G.-I. and Nishioka, I. (1989a) Tannins and related compounds LXXVII. Novel chalcan–flavan dimers, assamaicins A, B and C, and a new flavan-3-ol and proanthocyanidins from the fresh leaves of *Camellia sinensis* L. var. assamica Kitamura. *Chem. Pharm. Bull.*, **37**, 77–85.

Hashimoto, F., Nonaka, G.-I. and Nishioka, I. (1989b) Tannins and related compounds XC. 8C-ascorbyl-(−)-epigallocatechin-3-gallate and novel dimeric flavan-3-ols, oolonghomobisflavans A and B from oolong tea. *Chem. Pharm. Bull.*, **37**, 3255–3263.

Hashimoto, F., Nonaka, G.-I. and Nishioka, I. (1992) Tannins and related compounds. CXIV. Structures of novel fermentation products, theogallinin, theaflavonin and desgalloyltheaflavonin from black tea, and changes of tea leaf polyphenols during fermentation. *Chem. Pharm. Bull.*, **40**, 1383–1389.

Hatano, T., Shida, S., Han, L. et al. (1991) Tannins of Theaceous plants: III. Camelliatannins A and B, two new complex tannins from *Camellia japonica* L. *Chem. Phar. Bull.*, **39**, 876–880.

Henning, S.M., Niu, Y., Lee, N.H. et al. (2004) Bioavailability and antioxidant activity of tea flavanols after consumption of green tea, black tea, or a green tea extract supplement. *Am. J. Clin. Nutr.*, **80**, 1558–1564.

Henning, S.M., Niu, Y., Liu, Y. et al. (2005) Bioavailability and antioxidant effect of epigallocatechin gallate administered in purified form versus as green tea extract in healthy individuals. *J. Nutr. Biochem.*, **16**, 610–616.

Hodgson, J.M., Morton, L.W., Puddey, I.B. et al. (2000) Gallic acid metabolites are markers of black tea intake in humans. *J. Agric. Food Chem.*, **48**, 2276–2280.

Hodgson, J.M., Chan, S.Y., Puddey, I.B. et al. (2004) Phenolic acid metabolites as biomarkers for tea- and coffee-derived polyphenol exposure in human subjects. *Br. J. Nutr.*, **91**, 301–306.

Hollman, P.C.H., van het Hof, K., Tijburg, L. et al. (2001) Addition of milk does not affect the absorption of flavonols from tea in man. *Free Radic. Res.*, **34**, 297–300.

Hursel, R., Viechtbauer, W. and Westerterp-Plantenga M.S. (2009) The effects of green tea on weight loss and weight maintenance: a meta-analysis. *Int. J. Obes.*, **33**, 956–961.

IARC (1991) IARC monographs on the evaluation of carcinogenic risks to humans. *Coffee, Tea, Maté, Methylxanthines and Glyoxal*, Vol. 51. IARC, Lyon.

Ito, R., Yamamoto, A., Kodama, S. et al. (2003) A study on the change of enantiomeric purity of catechins in green tea infusion. *Food Chem.*, **83**, 563–568.

Iyer, L.V., Ho, M.N., Shinn, W.M. et al. (2003) Glucuronidation of 1′-hydroxyestragole (1′-HE) by human UDP-glucuronosyltransferases UGT2B7 and UGT1A9. *Toxicol. Sci.*, **73**, 36–43.

Jaganath, I.B., Mullen, W., Edwards, C.A. et al. (2006) The relative contribution of the small and large intestine to the absorption and metabolism of rutin in man. *Free Rad. Res.*, **40**, 1035–1046.

Jaiswal, R., Sovdat, T., Vivan, F. et al. (2010) Profiling and characterization by LC-MS^n of the chlorogenic acids and hydroxyycinnamoylshikimate esters in maté (*Ilex paraguariensis*). *J. Agric. Food Chem.*, **58**, 5471–5484.

Jöbstl, E., Fairclough, J.P.A., Davies, A.P. et al. (2005) Creaming in black tea. *J. Agric. Food Chem.*, **53**, 7997–8002.

Johnston, K., Sharp, P., Clifford, M.N. et al. (2005) Dietary polyphenols decrease glucose uptake by human intestinal Caco-2 cells. *FEBS Lett.* **579**, 1653–1657.

Joubert, E., Gelderblom, W.C.A., Louw, A. et al. (2008) South African herbal teas: *Aspalathus linearis, Cyclopia* spp. and *Athrixia phylicoides* – a review. *J. Ethnopharmacol.*, **119**, 376–412.

Joubert, E., Manley, M., Maicu, C. et al. (2010) Effect of pre-drying treatments and storage on color and phenolic composition of green honeybush (*Cyclopia subternata*) herbal tea. *J. Agric. Food Chem.*, **58**, 338–344.

Kamara, B.I., Brand, D.J, Brand E.V. et al. (2004) Phenolic metabolites from honeybust tea (Cyclopia subternata). *J. Agric. Food Chem.*, **52**, 5391–5395.

Kaundun, S.S. and Matsumoto, S. (2002) Heterologous nuclear and chloroplast microsatellite amplification and variation in tea, *Camellia sinensis*. *Genome*, **45**, 1041–1048.

Kida, T., Suzuki, Y., Matsumoto, N. et al. (2000) Identification of biliary metabolites of (−)-epigallocatechin gallate in rats. *J. Agric. Food Chem.*, **48**, 4151–4155.

Knight, S. (2004) *Metabolism of Dietary Polyphenols by Gut Flora.* Ph.D. University of Surrey, UK.

Kohri, T., Nanjo, F., Suziki, M. et al. (2001) Synthesis of (−)-[4–^{3}H]epigallocatechin gallate and its metabolic fate in rats after intravenous administration. *J. Agric. Food Chem.*, **49**, 1042–1048.

Kottra, G. and Daniel, H. (2007) Flavonoid glycosides are not transported by the human Na+/glucose transporter when expressed in Xenopus laevis oocytes, but effectively inhibit electrogenic glucose uptake. *J. Pharmacol. Exp.Ther.*, **322**, 829–835.

Krafczyk, N. and Glomb, M.A. (2008) Characterization of phenolic compounds in rooibos tea. *J. Agric. Food Chem.*, **56**, 3368–3376.

Krafczyk, N., Heinrich, T., Porzel, A. et al. (2009) Oxidation of the dihydrochalcone aspalathin leads to dimerization. *J. Agric. Food Chem.*, **57**, 6838–6843.

Kreuz, S., Joubert, E., Waldmann, K.H. et al. (2008) Aspalathin, a flavonoid in *Aspalathus linearis* (rooibos), is absorbed by pig intestine as a C-glycoside. *Nutr. Res.*, **28**, 690–701.

Krizman, M., Baricevic, D. and Prosek, M. (2007) Determination of phenolic compounds in fennel by HPLC and HPLC-MS using a monolithic reversed-phase column. *J. Pharm. Biomed. Anal.*, **43**, 481–485.

Kuhnert, N., Clifford, M.N. and Muller, A. (2010a) Oxidative cascade reactions yielding polyhydroxy-theaflavins and theacitrins in the formation of black tea thearubigins: evidence by tandem LC-MS. *Food and Funct.*, **1**, 180–199.

Kuhnert, N., Drynan, J.W., Obuchowicz, J. *et al.* (2010b) Mass spectrometric characterization of black tea thearubigins leading to an oxidative cascade hypothesis for thearubigin formation. *Rapid Commun. Mass Spectrom.*, **24**, 3387–3404.

Kyle, J.A., Morrice, P.C., McNeill, G. *et al.* (2007) Effects of infusion time and addition of milk on content and absorption of polyphenols from black tea. *J. Agric. Food Chem.*, **55**, 4889–4894.

Lakenbrink, C., Engelhardt, U.H. and Wray, V. (1999) Identification of two novel proanthocyanidins in green tea. *J. Agric. Food Chem.*, **47**, 4621–4624.

Lakenbrink, C., Lam, T.M.L., Engelhardt, U.H. *et al.* (2000a) New flavonol triglycosides from tea (*Camellia sinensis*). *Natl. Prod. Lett.*, **14**, 233–238.

Lakenbrink, C., Lapczynski, S., Maiwald, B. *et al.* (2000b) Flavonoids and other polyphenols in consumer brews of tea and other caffeinated beverages. *J. Agric. Food Chem.*, **48**, 2848–2852.

Li, C., Lee, M.J., Sheng, S. *et al.* (2000) Structural identification of two metabolites of catechins and their kinetics in human urine and blood after tea ingestion. *Chem. Res. Toxicol.*, **3**, 177–184.

Lin, J.-K., Lin, C.-L., Liang, Y.-C. *et al.* (1998) Survey of catechins, gallic acid, and methylxanthines in green, oolong, pu'erh, and black teas. *J. Agric. Food Chem.*, **46**, 3635–3642.

Lonac, M.C., Richards, J.C., Schweder, M.M. *et al.* (2011) Influence of short-term consumption of the caffeine-free, epigallocatechin-3-gallate supplement, Teavigo, on resting metabolism and the thermic effect of feeding. *Obesity (Silver.Spring)*, **19**, 298–304.

Luo, G. and Guenthner T.M. (1995) Metabolism of allylbenene 2′,3′-oxide and estragole 2′,3′-oxide in the isolated perfused rat liver. *J. Pharmacol. Exp. Ther.*, **272**, 588–596.

Luximon-Ramma, A., Bahorun, T., Crozier, A. *et al.* (2005) Characterization of the antioxidant functions of flavonoids and proanthocyanidins in Mauritian black teas. *Food Res. Int.*, **38**, 357–367.

Manach, C., Williamson, G., Morand, C. *et al.* (2005) Bioavailability and bioefficacy of polyphenols in humans. I. Review of 97 bioavailability studies. *Am. J. Clin. Nutr.*, **81**, 230S–242S.

McCall Smith, A. (1999) *The No.1 Ladies' Detective Agency*. Polygon, Edinburgh.

McCall Smith, A. (2000) *Tears of the Giraffe*. Polygon, Edinburgh.

McCall Smith, A. (2009) *Tea Time for the Traditionally Built*. Polygon, Edinburgh.

McKay, D.L. and Blumberg, J.B. (2007) A review of the bioactivity of South African herbal teas: rooibos (*Aspalathus linearis*) and honeybush (*Cyclopia intermedia*). *Phytother. Res.*, **21**, 1–16.

Menet, M.C., Sang, S., Yang, C.S. *et al.* (2004) Analysis of theaflavins and thearubigins from black tea extract by MALDI-TOF mass spectrometry. *J. Agric. Food Chem.*, **52**, 2455–2461.

Meng, X., Sang, S., Zhu, N. *et al.* (2002) Identification and characterization of methylated and ring-fission metabolites of tea catechins formed in humans, mice, and rats. *Chem. Res. Toxicol.*, **15**, 1042–1050.

Miguel, M.G., Cruz, C., Faleiro, L. *et al.* (2010) *Foeniculum vulgare* essential oils: chemical composition, antioxidant and antimicrobial activities. *Natl. Prod. Commun.*, **5**, 319–328.

Morton, J. F. (1989) Tannins as a carcinogen in bush tea, tea, maté and khat. In J.J. Hemmingway and J.J. Karchesy (eds), *Chemistry and Significance of Condensed Tannins*. Plenum Press, New York, pp. 403–415.

Mulder, T.P., van Platerink, C.J., Wijnand Schuyl, P.J. *et al.* (2001) Analysis of theaflavins in biological fluids using liquid chromatography–electrospray mass spectrometry. *J. Chromatogr. B Biomed. Sci. Appl.*, **760**, 271–279.

Mulder, T.P., Rietveld, A.G. and van Amelsvoort, J.M. (2005) Consumption of both black tea and green tea results in an increase in the excretion of hippuric acid into urine. *Am. J. Clin. Nutr.*, **81**, 256S–260S.

Mullen, W., Edwards, C.A. and Crozier, A. (2006) Absorption, excretion and metabolic profiling of methyl-, glucuronyl-, glucosyl and sulpho-conjugates of quercetin in human plasma and urine after ingestion of onions. *Br. J. Nutr.*, **96**, 107–116.

Mullen, W., Edwards, C.A. Serafini, M. *et al.* (2008a). Bioavailability of pelargonidin-3-glucoside and its metabolites in humans following the ingestion of strawberries with and without cream. *J. Agric. Food Chem.*, **56**, 713–719.

Mullen, W., Archeveque, M.-A., Edwards, C.A. *et al.* (2008b) Bioavailability and metabolism of orange juice flavanones in humans: impact of a full fat yogurt. *J. Agric. Food Chem.*, **56**, 11157–11164.

Mullen, W., Borges, G., Donovan, J.L. *et al.* (2009) Milk decreases urinary excretion but not plasma pharmacokinetics of cocoa flavan-3-ol metabolites in humans. *Am. J. Clin. Nutr.*, **89**, 1784–1791.

Nakashima, S., Matsuda, H., Oda, Y. *et al.* (2010) Melanogenesis inhibitors from the desert plant Anastatica hierochuntica in B16 melanoma cells. *Bioorg. Med. Chem.*, **18**, 2337–2345.

Nonaka, G.-I. and Nishioka, I. (1982) Tannins and related compounds. VII. Phenylpropanoid-substituted epicatechins, cinchonains from *Cinchona succirubra*. *Chem. Pharm. Bull.*, **30**, 4268–4276.

Nonaka, G.-I., Kawahara, O. and Nishioka, I. (1982) Tannins and related compounds. VIII. A new type of proanthocyanidin, cinchonains IIa and IIb from Cinchona succirubra. *Chem. Pharm. Bull.*, **30**, 4277–4282.

Nonaka, G.-I., Kawahara, O. and Nishioka, I. (1983) Tannins and related compounds. XV. A new class of dimeric flavan-3-ol gallates, theasinensins A and B, and proanthocyanidin gallates from green tea. *Chem. Pharm. Bull.*, **31**, 3906–3914.

Nonaka, G.-I., Sakai, R. and Nishioka, I. (1984) Hydrolysable tannins and proanthocyanidins from green tea. *Phytochemistry*, **23**, 1753–1755.

Nováková, L., Vildová, A., Mateus, J.P. *et al.* (2010) Development and application of UHPLC-MS/MS method for the determination of phenolic compounds in Chamomile flowers and Chamomile tea extracts. *Talanta*, **82**, 1271–1280.

Obuchowicz, J., Engelhardt, U.H. and Donnelly K. (2011) Flavanol database for green and black teas utilising ISO 14502-1 and ISO 14502-2 as analytical tools. *J. Food Comp. Anal.*, **24**, 411–417.

Omar, M.H., Mullen, W. and Crozier, A. (2011) Identification of proanthocyanidin dimers and trimers, flavones C-glycosides and antioxidants in Ficus deltoidea, a Malaysian herbal tea. *J. Agric. Food Chem.*, **54**, 1363–1369.

Opie, S.C., Clifford M.N. and Robertson, A. (1993) The role of (–)-epicatechin and polyphenol oxidase in the coupled oxidative breakdown of theaflavins. *J. Sci. Food Agric.*, **63**, 435–438.

Opie, S.C., Clifford M.N. and Robertson, A. (1995) The formation of thearubigin-like substances by in vitro polyphenol oxidase-mediated fermentation of individual flavan-3-ols. *J. Sci. Food Agric.*, **67**, 501–505.

Parejo, I., Jauregui, O., Sanchez-Rabaneda, F. *et al.* (2004) Separation and characterization of phenolic compounds in fennel (*Foeniculum vulgare*) using liquid chromatography-negative electrospray ionization tandem mass spectrometry. *J. Agric. Food Chem.*, **52**, 3679–3687.

Patel, K.R., Brown, V.A., Jones, D.J.L. *et al.* (2010) Clinical pharmacology of resveratrol and its metabolites in colorectal cancer patients. *Cancer Res.*, **70**, 7392–7399.

Price, K.R., Rhodes, M.J.C. and Barnes, K.A. (1998) Flavonol glycoside content and composition of tea infusions made from commercially available teas and tea products. *J. Agric. Food Chem.*, **46**, 2517–2522.

Raffo, A., Nicoli, S. and Leclercq, C. (2011) Quantification of estragole in fennel herbal teas: implications on the assessment of dietary exposure to estragole. *Food Chem. Toxicol.*, **49**, 370–375.

Record I.R. and Lane J.M. (2001) Simulated intestinal digestion of green and black teas. *Food Chem.*, **73**, 481–486.

Reeve, D.R. and Crozier, A. (1980) Quantitative analysis of plant hormones. In J. MacMillan (ed), *Hormonal Regulation of Development 1. Molecular Aspects of Plant Hormones.* Encyclopedia of Plant Physiology New Series, Vol. 9, Springer-Verlag, Heidelberg, pp. 203–280.

Rodríguez-Medina, I.C., Beltrán-Debón, R., Molina, V.M. *et al.* (2009) Direct characterization of aqueous extract of Hibiscus sabdariffa using HPLC with diode array detection coupled to ESI and ion trap MS. *J. Sep. Sci.*, **32**, 3441–3448.

Rolon, P.A., Castellsague, X., Benz, M. *et al.* (1995) Hot and cold maté drinking and esophageal cancer in Paraguay. *Cancer Epidemiol. Biomarkers Prev.*, **4**, 595–605.

Roowi, S., Mullen, W., Edwards, C.A. *et al.* (2009) Yoghurt impacts on the excretion of phenolic acids derived from colonic breakdown of orange juice flavanones in humans. *Mol. Nutr. Food Res.*, **53**, S44–S53.

Roowi, S., Stalmach, A., Mullen, W. *et al.* (2010) Green tea flavan-3-ols: colonic degradation and urinary excretion of catabolites by humans. *J. Agric. Food Chem.*, **58**, 1296–1304.

Sakamoto, Y. (1967) Flavones in green tea. Part I. Isolation and structure of flavones occurring in green tea infusion. *Agric. Biol. Chem.*, **31**, 1029–1034.

Sakamoto, Y. (1969) Flavones in green tea. Part II. Identification of isovitexin and saponarin. *Agric. Biol. Chem.*, **33**, 959–961.

Sakamoto, Y. (1970) Flavones in green tea. Part III. Structure of pigments IIIa and IIIb. *Agric. Biol. Chem.*, **34**, 919–925.

Sang, S., Tian, W., Meng, X. *et al.* (2002) Theadibenztropolone A, a new type pigment from enzymatic oxidation of (−)-epicatechin and (−)-epigallocatechin gallate and characterized from black tea using LC/MS/MS. *Tetrahedron. Lett.*, **43**, 7129–7133.

Sang, S., Tian, S., Stark, R.E. *et al.* (2004) New dibenzotropolone derivatives characterized from black tea using LC/MS/MS. *Bioorg. Med. Chem.*, **12**, 3009–3017.

Sang, S., Lee, M.J., Yang, I. *et al.* (2008) Human urinary metabolite profile of tea polyphenols analyzed by liquid chromatography/electrospray ionization tandem mass spectrometry with data-dependent acquisition. *Rapid Comm. Mass Spectrom.*, **22**, 1567–1578.

Sanugul, K., Akao, T., Li, Y. *et al.* (2005) Isolation of human intestinal bacterium that transforms mangiferin to norathyriol and inducibility of the enzyme that cleaves a C-glucosyl bond. *Biol. Pharm. Bull.*, **28**, 1672–1678.

Segura-Carretero, A., Puertas-Mejía, M.A., Cotracero-Ramírez, S. *et al.* (2008) Selective extraction, separation, and identification of anthocyanins from *Hibiscus sabdariffa* L. using solid phase extraction-capillary electrophoresis-mass spectrometry (time-of-flight/ion trap). *Electrophoresis*, **29**, 2852–2861.

Self, H.L., Brown, R.R. and Price, J.M. (1960) Quantitative studies on the metabolites of tryptophan in the urine of swine. *J. Nutr.*, **70**, 21–25.

Serafini, M., Ghiselli, A. and Ferro-Luzzi, A. (1996) In vivo antioxidant effect of green and black tea in man. *Eur. J. Clin. Nutr.*, **50**, 28–32.

Sewram, V., De Stefani, E., Brennan, P. *et al.* (2003) Maté consumption and the risk of squamous cell esophageal cancer in Uruguay. *Cancer Epidemiol. Biomarkers Prev.*, **2**, 508–513.

Shahrzad, S and Bitsch, I. (1998) Determination of gallic acid and its metabolites in human plasma and urine by high-performance liquid chromatography. *J. Chromatogr. B*, **705**, 87–95.

Shahrzad, S., Aoyagi, K., Winter, A. *et al.* (2001) Pharmacokinetics of gallic acid and its relative bioavailability from tea in healthy humans. *J. Nutr.*, **131**, 1207–1210.

Shao, W., Powell, C. and Clifford M.N. (1995) The analysis by HPLC of green, black and Pu'er teas produced in Yunnan. *J. Sci. Food Agric.*, **69**, 535–540.

Spiro, M. and Jaganyi, D. (1993) What causes scum on tea. *Nature*, **364**, 581.

Stalmach, A., Mullen, W., Barron, D. *et al.* (2009a) Metabolite profiling of hydroxycinnamate derivatives in plasma and urine following the ingestion of coffee by humans: identification of biomarkers of coffee consumption. *Drug Metab. Disp.*, **37**, 1759–1768.

Stalmach, A., Mullen, W., Pecorari, M. *et al.* (2009b) Bioavailability of C-linked dihydrochalcone and flavanone glucosides in humans following ingestion of unfermented and fermented rooibos teas. *J. Agric. Food Chem.*, **57**, 7104–7111.

Stalmach, A., Troufflard, S., Serafini M. *et al.* (2009c) Absorption, metabolism and excretion of Choladi green tea flavan-3-ols by humans. *Mol. Nutr. Food Res.*, **53**, S44–S53.

Stalmach, A., Mullen, W., Steiling, H. *et al.* (2010a) Absorption, metabolism, efflux and excretion of green tea flavan-3-ols in humans with an ileostomy. *Mol. Nutr. Food Res.*, **54**, 323–334.

Stalmach, A., Steiling, H., Williamson, G. *et al.* (2010b) Bioavailability of chlorogenic acids following acute ingestion of coffee by humans with an ileostomy. *Arch. Biochem. Biophys.*, **501**, 98–105.

Stendell-Hollis, N.R., Thomson, C.A., Thompson, P.A. *et al.* (2010) Green tea improves metabolic biomarkers, not weight or body composition: a pilot study in overweight breast cancer survivors. *J. Hum. Nutr. Diet.*, **23**, 590–600.

Stewart, A.J., Mullen, W. and Crozier, A. (2005) On-line high-performance liquid chromatography analysis of the antioxidant activity of phenolic compounds in green and black tea. *Mol. Nutr. Food Res.*, **49**. 52–60.

Stoupi, S. (2010) *In Vitro and In Vivo Metabolic Studies of Dietary Flavan-3-ols*. Ph.D. University of Surrey, UK.

Stoupi, S., Williamson, G., Drynan, J.W. *et al.* (2010a) Procyanidin B2 catabolism by human fecal microflora: partial characterization of 'dimeric' intermediates. *Arch. Biochem. Biophys.*, **501**, 73–78.

Stoupi, S., Williamson, G., Drynan, J.W. *et al.* (2010b) A comparison of the in vitro biotransformation of (−)-epicatechin and procyanidin B2 by human faecal microbiota. *Mol. Nutr. Food Res.*, **54**, 747–759.

Subramanian, R., Venkatesh, P., Ganguli, S. *et al.* (1999) Role of polyphenol oxidase and peroxidase in the generation of black tea theaflavins. *J. Agric. Food Chem.*, **47**, 2571–2578.

Takeo, T. (1992) Green and semi-fermented teas. In K.C. Willson and M.N. Clifford (eds), *Tea: Cultivation to Consumption*. Chapman and Hall, London, pp. 413–457.

Tanaka, T., Betsumiya, Y., Mine, C. *et al.* (2000) Theanaphthoquinone, a novel pigment oxidatively derived from theaflavin during tea fermentation. *Chem. Commun.*, 1365–1366.

Tanaka, T., Inoue, K., Betsumiya, Y. *et al.* (2001) Two types of oxidative dimerization of the black tea polyphenol theaflavins. *J. Agric. Food Chem.*, **49**, 5785–5789.

Tanaka, T., Mine, C., Watarumi, S. *et al.* (2002a) Accumulation of epigallocatechin quinone dimers during tea fermentation and formation of theasinensins. *J. Natl. Prod.*, **65**, 1582–1587.

Tanaka, T., Mine, C., Inoue, K. et al. (2002b) Synthesis of theaflavin from epicatechin and epigallocatechin by plant homogenates and role of epicatechin quinone in the synthesis and degradation of theaflavins. *J. Agric. Food Chem.*, **50**, 2142–2148.

Tanaka, T., Mine, C. and Kuono, I. (2002c) Structures of two new oxidation products of green tea polyphenols generated by model tea fermentation. *Tetrahedron*, **58**, 8851–8856.

Tanaka, T., Watarumi, S., Matsuo, Y. et al. (2003) Production of theasinensins A and D, epigallocatechin gallate dimers of black tea, by oxidation-reduction dismutation of dehydrotheasinensin A. *Tetrahedron*, **59**, 7939–7947.

Tanaka, T., Matsuo, Y. and Kouno, I. (2005a) A novel black tea pigment and two new oxidation products of epigallocatechin-3-O-gallate *J. Agric. Food Chem.*, **53**, 7571–7578.

Tanaka, T., Watarumi, S., Fujieda, M. et al. (2005b) New black tea polyphenol having N-ethyl-2-pyrrolidinone moiety derived from tea amino acid theanine: isolation, characterization and partial synthesis. *Food Chem.*, **93**, 81–87.

Tschiggerl, C. and Bucar, F. (2010) Volatile fraction of lavender and bitter fennel infusion extracts. *Natl. Prod. Commun.*, **5**, 1431–1436.

Unno, T., Kondo, K., Itakura, H. et al. (1996) Analysis of (–)-epigallocatechin gallate in human serum obtained after ingesting green tea. *Biosci. Biotechnol. Biochem.*, **60**, 2066–2068.

van het Hof, K.H., Kivits, G.A., Weststrate, J.A. et al. (1998) Bioavailability of catechins from tea: the effect of milk. *Eur. J. Clin. Nutr.*, **52**, 356–359.

Verzelloni, E., Pellacani, C., Tagliazucchi, D. et al. (2011) Antiglycative and neuroprotective activity of colon-derived polyphenol catabolites. *Mol. Nutr. Food Res.*, **55**.

Walle, T, Hsieh, F., De Legge, M.H. et al. (2004) High absorption but very low bioavailability of oral resveratrol in humans. *Drug Met. Disp.*, **32**, 1377–1382.

Wan, X.C., Nursten, H.E., Cai, Y. et al. (1997) A new type of tea pigment from the chemical oxidation of epicatechin gallate and isolated from tea. *J. Sci. Food Agric.*, **74**, 401–408.

Wang, H. and Helliwell, K. (2000) Epimerisation of catechins in green tea infusions. *Food Chem.*, **70**, 337–341.

Williamson, G. and Clifford, M.N. (2010) Colonic metabolites of berry polyphenolics: the missing link to biological activity? *Br. J. Nutr.*, **104**, S48–S66.

Willson, K.C. (1999) *Coffee, Cocoa and Tea*. CABI Publishing, Wallingford, UK.

Wiseman, R.W., Miller, E.C., Miller, J.A. et al. (1987) Structure-activity studies of the hepatocarcinogenicities of alkylbenzene derivatives related to estragole and safrol on administration to preweanling male C75BL/6J x C3H/HeJ F1 mice. *Cancer Res.*, **47**, 2275–2283.

Yoshida, T., Nakazawa, T., Hatano, T. et al. (1994) Tannins from theaceous plants. 6. A dimeric hydrolyzable tannin from *Camellia oleifera*. *Phytochemistry*, **37**, 241–244.

Yoshikawa, M., Xu, F., Morikawa, T. et al. (2003) Anastatins A and B, new skeletal flavonoids with hepatoprotective activities from the desert plant *Anastatica hierochuntica*. *Bioorg. Med. Chem. Lett.*, **13**, 1045–1049.

Yoshino, K., Suzuki, M., Sasaki, K. et al. (1999) Formation of antioxidants from (–)-epigallocatechin gallate in mild alkaline fluids, such as authentic intestinal juice and mouse plasma. *J. Nutr. Biochem.*, **10**, 223–229.

Zeller, A. and Rychlik, M. (2006) Character impact odorants of fennel fruits and fennel tea. *J. Agric. Food Chem.*, **54**, 3686–3692.

Zeller, A., Horst, K. and Rychlik, M. (2009) Study of the metabolism of estragole in humans consuming fennel tea. *Chem. Res. Toxicol.*, **22**, 1929–1937.

Zhao, Y., Chen, P., Lin, L. *et al.* (2011) Tentative identification, quantitation, and principal component analysis of green, pu-erh, green and white teas using UPLC/DAD/MS. *Food Chem.*, **126**, 1268–1277.

Zhou, Z.H., Zhang, Y.J., Xu, M. *et al.* (2005) Puerins A and B, two new 8-C-substituted flavan-3-ols from pu'er tea. *J. Agric. Food Chem.*, **53**, 8614–8617.

Zhu, Q.A., Zhang, A. Tsang, D. *et al.* (1997) Stability of green tea catechins. *J. Agric. Food Chem.*, **45**, 4624–4628.

Zhu, B.T., Patel, U.K., Cai, M.X. *et al.* (2001) Rapid conversion of tea catechins to monomethylated products by rat liver cytosolic catechol-O-methyltransferase. *Xenobiotica*, **31**, 879–890.

Zidorn, C., Johrer, K., Ganzera, M. *et al.* (2005) Polyacetylenes from the Apiaceae vegetables carrot, celery, fennel, parsley, and parsnip and their cytotoxic activities. *J. Agric. Food Chem.*, **53**, 2518–2523.

Chapter 4
Teas, Tisanes and Health

Diane L. McKay[1], Marshall G. Miller[2] and Jeffrey B. Blumberg[1]

[1]Jean Mayer USDA Human Nutrition Research Center on Aging, Tufts University, 711 Washington Street, Boston, MA 02111, USA
[2]Department of Psychology, Tufts University, 490 Boston Avenue, Medford, MA 02155, USA

4.1 Introduction

The observed health benefits of plant-based foods and beverages are not simply due to their macro- and micronutrient content alone but also due to the presence of phytochemicals (Hertog *et al.* 1995; Kromhout *et al.* 1996). The most numerous and widely distributed class of phytochemicals is the polyphenols, which includes the chromones, coumarins, lignans, stilbenes, xanthones and the ubiquitous flavonoids. Within the past decade, many polyphenols, particularly the flavonoids, have been reported to possess antioxidant, antiatherosclerotic, anti-inflammatory, antimutagenic, antitumour and antiviral activities (Nijveldt *et al.* 2001). While further research is necessary to elucidate and quantify the specific contributions of phytochemicals to health promotion and disease prevention, virtually all dietary guidelines created by regulatory agencies and healthcare organisations include recommendations for generous intakes of plant foods, including fruits, vegetables and whole grain cereals. Interestingly, recommendations for the consumption of plant-based beverages such as tea (*Camellia sinensis*) and tisanes (herbal teas) are absent despite their being particularly rich sources of phytochemicals, especially polyphenols (McKay *et al.* 2010).

People have been brewing tea made from the leaves of the *C. sinensis* plant for almost 50 centuries. Although health benefits have been attributed to tea consumption since the beginning of its history, scientific investigations of this beverage and its constituents have been under way for less than three decades. Observational studies have associated tea drinking with a reduced risk of cardiovascular diseases (CVD) and cancer. Within the last few years, clinical studies have revealed several physiological responses to tea, which may be relevant to the promotion of health and the prevention or treatment of some chronic diseases (McKay and Blumberg 2002).

Teas, Cocoa and Coffee: Plant Secondary Metabolites and Health, First Edition.
Edited by Professor Alan Crozier, Professor Hiroshi Ashihara and Professor F. Tomás Barberán.
© 2012 Blackwell Publishing Ltd. Published 2012 by Blackwell Publishing Ltd.

The use of herbal preparations, typically prepared by steeping or heating the crude plant material, has prevailed for centuries, and healthcare providers in Europe and Asia even today frequently prescribe herbal teas. However, such practices are largely based on folklore or schools of traditional medicine rather than evidence-based research. In many cases, the bioactivity of these plants appears to be derived from plant secondary metabolites, such as the polyphenols (Huffman 2003).

While there is an extensive and growing body of literature suggesting health benefits associated with drinking teas prepared with *C. sinensis* (i.e. black, green and oolong teas), evidence-based information regarding the effects of most herbal teas, or tisanes, is quite limited. We review here the available scientific literature directly related to the potential health benefits of infusions prepared with *C. sinensis* (black, green and oolong teas), *Ilex paraguariensis* (yerba maté), *Polygonum cuspidatum* (Itadori tea), *Chamomilla recutita* (chamomile tea), *Hibiscus sabdariffa* (hibiscus tea), *Aspalathus linearis* (rooibos tea) and *Cyclopia intermedia* (honeybush tea). Chapter 3 contains information on the principal phytochemicals in these teas and their bioavailability. When evidence from human studies is limited, animal and *in vitro* studies are presented. The relevant literature was identified principally from the PubMed and Medline databases, with literature searches employing both common and Latin names for each plant. In general, articles available in a language other than English were considered if the abstract was in English and/or if the full-text article pertained to a human study and was completed or referenced elsewhere within the past 20 years – except for *C. sinensis* where our search was focused on the last decade. Reviews of earlier studies on *C. sinensis* are available, including Yang and Landau (2000), McKay and Blumberg (2002), and Cabrera *et al.* (2006).

4.2 Black, oolong and green tea (*C. sinensis*)

After water, tea is the most popularly consumed beverage worldwide with a per capita consumption of ~120 mL per day (McKay and Blumberg 2002). All tea is produced from the leaves of the tropical evergreen *C. sinensis*. There are three main types of tea with black tea made via a post-harvest 'fermentation', an auto-oxidation catalysed by polyphenol oxidase. After picking, leaves for green tea are steamed to inactivate polyphenol oxidase prior to drying. Oolong tea is produced by a partial oxidation of the leaf, intermediate between the process for green and black tea. Recent human studies, i.e. observational studies and clinical trials, examining the health effects of black, green and oolong tea consumption are presented. Intervention studies are summarised in Table 4.1.

4.2.1 Black tea

4.2.1.1 Cancer

With the exception of breast and ovarian cancers, few studies have examined the specific association between black tea consumption and cancer risk. With regard to ovarian cancer, the evidence is conflicting. In a retrospective study of 696 Australian women

Table 4.1 Summary of human intervention studies (2001–2009) with *Camellia sinensis* teas

Reference	Dosage form	Subjects	Dose	Duration	Outcome
Black tea					
Fujita and Yamagami (2008)	Tablet	47 hypercholesterolaemics (40–70 years)	999 mg/day black tea extract or placebo with meals	3 months	Reduced total and LDL cholesterol and TG relative to baseline; effects extended 1 month following treatment termination
Bryans et al. (2007)	Beverage	16 healthy adults (35.5 ± 1.5 years)	1 g instant black tea + 75 g glucose + 250 mL water; 0.052 g caffeine + 75 g glucose + 250 mL water; or 250 mL water + 75 g glucose	Acute	Increased plasma insulin and decreased plasma glucose levels
Mukamal et al. (2007)	Beverage	28 adults with diabetes or 2 other CVD risk factors (>55 years)	6 g black tea powder/750 mL vs. water daily	6 months	No effect on cardiovascular biomarkers
Steptoe et al. (2007)	Liquid	75 healthy men	Black tea extract (≈1 cup strong black tea) or placebo, 4/day, counterbalanced to lemon or apple flavouring	1.5 months	Reduced cortisol levels and platelet activation; increased self-reports of relaxation
Hirata et al. (2004)	Liquid	10 men (26 ± 3 years)	2.1 g tea solids vs. 120 mg caffeine in 450 mL water	Acute	Increased coronary flow velocity reserve after 2 h relative to control
Oolong tea					
Shimada et al. (2004)	Beverage	10 patients with angina pectoris, 12 with former myocardial infarction	6 g oolong tea leaves/L or 1 L water	1 month	Increased plasma adiponectin and LDL particle size; decreased haemoglobin A_{1c}

(*continued*)

Table 4.1 (Continued)

Reference	Dosage form	Subjects	Dose	Duration	Outcome
Hosoda et al. (2003)	Beverage	20 diabetics	15 g oolong tea leaves/1.5 L or 1.5 L water	2 weeks	Reduced plasma glucose and fructosamine levels relative to water
Komatsu et al. (2003)	Beverage	11 healthy women (20 ± 1 years)	15 g oolong tea leaves + 300 mL water, 2.5 g powdered green tea + 300 mL water, or 300 mL water	Acute	Increased energy expenditure 10% (oolong) and 4% (green) relative to water
Rumpler et al. (2001)	Beverage	12 healthy men (25–60 years)	15 g oolong tea leaves/1.5 L water; 7.5 g oolong tea/1.5 L water, 270 mg caffeine/1.5 L water or 1.5 L water	Acute	15 g (and caffeine) increased energy expenditure and fat oxidation relative to water
Green tea					
Belza et al. (2009)	Tablet	12 healthy men (23.7 ± 2.6 years)	500 mg green tea extract, 400 mg tyrosine, 50 mg caffeine or placebo	Acute	No thermogenic effect was observed
Janjua et al. (2009)	Capsule	56 women (25–75 years)	2 green tea capsules (≈14 cups green tea) or placebo	2 year	Decreased in solar damage, reduced erythema and telangiectasias; fewer brown and UV spots relative to baseline
Maki et al. (2009)	Liquid	107 sedentary adults (21–65 years; BMI 25–40)	Green tea extract (≈625 mg catechins) in citrus beverage or citrus beverage	12 weeks	Reduced in abdominal fat area, subcutaneous abdominal fat area and fasting serum triglycerides relative to placebo; no decrease in body weight
Reinback et al. (2009)	Mixed	27 adults	510 mg cayenne capsule, 3.5 dL green tea drink (598.5 mg catechins, 77 mg caffeine), 2.3 mg CH-19 sweet pepper capsule, 3.5 dL green tea drink + 510 mg cayenne capsule, placebo	Acute	Green tea reduced desire to eat fatty, salty and hot foods; no effect on body weight or heart rate

Reference	Form	Subjects	Dose	Duration	Outcome
Alexopoulos et al. (2008)	Beverage	7 healthy smokers, 7 healthy non-smokers (30 ± 3 years)	6 g green tea, 235 mg caffeine or water	Acute	Increased flow-mediated dilation
Auvichayapat et al. (2008)	Capsule	60 adults (40–60 years; BMI >25)	750 mg green tea extract or placebo	12 weeks	Increased weight loss and resting energy expenditure
Fukino et al. (2008)	Liquid	60 borderline diabetics	Green tea extract (544 mg polyphenols)	2 months	Reduced haemoglobin A_{1c}
Hsu et al. (2008)	Capsule	78 women (BMI >27)	1.2 g green tea extract or placebo	12 weeks	No effect on body weight or composition
Matsuyama et al. (2008)	Beverage	38 children (6–16 years; BMI ≥28)	9 g green tea leaves + 1 L water + extract (576 mg / 340 mL catechins) or 9 g green tea leaves + 1 L water control (75 mg /340 mL catechins)	24 weeks	Greater decease in waist circumference, SBP and LDL cholesterol among those with above-median baseline body weight relative to control
Venables et al. (2008)	Capsule	12 healthy men	340 mg green tea polyphenols or placebo	Acute	Increased fat oxidation relative to placebo
Nagao et al. (2007)	Beverage	240 adults with visceral fat-type obesity (age 41.7 ± 9.9 years; BMI 26.8 ± 2.0)	340 mL green tea + green tea extract (583 mg catechins) or green tea control (96 mg catechins)	12 weeks	Increased magnitude of change for LDL cholesterol, SBP, body fat, waist and hip circumference, total and visceral and subcutaneous fat area relative to control
Chan et al. (2006)	Capsule	34 obese women with polycystic ovary syndrome	Green tea extract or placebo	3 months	No effect on body weight or glucose or lipid metabolism

(continued)

Table 4.1 (Continued)

Reference	Dosage form	Subjects	Dose	Duration	Outcome
Diepvens et al. (2006)	Capsule	46 women (BMI 27.7 ± 1.8)	Green tea extract (1206.9 mg) or placebo	83 days	No effect on blood parameters relating to energy metabolism
Vlachopoulos et al. (2006)	Beverage	13 healthy adults (26–34 years)	6 g green tea leaves/450 mL water, 6 g black tea/450 mL water, 125 mg caffeine/450 mL water, or 450 mL water	Acute	Green tea increased augmented aortic pressure and augmentation index relative to water; black tea increased pulse wave velocity, augmentation index, and peripheral and central SBP relative to hot water
Diepvens et al. (2005)	Capsule	46 women (BMI 27.6 ± 1.8)	Green tea extract (1125 mg catechins) + 225 mg caffeine or placebo	83 days	No effect on body weight; increased hunger and prospective food consumption; decreased thirst
Fukino et al. (2005)	Liquid	66 borderline diabetics (32–73 years)	Extract (544 mg polyphenols) + water or nothing	2 months	Insulin concentration correlated with polyphenol intake
Kovacs et al. (2004)	Capsule	104 adults (18–60 years; BMI 25–35)	450 mg green tea extract or placebo	13 weeks	No effect on body weight maintenance after weight loss
Nagaya et al. (2004)	Beverage	20 healthy male smokers, 7 age-matched, non-smoking, controls	400 mL green tea or water	acute	Increased blood flow in smokers relative to water

(aged 18–79 years) newly diagnosed with epithelial ovarian cancer and 786 healthy control subjects, Jordan *et al.* (2004) found no association with black tea consumption. Similarly, Steevens *et al.* (2007) found no association between black tea and ovarian cancer risk in their analysis of 2083 Dutch women. Conversely, Larson and Wolk (2005) reported a 46% decrease in ovarian cancer risk with the consumption of ≥ 2 cups per day, compared to non-drinkers, in a study of 61,057 Swedish women (40–76 years) followed up over a 14-year period (Odds ratio [OR] $= 0.54$, 95% confidence interval [CI]: 0.31–0.91, $P = 0.03$). These findings were supported by Baker *et al.* (2007) in a study of American women (114 patients *vs.* 868 age-matched controls) in which the consumption of ≥ 2 cups per day black tea was associated with a 30% reduction in ovarian cancer risk (OR $= 0.70$, 95% CI: 0.51–0.97, $P = 0.03$).

Only one study examined the association between black tea and breast cancer. Baker *et al.* (2006) compared 1932 breast cancer patients with 1895 hospitalised, non-neoplastic control patients. Although black tea was not associated with overall breast cancer risk, it was inversely associated with the risk for lobular-type breast cancer among premenopausal women relative to non-tea drinkers (OR $= 0.52$, 95% CI: 0.28–0.97).

Il'Yasova *et al.* (2003) compared 663 men and women diagnosed with rectal cancer (aged 40–79 years) with 323 controls, and their analysis revealed an inverse relationship between black tea consumption and rectal cancer risk among women (OR $= 0.40$, 95% CI: 0.23–0.70) and men (OR $= 0.77$, 95% CI: 0.42–1.43) when high consumers (>160 g dry leaves per month) were compared to low consumers (<80 g dry leaves per month). However, the epidemiological evidence for an association between black tea consumption and colorectal cancer risk is inconsistent (Arab and Il'Yasova 2003; Gardner *et al.* 2003).

4.2.1.2 Cardiovascular disease

The clinical evidence supporting an effect of black tea on cardiovascular health is conflicting. In a crossover study by Hirata *et al.* (2004), ten young, healthy males aged 26 ± 3 years were asked to consume 450 mL of a beverage containing either a black tea extract or caffeine (120 mg) on one of two occasions following a flavonoid-deficient diet for 24 hours, with a 1-week washout period between treatments. Their results showed that the black tea treatment significantly increased coronary flow velocity reserve at 2 hours relative to the control ($P < 0.0001$), suggesting it may improve blood vessel function. In a study by Fujita and Yamagami (2008), hypercholesterolaemic patients aged 40–70 years consumed either a Chinese black tea (Pu-ehr) extract ($n = 25$, 1 g per day) or placebo tablets ($n = 22$) for 3 months. Following treatment, subjects in the black tea group had significantly lower total cholesterol, low-density lipoprotein (LDL) cholesterol and triglyceride concentrations compared with the placebo group ($P < 0.05$). These reductions were maintained for 1 month following treatment termination ($P < 0.05$).

In contrast, Vlachopoulos *et al.* (2006) showed that black tea consumption acutely increased blood pressure (BP) and aortic stiffness. In their crossover study, 16 healthy volunteers consumed one 450 mL serving of black tea (6 g), caffeine (175 mg) or hot water. Compared to hot water, black tea increased pulse wave velocity at 30 minutes

($P < 0.05$), but attenuated it during the remaining time points up to 180 minutes. Black tea also increased the augmentation index (peripheral pressure/aortic pulse pressure ratio), indicating higher aortic pressure due to aortic stiffness, as well as peripheral and central systolic blood pressure (SBP) relative to hot water alone ($P < 0.01$).

Interestingly, Mukamal *et al.* (2007) found no effect of black tea on several cardiovascular biomarkers, including lipids, inflammatory markers, haemoglobin, adhesion molecules, prothrombotic and fibrinolytic parameters, and lipoprotein oxidisability, in adults (≥ 55 years) with elevated CVD risk following the consumption of black tea (6 g powder per 3 glasses water per day; $n = 14$) *vs.* water alone (3 glasses per day; $n = 14$) after 6 months. However, in a previous study of 1900 men and women, Mukamal *et al.* (2002) found that moderate (<14 cups per week) and heavy (≥ 14 cups per week) tea drinkers had a lower risk of mortality (28% and 44%, respectively) following non-fatal myocardial infarct relative to non-drinkers ($P < 0.001$).

4.2.1.3 Cognition and neurodegenerative disease

Black tea was inversely associated with risk for depression among 2011 subjects in the Kuopio Depression Study (Hintikka *et al.* 2005). Depression, assessed with the Beck Depression Inventory, was 53% lower among subjects who consumed black tea daily relative to non-tea drinkers (OR = 0.47, 95% CI: 0.27–0.83, no *P*-value reported). Coffee was not associated with depression risk, suggesting the observed effect was not due to higher caffeine consumption.

As part of the Singapore Chinese Health Study, Tan *et al.* (2008) examined the effects of black tea consumption on Parkinson's disease among 63,257 Chinese adults. After ~ 7 year of follow-up, they found black tea was associated with a 71% lower risk of Parkinson's disease when the highest (>23 cups per month) and lowest (<5 cups per month) tertiles of consumption were compared (relative risk [RR] = 0.29, 95% CI: 0.13–0.67, $P = 0.0006$). No risk reduction was reported with green tea.

Experimental evidence also suggests that black tea may affect mood and cognition. Steptoe *et al.* (2007) randomised 75 healthy men to receive either a black tea beverage, equivalent to 4 cups per day strong tea, or a caffeine-equivalent placebo beverage for 6 weeks following a 4-week run-in period during which dietary polyphenols were restricted. Cardiovascular measures (BP and heart rate) were obtained before, during and after two challenging behavioural tasks. Salivary cortisol, platelet activation (platelet–monocyte, –neutrophil and –leucocyte aggregates), and subjective stress measures were assessed before and after tasks. BP, heart rate and stress measures did not differ between the two groups. However, platelet activation ($P < 0.005$), post-task cortisol levels ($P = 0.032$) and subjective relaxation measures ($P = 0.036$) were all lower following tea treatment compared with the control beverage, suggesting black tea may aid stress recovery.

4.2.1.4 Diabetes

Bryans *et al.* (2007) evaluated the effects of instant black tea on fasting blood glucose and insulin response in 16 healthy adults aged 35.5 ± 1.5 years using a three-way crossover design. After a 12-hour fast, subjects were asked to consume 250 mL water either a placebo (75 g glucose), caffeine (0.052 g + 75 g glucose) or commercially available instant black tea beverage (1 g + 75 g glucose). A group assigned to receive a higher dose of black tea (3 g) was dropped from the study due to a high incidence of reported gastrointestinal symptoms. Among the remaining groups, plasma insulin levels increased significantly after 90 minutes with black tea relative to caffeine and at 150 minutes relative to caffeine only ($P < 0.01$), while plasma glucose levels were significantly lower after 120 minutes with black tea relative to caffeine and water ($P < 0.01$). The results of this study suggest that black tea can modulate plasma insulin and glucose levels acutely.

4.2.1.5 Bone health

Observational evidence suggests that black tea may prevent bone loss among older women. During a 5-year prospective calcium supplementation study in 1500 Australian women aged 70–85 years, Devine et al. (2007) interviewed subjects regarding their dietary habits, including tea consumption, and assessed bone mineral density at 1 and 5 years. After 4 years of follow-up, black tea drinkers ($n = 122$) lost significantly less total hip ($P < 0.01$), trochanter ($P < 0.05$) and intertrochanter ($P < 0.05$) bone mineral density relative to non-drinkers ($n = 42$). They also observed a significant dose–response effect of black tea on degree of trochanter bone loss ($P < 0.05$). A cross-sectional analysis conducted at the end of the study ($n = 1027$) indicated that black tea drinkers had significantly higher hip ($P < 0.05$) and trochanter ($P < 0.01$) bone mineral densities relative to non-drinkers.

4.2.2 Oolong tea

4.2.2.1 Cancer

Oolong tea was associated with an increased risk of bladder cancer in a small cohort of newly diagnosed Chinese patients by Lu et al. (1999), in which 40 patients were compared with 160 sex- and age-matched non-neoplastic controls. Their analysis of structured interviews, including questions regarding demographics, food and beverage frequencies, smoking, etc., showed that drinking oolong tea was associated with a threefold higher increase in bladder cancer risk relative to non-drinkers (OR = 3.0, 95% CI: 1.20–7.47, no P-value reported). They also found that consuming non-oolong tea, described as black and/or other green teas, was associated with a much higher risk of bladder cancer (OR = 14.86, 95% CI: 2.13–103.83, no P-value reported).

4.2.2.2 Cardiovascular disease

Oolong tea was associated with a decreased risk of ischaemic stroke in a study by Liang et al. (2009), who compared 374 Chinese stroke patients with 464 controls and found a 79% lower risk among those who consumed ≥ 2 cups per day of oolong tea relative to those who drank <1 cup per day (OR = 0.21, 95% CI: 0.09–0.50, $P < 0.001$) and a 75% lower risk with 1–2 cups per day (OR = 0.25, 95% CI: 0.11–0.56, $P < 0.001$). They observed a similar association for green, but not black, tea.

Clinical evidence suggests that oolong tea may confer health benefits in CVD patients. Shimada et al. (2004) investigated the effects of oolong tea on several CVD biomarkers, including total, high-density lipoprotein (HDL) and LDL cholesterol, serum triglycerides, plasma glucose and adiponectin, in 22 patients with angina pectoris or a previous myocardial infarction with a randomised, placebo-controlled, crossover design. Subjects consumed either water (1 L) or the same volume of oolong tea (6 g tea leaves) daily for 1 month. Among those who presented with adiponectin levels <7 µg/mL at baseline, oolong tea significantly increased levels relative to baseline ($P < 0.01$). Increased plasma LDL particle size ($P < 0.01$) and decreased haemoglobin A_{1c} levels ($P < 0.05$) were also observed with oolong tea, but not with water.

4.2.2.3 Diabetes

Oolong tea may be helpful in controlling blood glucose levels among diabetic patients. In a placebo-controlled, crossover study, 20 Taiwanese diabetics (10 men and 10 women) who were taking antihyperglycaemic drugs were asked to consume either 1.5 L oolong tea (15 g tea leaves) or water daily for 4 weeks each (Hosoda et al. 2003). Each treatment period was preceded by a 2-week washout period during which water was the only beverage consumed. Both plasma glucose levels and fructosamine were significantly reduced from baseline levels following oolong tea consumption ($P < 0.01$), and no significant changes were observed with water. However, no between-group differences were reported.

4.2.2.4 Obesity

Two small clinical studies reported the potential effects of oolong tea on obesity. Rumpler et al. (2001) examined the effects of oolong consumption on energy expenditure in 12 healthy men aged 25–60 years, using a four-way crossover study design. After a 4-day abstinence from caffeine and flavonoid-containing foods, participants were asked to consume 1.5 L of water, caffeinated water (270 mg), full-strength tea (15 g) or half-strength tea (7.5 g) daily for 3 days along with a standard American diet. Energy expenditure was measured indirectly for 24 hours using a room calorimeter. In their analysis, the beverages containing full-strength tea and caffeine raised subjects' 24-hour energy expenditure by 2.9% and 3.4%, respectively, relative to baseline ($P < 0.05$). Fat oxidation also increased with these beverages (12% and 8%, respectively) relative to baseline ($P < 0.05$). No between-group differences were reported. These beverages also elevated plasma caffeine levels similarly 8 hours following the last dose, suggesting the observed effects may have been due to the caffeine content of the beverages, rather than tea polyphenols alone.

Komatsu et al. (2003) also used a crossover design to compare the effects of oolong tea in 11 healthy Japanese women aged 20 ± 1 years. After consuming 300 mL of water, oolong tea (15 g leaves) or green tea (2.5 g powdered tea), indirect calorimetry revealed a cumulative 10% increase in energy expenditure within 120 minutes of oolong tea consumption, with levels attained at 60 and 90 minutes being significantly higher relative to water ($P < 0.05$). Green tea showed a cumulative 4% increase in energy expenditure within 120 minutes of consumption. The authors suggest that the highly polymerised polyphenol content of oolong tea (twofold higher than green tea) may be responsible for the greater observed increase in energy expenditure rather than the caffeine content as green tea contained more caffeine than oolong tea (77 vs. 161 mg).

4.2.3 Green tea

4.2.3.1 Cancer

Several recent observational studies suggest that green tea consumption may protect against oesophageal cancer. In a European, multicenter, case–control study, Lagiou *et al.* (2009) compared 2304 patients with upper aerodigestive tract cancer with 2227 sex-, age- and location-matched controls. Their results showed that frequent consumption of green tea (above *vs.* below median intake levels) was associated with reduced disease risk when controlled for coffee consumption (OR = 0.82, 95% CI: 0.69–0.98, no *P*-value reported). Three smaller case–control studies conducted in Asian populations similarly showed a protective effect (Wang *et al.* 2006a, 2007; Chen *et al.* 2009). However, it should be noted that a recent systematic review of 59 studies concluded that the consumption of very hot foods and beverages, such as green tea, may independently increase oesophageal cancer risk (Islami *et al.* 2009).

The association between green tea and gastric cancers is less clear. In one study, Mu *et al.* (2005) compared 206 Chinese patients newly diagnosed with stomach cancer (>90% adenocarcinoma of the distal stomach) with 415 healthy, randomly selected controls. Subjects who consumed ≥1 cup per day for ≥6 months had a reduced risk of stomach cancer (OR = 0.59, 95% CI: 0.34–1.01, no *P*-value reported). In a prospective cohort study of 72,943 Japanese participants including 892 gastric cancer cases, Sasazuki *et al.* (2004) found no association between green tea consumption and gastric cancer risk. In their analysis, a reduced risk was observed among women with cancer in the most distal subsite who reported consuming the highest amount of green tea (≥5 cups per day; RR = 0.51, 95% CI: 0.30–0.86, $P = 0.01$). In a case–control study by Hoshiyama *et al.* (2004), no association between green tea and gastric cancer was reported. Similarly, no association was reported for risk of liver cancer (Montella *et al.* 2007), pancreatic cancer (Luo *et al.* 2007) or pancreatic cancer mortality (Lin *et al.* 2008) in other observational studies.

The association between green tea consumption and colorectal cancer is also unclear. In a large cohort of Chinese women, aged 40–70 years, followed up for 6 years, Yang *et al.* (2007) showed that patients who regularly consumed green tea had a 54% lower risk of developing colon (RR = 0.66, 95% CI: 0.43–1.01, no *P*-value reported) and rectal cancers (RR = 0.58, 95% CI: 0.35–0.98, no *P*-value reported). An additional reduction in risk was observed with increased quantity ($P = 0.01$) and duration ($P = 0.006$) of tea consumption. Sun *et al.* (2007) followed another large cohort of Chinese men and women, aged 45–74 years, for an average of 8.9 years and found a non-significant increase in colorectal cancer risk (RR = 1.12, 95% CI: 0.97–1.29) among green tea drinkers. This increase appears to have been in men only (RR = 1.31,

95% CI: 1.08–1.58) rather than in women (RR = 0.89, 95% CI: 0.71–1.12), and was high particularly among men with advanced stages of disease who consumed green tea monthly (RR = 1.38, 95% CI: 0.92–2.07), weekly (RR = 1.35, 95% CI: 0.96–1.89) and daily (RR = 1.85, 95% CI: 1.34–2.54; $P = 0.0002$ for trend).

Green tea was associated with a reduced risk of haematological cancer in a large Japanese cohort study by Naganuma et al. (2009). After a 9-year follow-up with 157 cases of haematological malignancies (119 lymphoid neoplasms and 36 myeloid neoplasms) identified, the consumption of ≥5 cup per day green tea was associated with a reduced risk of haematological malignancy relative to those who consumed <1 cup per day (hazard ratio (HR) = 0.58, 95% CI: 0.37–0.89). Risk estimates for lymphoid and myeloid neoplasms were similarly reduced (HR = 0.52, 95% CI: 0.31–0.87 and 0.76, 95% CI: 0.32–1.78, respectively). In another study, Kuo et al. (2009) compared 252 young Taiwanese leukaemia patients (<30 years of age) with 637 age- and sex-matched controls. Among the older patients in this cohort, aged 16–29 years, consumption of green tea in higher cumulative quantities (>550 cups from birth to diagnosis) was associated with a 53% lower risk (OR = 0.47, 95% CI: 0.23–0.97). A smaller case–control study compared 107 Chinese adult leukaemia patients with 110 controls and showed similar results (Zhang et al. 2008).

Evidence suggests a weak association between green tea and the risk of developing breast cancer. In a recent case–control study by Shrubsole et al. (2009), 3454 Chinese breast cancer patients aged 20–74 years were compared with 3474 controls. Green tea consumption was associated with a 12% lower risk for breast cancer relative to those who consumed none (OR = 0.88, 95% CI: 0.79–0.98, no P-value reported). However, this reduction was no longer significant when postmenopausal women alone were considered (OR = 0.88, 95% CI: 0.74–1.04, no P-value reported). A follow-up study conducted in a cohort of 74,942 Chinese women found that those who regularly consumed green tea starting at age ≤25 years had a 31% lower risk of developing premenopausal breast cancer relative to non-drinkers (HR = 0.69, 95% CI: 0.41–1.17), yet a 1.6-fold higher increase in the risk of developing postmenopausal breast cancer (HR = 1.61, 95% CI: 1.18–2.20), suggesting that green tea may delay the onset of breast cancer (Dai et al. 2010). These findings are supported by the results of a previous study by Zhang et al. (2007) who compared 1009 Chinese women with breast cancer (aged 20–87 years) with age-matched controls and found a significant dose–response effect ($P < 0.001$ for trend); the ORs for women consuming dried green tea leaves at 1–249, 250–499, 500–749, and ≥750 g per year were 0.87 (0.73–1.04), 0.68 (0.54–0.86), 0.59 (0.45–0.77) and 0.61 (0.48–0.78), respectively. Similar dose–response relationships were observed for duration, number of cups consumed and new batches prepared per day.

Inoue et al. (2008) compared 380 Chinese women with 662 controls, and while they failed to find an overall association between green tea and breast cancer risk, they did find that women with folate intake <133.4 µg per day who also consumed ≥1 cup per week green tea had a lower risk relative to those who consumed <1 cup per week (OR = 0.45, 95% CI: 0.26–0.79, $P = 0.02$). Interestingly, the women in this study with methylenetetrahydrofolate reductase and thymidylate synthase 0–1 variant allele genotypes, who also consumed ≥1 cup per week green tea, appeared to have a reduced risk relative to those who consumed <1 cup per week (OR = 0.66, 95% CI: 0.45–0.98), suggesting that green tea may act via folate pathway inhibition to protect against breast cancer.

The evidence for an association between green tea and the risk of endometrial and ovarian cancers is equivocal. In a study by Kakuta *et al.* (2009), 152 Japanese women with confirmed endometrial endometrioid adenocarcinoma were compared with 285 healthy age- and location-matched controls. Those in the highest quartile of green tea intake (>4 cups per day) had a 67% lower risk compared to the lowest quartile (OR = 0.33, 95% CI: 0.15–0.75, $P = 0.007$). However, a 15-year prospective study of 53,724 Japanese women (aged 40–69 years) by Shimazu *et al.* (2008), which identified 117 cases of endometrial cancer, failed to show an effect of green tea ($P = 0.22$). Another study followed 254 Chinese patients diagnosed with ovarian cancer for a minimum of 3 years and found green tea was associated with a 45% reduction in mortality risk (HR = 0.55, 95% CI: 0.34–0.90, no P-value reported) (Zhang *et al.* 2004).

Green tea may reduce the risk of prostate cancer, according to a study by Kurahashi *et al.* (2008a). In their cohort of 49,920 Japanese men aged 40–69 years, 404 new cases of prostate cancer were identified during 11–14 years of follow-up. After that time, they reported a dose-dependent inverse association between green tea consumption and advanced prostate cancer (≥ 5 cups per day; OR = 0.52, 95% CI: 0.28–0.96, $P = 0.01$) relative to low consumers (<1 cup per day). These findings are supported by a case–control study by Jian *et al.* (2004), in which 130 Chinese patients with confirmed adenocarcinoma of the prostate were compared with 274 age-matched, non-malignant controls. In this study, green tea was inversely associated with prostate cancer risk relative to non-drinkers (OR = 0.28, 95% CI: 0.17–0.47, no P-value reported), and this risk was further reduced among those who consumed >3 cups per day (OR = 0.27, 95% CI: 0.15–0.48) and for >40 years (OR = 0.12, 95% CI: 0.06–0.26) relative to seldom and never drinkers. However, a study of 19,561 Japanese men, including 110 cases of prostate cancer by Kikuchi *et al.* (2006) reported no association between green tea intake and prostate cancer risk ($P = 0.81$).

The effects of green tea on other cancer types have also been investigated. Rees *et al.* (2007) found an inverse association between green tea and risk of squamous cell ($n = 696$; OR = 0.70, 95% CI: 0.53–0.92) and basal cell carcinomas ($n = 770$; OR = 0.79, 95% CI: 0.63–0.98) relative to age- and sex-matched controls ($n = 715$). Conversely, green tea consumption was not associated with a reduced risk of either lung (Li *et al.* 2008) or oral cancers (Ide *et al.* 2007), but ≥ 5 cups per day was associated with an increased risk of bladder cancer among women (HR = 2.29, 95% CI: 1.06–4.92, $P = 0.03$) in a study by Kurahashi *et al.* (2008b).

To date, few clinical trials have assessed the anticancer effects of green tea consumption. A phase I clinical trial conducted by Laurie *et al.* (2005) investigated the maximum tolerated dose of green tea extract among 17 patients with advanced, incurable lung cancer (≥ 3 weeks following chemotherapy). Patients were administered a once-daily oral dose of green tea extract starting at 0.5 g/m^2 during 4-week cycles. According to their dose escalation scheme, an additional participant was recruited at each dose level, though individual patients did not experience increased doses, with additional recruitment pending toxicity symptoms. Toxicity was assessed according to the National Cancer Institute Common Toxicity Criteria (V.2). Dosing escalated as high as 8 g/m^2, then de-escalated to 3 g/m^2, which was later determined to be the maximum tolerated dose. A previous phase I study conducted by Pisters *et al.* (2001) reported a slightly higher maximum tolerated dose of 4.2 g/m^2.

A study of decaffeinated green tea extract (2–250 mg capsules per day) in 15 patients aged 61–84 years with hormone refractory prostate cancer by Choan *et al.* (2005) was concluded prematurely based on the null findings published by Jatoi *et al.* (2003). Both studies failed to find evidence of an anticancer effect of green tea despite the reported bioavailability of tea polyphenols in prostate by Henning *et al.* (2006).

In a pilot study of 125 patients (20–80 years), Shimizu *et al.* (2008) investigated the effect of green tea extract on colorectal adenomas. Adenomas were initially removed 1 year prior. Patients ($n = 125$) were then randomised to receive either green tea extract tablets (3 per day), equivalent to 6 cups of green tea, or no extract for 1 year. No placebo was used as patients already consumed ≈6 cups per day green tea, and the authors were interested in the effects of increased consumption. After this time, half as many patients in the green tea group were diagnosed with new colorectal adenomas as the placebo group ($n = 9$ vs. 20; RR = 0.49, 95% CI: 0.24–0.99, $P < 0.05$). Relapsed adenomas were also significantly smaller in the green tea group relative to placebo ($P = 0.001$).

4.2.3.2 Cardiovascular disease

Some observational evidence suggests that green tea consumption may be associated with a lower risk for CVD. Sano *et al.* (2004) compared 109 Japanese patients with coronary stenosis aged 60 ± 0.9 years with 94 control patients and reported an inverse association between green tea and coronary artery disease (OR = 0.84, 95% CI: 0.76–0.91, $P = 0.001$). However, no associations with cardio- or cerebrovascular events were found within the follow-up period of 4.9 ± 0.1 years. In a cohort of 51,225 Japanese adults (aged 40–79 years), Kuriyama *et al.* (2006a) found a dose-dependent decrease in all-cause mortality with >4 cups per day vs. <1 cup per day (HR = 0.84, 95% CI: 0.77–0.92, $P < 0.001$). During the 11-year follow-up, the risk for women was decreased further (interaction with sex, $P = 0.03$). They also observed a dose-dependent decrease in CVD mortality with consumption of >4 vs. <1 cup per day (HR = 0.74, 95% CI: 0.62–0.89, $P < 0.001$), with a more pronounced effect among women during the 7-year follow-up.

Green tea was associated with reduced hypertension in a population of 1507 newly diagnosed Taiwanese adults, aged ≥20 years (Yang *et al.* 2004). Compared to non-drinkers, moderate (120–599 mL per day) and high (>600 mL per day) consumption was associated with an 81–87% risk reduction (OR = 0.19, 95% CI: 0.04–0.85 and 0.13, 95% CI: 0.03–0.61, respectively, no *P*-values reported).

Green tea drinking was also associated with a reduced risk of stroke in a case–control study of 402 Japanese patients with subarachnoid haemorrhage and age- and sex-matched controls (Okomoto 2006). Consumption of >1 cup per day was associated with a 62% lower risk of haemorrhage among those with a positive family history (OR = 0.38, 95% CI: 0.32–0.95, $P = 0.02$). Another study with 6358 Japanese adults found moderate (>2–<5 cups per day) to high (≥5 cups per day) consumption of green tea was associated with a 67–69% reduced risk (HR = 0.43, 95% CI: 0.25–0.74 and 0.41, 0.24–0.70, respectively) (Tanabe *et al.* 2008). A recent study by Liang *et al.*

(2009) compared 374 ischaemic stroke patients with 464 stroke-free controls and found an 88% lower risk of ischaemic stroke with 1–2 cups per day (OR = 0.12, 95% CI: 0.05–0.29, $P < 0.001$) and a 72% reduction with ≥2 cups per day (OR = 0.28, 95% CI: 0.15–0.53, $P < 0.001$) relative to those who consumed <1 cup per day. Similar results were found for oolong, but not black, tea.

Clinical trials of green tea have also shown beneficial effects on some CVD risk factors, but not on others. The effects of green tea on acute endothelial function were evaluated in a randomised, crossover study by Nagaya et al. (2004) in 20 healthy male smokers and 7 non-smoking, age-matched controls who consumed either 400 mL of green tea or hot water. At baseline, the forearm blood flow of smokers was significantly lower than the non-smokers ($P < 0.05$). After consuming green tea, the blood flow in smokers was significantly increased relative to hot water ($P < 0.001$), thus attenuating endothelial dysfunction in this high-risk group. Alexopoulos et al. (2008) found that green tea (6 g), but not caffeine (125 mg) or hot water, significantly increased flow-mediated dilation by 3.69% ($P = 0.02$) among 14 healthy smoking and non-smoking adults (50% smokers; 30 ± 3 years).

Vlachopoulos et al. (2006) examined the effects of green (and black) tea on aortic stiffness and wave reflections in a crossover study of 13 healthy volunteers (7 men and 5 smokers), aged 26–34 years, who consumed 450 mL of green tea (6 g leaves), caffeine (125 mg) or hot water. At baseline and for 3 hours following ingestion, aortic stiffness was assessed by means of carotid–femoral pulse wave velocity and wave reflection. The administration of green tea increased the aortic pressure ($P < 0.05$) relative to water.

4.2.3.3 Cognition and neurodegenerative disease

Although an early study of green tea consumption and mental health failed to detect a statistically significant association (Shimbo et al. 2005), subsequent studies have reported a decreased risk for depression and psychological distress. Kuriyama et al. (2006b) analysed the Mini-Mental State Examination (MMSE) data from 1003 Japanese adults, aged ≥70 years, in a nested community-based cohort. They found green tea consumption was associated with increased MMSE scores, indicating increased cognitive function, with 4–6 cups per week (OR = 0.62, 95% CI: 0.33–1.19) and ≥2 cups per day (OR = 0.62, 95% CI: 0.30–0.72, $P = 0.0006$) relative to those who consumed <3 cups per week. No significant associations were observed with black or oolong tea ($P = 0.33$ for trend) or coffee ($P = 0.70$ for trend).

Niu et al. (2009) investigated green tea and depression using the 30-item Geriatric Depression Scale (GDS) in 1058 Japanese adults aged ≥70 years. They showed that green tea consumption was inversely associated in a dose-dependent manner with the incidence of mild depression; i.e. less severe symptoms were observed with 2–3 cups per day (adjusted OR = 0.96; 95% CI: 0.66–1.42) and ≥4 cups per day (adjusted OR = 0.56; 95% CI: 0.39–0.81, $P = 0.001$ for trend). Similar results were observed for severe depression. No significant associations were observed for black and oolong teas ($P = 0.06$ for trend) or coffee ($P = 0.49$ for trend).

Another Japanese study by Hozawa et al. (2009) investigated the incidence of psychological distress among 42,093 Japanese adults (aged ≥40 years) using the Kessler 6-Item Psychological Distress Scale. Their analysis revealed an association between green tea intake and a reduced risk for psychological distress with a 5% risk reduction for 1–2 cups per day (OR = 0.95, 95% CI: 0.86–1.06), 11% for 3–4 cups per day (OR = 0.89, 95% CI: 0.79–1.00) and 20% for ≥5 cups per day (OR = 0.80, 95% CI: 0.70–0.90, $P < 0.001$) relative to non-drinkers. However, the consumption of black tea was not associated with a reduced risk among those consuming 1–2 cups per day (OR = 1.14, 95% CI: 0.95–1.36) or ≥3 cups per day (OR = 1.11, 95% CI: 0.78–1.58) relative to non-drinkers. The authors suggest that this absence of an association may be due to the less frequent consumption of black tea in Japan.

4.2.3.4 Diabetes

There is some evidence that green tea drinking is associated with a decreased risk of developing type 2 diabetes. Iso et al. (2006) followed a cohort of 17,413 healthy Japanese adults aged 40–65 years for a 5-year period and found a 33% reduced risk with ≥6 cups per day compared to <1 cup per day (OR = 0.67, 95% CI: 0.47–0.94, $P = 0.002$). However, a study by Odegaard et al. (2009) followed 36,908 Chinese adults (aged 45–74 years) for 5 years and observed no association between green tea and risk of type 2 diabetes.

Few clinical trials have investigated the effects of green tea on diabetes. An initial study failed to observe an effect of green tea extract (544 mg polyphenols per day) among 66 borderline diabetics (Fukino et al. 2005), although insulin levels were correlated with polyphenol intake. A subsequent study by Fukino et al. (2008) examined the effects of a powdered green tea extract (544 mg polyphenols per day for 2 months) in 60 Japanese patients (32–73 years) who were considered borderline diabetic (blood glucose >6.1 mmol/L fasted or >7.8 mmol/L non-fasted) and found significantly lower levels of haemoglobin A_{1c} over time ($P = 0.03$). Although these results suggest a possible effect of green tea on blood glucose control, this study was not placebo controlled.

4.2.3.5 Obesity

The evidence supporting a clinically significant antiobesity effect of green tea is not yet convincing. Many studies have failed to show an effect in humans (Kovacs et al. 2004; Diepvens et al. 2005; Chan et al. 2006; Diepvens et al. 2006; Hsu et al. 2008; Belza et al. 2009; Reinback et al. 2009), while others have reported significant effects. In a multicentre trial by Nagao et al. (2007), 240 adults aged 25–55 years were asked to consume 340 mL per day of green tea plus green tea extract or a control beverage for 12 weeks. Cardiovascular risk factors and anthropometric measures were assessed. There were no main effects with green tea treatment, although the magnitude of change for LDL cholesterol (−0.09 vs. 0.04 mmol/L), SBP (−2.7 vs. 0 mm Hg), body fat (−2.5 vs. −0.7%), waist (−2.5 vs. 0 cm) and hip circumference (−2.3 vs. −0.1 cm), total fat area (−16 vs. 0.1 cm^2), visceral fat area (−10.3 vs. −3.9 cm^2) and subcutaneous fat area (−5.7 vs. 4.0 cm^2) were all greater in the green tea group compared to the control group ($P > 0.05$ for all).

Matsuyama et al. (2008) conducted a randomised, controlled trial with 38 obese Japanese children, aged 6–16 years. Subjects received either green tea (9 g leaves + 1 L hot water; 75 mg/340 mL catechins) or green tea containing green tea extract (9 g leaves + 1 L hot water + extract; 576 mg/340 mL catechins). After 24 weeks, only subjects with above-median weight at baseline were compared. Those in the green tea plus extract group showed a greater decrease in waist circumference (−0.2 vs. 1.4 cm), SBP (−2.1 vs. 7.8 mm Hg) and LDL cholesterol (−0.29 vs. −0.01 mM) from 0 to 24 weeks than did those in the green tea-only group.

A randomised, controlled trial of 60 obese Thai adults (aged 40–60 years) by Auvichayapat et al. (2008) compared the effects of green tea extract (three 250 mg capsules per day) with a placebo on anthropometrics, body composition and energy expenditure. Dietary intake was controlled by providing subjects with three standardised meals daily. After 8 and 12 weeks, the green tea group lost more body weight (−4.44 vs. −1.93 kg and −2.7 vs. 2 kg, respectively, $P < 0.05$). At 8 weeks, when weight loss was most different from controls, resting energy expenditure was significantly higher (8351.12 vs. 8167.74 kJ per day, $P < 0.01$) and might account for the increased weight loss among those in the green tea extract group.

Venables et al. (2008) assessed the effects of green tea on energy expenditure during exercise in a crossover study of 12 healthy men. Subjects cycled at 60% VO_{2max} for 30 minutes before and after supplementation with three capsules of green tea extract (340 mg polyphenols) or placebo. Fat oxidation rates were 17% higher following green tea extract relative to placebo ($P < 0.05$); fat oxidation, relative to total energy expenditure, was similarly increased.

A randomised, placebo-controlled trial of 107 sedentary adults (aged 21–65 years, BMI 25–40 kg/m^2) by Maki et al. (2009) evaluated the effect of a green tea extract on body composition and weight. Participants consumed 500 mL per day of a control beverage with or without green tea extract ($\approx$625 mg catechins). After 12 weeks, body weight did not change in either group; however, reductions in total abdominal fat area ($P = 0.013$), subcutaneous abdominal fat area ($P = 0.019$) and fasting serum triglycerides (−11.2 vs. 1.9%, $P = 0.023$) were greater among those who consumed the green tea extract compared with the placebo.

4.2.3.6 Dental health

Green tea drinking was associated with better dental health in a cohort of 25,078 Japanese participants aged 40–64 years. In their analysis, Koyama et al. (2010) found that drinking green tea was associated with less tooth loss among men and women at 1–2 cups per day (OR = 0.82, 95% CI: 0.74–0.91 and 0.87, 95% CI: 0.78–0.97, respectively), 3–4 cups per day (OR = 0.82, 95% CI: 0.73–0.92 and 0.87, 95% CI: 0.77–0.98, respectively), and ≥5 cups per day intakes (OR = 0.77, 95% CI: 0.66–0.89 and 0.89, 95% CI: 0.78–1.01, respectively) relative to consuming <1 cups day ($P < 0.0001$ and $P < 0.011$, respectively). Oolong tea did not appear to reduce the risk of tooth loss.

4.2.3.7 Dermatological health

While green tea extract is a popular additive to many skin care products, only one recent study has examined the effect of ingested green tea extract on skin. Janjua et al. (2009) randomly assigned 56 female participants aged 25–75 years to receive either a placebo or two green tea capsules daily (≈14 cups tea) for 2 years. Digital photographs of patients' faces were examined at baseline, 6, 12 and 24 months by a blinded, board-certified dermatologist and assessed for photo damage. At 6 months, participants in the green tea group showed reduced solar damage relative to baseline ($P = 0.02$). Erythema and telangiectasias were also reduced in the green tea group at 6 and 12 months relative to baseline ($P = 0.05$ and 0.02, respectively). The green tea group showed fewer brown and UV spots relative to baseline ($P = 0.02$ and 0.03, respectively) at 12 months, but not at 24 months. The placebo group had no significant improvements in any measures tested relative to baseline. Although these findings suggest that green tea reduces photo damage to skin, the changes in this group were not significantly different from those observed in the placebo group.

4.3 Other teas and tisanes

4.3.1 *Yerba maté* (Ilex paraguariensis)

Yerba maté tea is made with the dried leaves of *I. paraguariensis*, a subtropical evergreen tree native to South America. Fresh maté leaves undergo several stages of processing including blanching, drying and ageing prior to packaging (Heck and De Mejia 2007). During blanching, the maté leaves are flash-heated over an open flame to deactivate polyphenol oxidase and inhibit oxidation. The leaves are then dried very slowly, often using wood smoke, and aged for up to 12 months for flavour development. This herbal tea is traditionally prepared by packing the dried leaves (about 50 g) into a calabash gourd and pouring hot water (0.5–1.0 L) over them before the brew is shared with others through a straw (bombilla), typically in a social setting. The name 'maté' is derived from the Quechua word 'maté', meaning a cup or vessel used for drinking. Other names for beverages made with *I. paraguariensis* include Jesuit's tea, Paraguayan tea, cimarrón and chimarraō. In the United States, maté is commercially packed in individual tea bags (1–2 g) or as a tea concentrate for use in the food and dietary supplement industries.

Maté has long been used by the indigenous peoples of South America, especially the Guarani Indians. In the nineteenth century, South American gauchos (cowboys) relied on maté not only as a stimulant but also as a vital part of their diet (Goldenberg 2002). In folk medicine, maté is used as a central nervous system stimulant, diuretic and antirheumatic (Gosmann *et al.* 1989). The German Commission E approved the internal use of maté for mental and physical fatigue (Klein and Riggins 1998a).

In vitro, water extracts of *I. paraguensis* have been shown to inhibit the formation of advance glycation end products (Lunceford and Gugliucci 2005), an action most likely due to chlorogenic and caffeic acids (Gugliucci *et al.* 2009); inhibit protein nitration and enhance survival following peroxynitrite-induced cytotoxicity (Bixby *et al.* 2005); inhibit topoisomerase II activity and squamous cancer cell proliferation (Ramirez-Mares *et al.* 2004; De Mejia *et al.* 2005); and protect against H_2O_2-induced DNA damage (Bracesco *et al.* 2003; Miranda *et al.* 2008). Yerba maté beverages have also been shown to protect human LDL against peroxidation *in vitro* (Gugliucci and Stahl 1995) and *ex vivo* (Gugliucci 1996).

Mosimann et al. (2006) found yerba maté exhibited an antiatherosclerotic effect in hypercholesterolaemic rabbits. Although 400 mL per day of *I. paraguariensis* for 60 days inhibited the progression of atherosclerosis in cholesterol-fed rabbits, no reductions in serum cholesterol, changes in aortic lipid peroxidation or antioxidant enzyme activities were observed. In a mouse model, Lanzetti et al. (2008) found that maté reduced lung damage induced by short-term exposure to cigarette smoke. In their experiment, mice were given either distilled water (control), 150 mg/kg of reconstituted lyophilised maté by oral gavage or i.p. prior to exposing them to six commercial filtered cigarettes daily for 5 consecutive days. Compared to controls, the group given maté orally had fewer alveolar macrophages and neutrophils ($P < 0.001$), reduced lipid peroxidation ($P < 0.01$), tumour necrosis factor α ($P < 0.05$) and matrix metalloproteinase-9 activity ($P < 0.001$) in their lung extracts, suggesting an acute anti-inflammatory effect.

Although *in vitro* and animal studies suggest a potential anticarcinogenic effect of maté, the available epidemiological evidence does not. Several case–control studies conducted in South America reported an association between the consumption of yerba maté tea and an increased risk of oral (Pintos et al. 1994; Goldenberg 2002), laryngeal (Pintos et al. 1994), oesophageal (Sewram et al. 2003), lung (De Stefani et al. 1996), kidney (De Stefani et al. 1998) and bladder cancers (De Stefani et al. 1991, 2007; Bates et al. 2007). Tobacco use, the consumption of alcohol and high-temperature beverages are confounding factors associated with both maté drinking and the development of many, but not all, of these cancers. The *N*-nitroso compounds and polycyclic aromatic hydrocarbons (PAH) present in maté are suspected carcinogens, and may also play a role in its procarcinogenic action (Goldenberg et al. 2003). In a cross-sectional study of 200 Brazilians (100 males and 100 females; 50 of each group were current smokers and 50 non-smokers), Fagundes et al. (2006) found that any maté consumption increased urine concentrations of 1-hydroxypyrene glucuronide (1-OHPG), a PAH metabolite ($P = 0.0053$), and observed a stepwise increase in 1-OHPG concentration with volume of maté consumed. As PAH are introduced by wood smoke during the drying process, the contamination of maté leaves with PAH presents a plausible explanation for increased rates of cancer with maté consumption.

Few clinical trials of maté have been published. In an uncontrolled trial of 15 healthy females with a mean age of 25 years, Matsumoto et al. (2009) reported a decrease in lipid peroxidation and a concomitant increase in plasma resistance to Cu^{+2}-induced oxidation with both acute (1-hour) and prolonged (1-week) consumption of instant maté tea (5 g/500 mL cold water) ($P < 0.05$ for all). After 1 week, plasma total antioxidant status and gene expression of the antioxidant enzymes glutathione peroxidase, superoxide dismutase and catalase also increased ($P < 0.01$ for all). In a non-placebo-controlled trial, De Morais et al. (2009) reported significant reductions in LDL cholesterol compared with baseline levels among normolipidaemic ($n = 15$), dyslipidaemic ($n = 57$) and hypercholesterolaemic subjects on long-term statin therapy ($n = 30$) after consuming 330 mL of green or roasted yerba maté tea three times per day for 40 days ($P < 0.05$ for all). No changes in triglyceride levels were observed.

Two randomised, controlled trials of a weight-loss supplement containing an unknown quantity of yerba maté in combination with other herbal ingredients failed to show clinically significant changes in weight within 10–12 weeks (Andersen and Fogh 2001; Opala *et al.* 2006). Andersen and Fogh (2001) did report a mean 5.1 kg loss in the supplemented group at 45 days (*vs.* 0.3 kg with placebo; no *P*-value reported). However, since the subjects' diets were self-selected and their food intake was not assessed, the observed effect cannot be attributed solely to the yerba maté-containing supplement.

4.3.2 *Itadori* (Polygonum cuspidatum)

Japanese knotweed (*P. cuspidatum*) is a shrub-like, herbaceous perennial with thick, hollow stems resembling bamboo, and can reach a height of 2–3 m. Clusters of the plant arise from fibrous roots and produce a spreading rhizome system that may extend to more than 8 m. Native to Japan and eastern Asia, *P. cuspidatum* is considered an invasive weed in the United Kingdom and North America. In Asia, the root of the plant is dried and infused into a tea called Itadori, which means 'well-being' in Japanese (Burns *et al.* 2002).

The roots of *P. cuspitadum* constitute one of the, if not the richest, sources of resveratrol, particularly *trans*-resveratrol and its glucoside (Vastano *et al.* 2000). In plants, resveratrol is considered a phytoalexin, produced in response to infection and stress. Although this stilbene has been identified in at least 72 plant species, common food sources are limited to wine, grapes and grape juice, cranberries and cranberry juice, peanuts and peanut products, as well as dark chocolate and chocolate liquor (Counet *et al.* 2006), and the resveratrol content of these products is, at best, minimal (Crozier *et al.* 2009). During the last decade, there have been numerous reports of resveratrol, and its *in vitro* and *in vivo* effects relevant to its actions to reduce the pathogenesis of CVD, breast cancer, colorectal cancer, prostate cancer, inflammation and ageing (Ragab *et al.* 2006). However, much of the epidemiological and clinical evidence for disease prevention associated with resveratrol comes from studies on the consumption of red wine, grapes and grape juice rather than Itadori tea (Cassidy *et al.* 2000).

Itadori tea has been used for centuries in Japan and China as a traditional herbal remedy for many diseases including heart disease and stroke (Kimura *et al.* 1985), yet no clinical trials or human studies of Itadori tea have been published.

In vitro, water extracts of *P. cuspitadum* have been shown to inhibit xanthine oxidase activity (Kong *et al.* 2000), the expression of inducible nitric oxide synthase (iNOS) and cyclooxygenase 2 (COX-2) gene expression in macrophages (Kim *et al.* 2007), and cholesterol ester formation in human hepatocytes (Park *et al.* 2004). In a study of ethanol extracts (70%) from 32 traditional Chinese medicinal plants used to treat menopausal symptoms, *P. cuspitadum* was found to have the highest oestrogenic relative potency *in vitro* (Zhang *et al.* 2005).

In animal models, components of *P. cuspitadum* have been shown to inhibit lipid peroxidation in rats fed peroxidised oil (Kimura *et al.* 1983), lower cholesterol in hyperlipidaemic rabbits fed a diet high in fat and cholesterol (Xing *et al.* 2009), exert a neuroprotective effect on cerebral injury induced by ischaemia/reperfusion in rats (Cheng *et al.* 2006; Wang *et al.* 2007b) and improve post-ischaemic left ventricular functions and aortic flow in rats (Sato *et al.* 2000).

4.3.3 *Chamomile* (Chamomilla recutita *L.*)

Chamomile (*C. recutita* L., synonymous with *Matricaria recutita* L. Rauschert, and *Matricaria chamomilla*), an annual plant indigenous to Europe and Western Asia, is cultivated in Germany, Hungary, Russia and elsewhere in southern and eastern European countries for the flower heads. Infusions and essential oils from fresh or dried flower heads have aromatic, flavouring and colouring properties; both are used in a number of commercial products including soaps, detergents, perfumes, lotions, ointments, hair products, baked goods, confections, alcoholic beverages and herbal teas (McKay and Blumberg 2006). Other common names for this plant include German chamomile, Hungarian chamomile, mayweed, sweet false chamomile or wild chamomile. Over 120 constituents have been identified in chamomile flowers including several flavonoids and other phenolic compounds (Mann and Staba 1986). The major flavonoids present in the flowers of this plant are apigenin, quercetin, patuletin and luteolin. Interestingly, chamomile is one of the richest natural sources of apigenin at 840 mg/100 g dried material. Teas brewed from chamomile contain 10–15% of the essential oil available in the flower, the main constituents of which include the terpenoids α-bisabolol and its oxides and azulenes, including chamazulene. Qualitative and quantitative differences in the essential oil can vary significantly between growing regions, in cultivated *vs.* wild plant populations, and with different processing conditions (McKay and Blumberg 2006).

In the United States, chamomile is one of the most widely used herbal tea ingredients. The German Commission E approved the internal use of *C. recutita* for gastrointestinal spasms and inflammatory diseases of the gastrointestinal tract (Klein and Riggins 1998b). Consuming an infusion of 3 g in 150 mL water three to four times daily is recommended for gastrointestinal complaints, while its use as a mouthwash or gargle is recommended for inflammation of the mucous membranes of the mouth and throat (Wichtl and Bisset 1994; Klein and Riggins 1998b).

In animal models, chamomile extracts have demonstrated anti-inflammatory (Shipochliev *et al.* 1981; Al-Hindawi *et al.* 1989), antipruritic (Kobayashi *et al.* 2003, 2005), antispasmodic (Achterrath-Tuckermann *et al.* 1980; Forster *et al.* 1980) and hypocholesterolaemic effects (Al-Jubouri *et al.* 1990). More recently, the antidiabetic effects of chamomile have been examined in streptozotocin-induced diabetic rats (Cemek *et al.* 2008; Kato *et al.* 2008). Kato *et al.* (2008) showed the administration of a hot-water chamomile extract for 21 days suppressed blood glucose levels, enhanced liver glycogen levels and inhibited aldose reductase. After 14 days, Cemek *et al.* (2008) found a dose-dependent reduction in postprandial hyperglycaemia and oxidative stress with administration of an ethanol extract of chamomile. To date, none of these reported effects have been demonstrated in humans.

Despite the popular use of chamomile tea as a mild sedative and sleep aid, the German Commission E did not grant approval for such use due to the lack of published research in this area. In mice, extracts of *C. recutita* have been shown to potentiate hexobarbital-induced sleep (Della Loggia *et al.* 1982) and delay the onset of picrotoxin-induced seizure (Heidari *et al.* 2009). However, the only evidence supporting a potential effect on sleep comes from a small, uncontrolled study conducted by Gould *et al.* (1973) in which 10 of 12 heart disease patients, hospitalised for cardiac catheterisation, reportedly fell into a deep sleep within 10 minutes of consuming two cups of chamomile tea. More recently, the results of a randomised, double-blind, placebo-controlled trial conducted by Amsterdam et al. (2009) suggest a modest anxiolytic effect of chamomile in patients with generalised anxiety disorder. Subjects were given one capsule daily (chamomile extract standardised to 1.2% apigenin or placebo) during the first week with weekly dose escalation, up to a maximum of five capsules per day, among subjects who reported a reduction of $\leq$50% in total Hamilton Anxiety Rating (HAM-A) score. After 8 weeks, a significant reduction in HAM-A score was observed in the chamomile extract group compared to the placebo group ($P = 0.047$). Although these studies are suggestive of some benefit, the calmative effects of chamomile tea in humans have yet to be confirmed.

No clinical trials have examined the gastrointestinal effects of chamomile tea alone. However, a herbal mixture including chamomile has been shown to effectively reduce the symptoms of colic in infants within 1 week (Weizman et al. 1993; Savino et al. 2005), and a chamomile/apple pectin preparation has been shown to shorten the duration of non-specific diarrhoea in young children (De la Motte et al. 1997; Becker et al. 2006) when compared with placebo treatment. A summary of human studies with C. recutita is presented in Table 4.2.

No long-term adverse effects of chamomile tea consumption have been reported. There have been a few case reports of atopic and contact dermatitis with chamomile use (Pereira et al. 1997; Rodriguez-Serna et al. 1998; Rycroft 2003). In rare cases, chamomile may cause severe allergic reactions (Subiza et al. 1989; Reider et al. 2000), particularly among those who also have allergies to other plants in the daisy family (*Asteraceae* or *Compositae*), including mugwort (*Artemisia vulgaris*) (De la Torre Morin et al. 2001) and ragweed (*Ambrosia trifida*) (Subiza et al. 1989). Potential drug interactions with warfarin have been suggested, but not demonstrated (Miller 1998; Heck et al. 2000; Abebe 2002). Similarly, the additive effects of chamomile in combination with sedatives, such as opioid analgesics, benzodiazepines or alcohol, is theoretical (O'Hara et al. 1998; Abebe 2002; Larzelere and Wiseman 2002). Bianco et al. (2008) reported a higher prevalence of *Clostridium botulinum* spores in unwrapped chamomile sold by weight (in bulk) compared with chamomile sold in tea bags ($P < 0.01$). Therefore, dried chamomile flowers intended for use in infants or other vulnerable populations should be packaged appropriately.

4.3.4 *Hibiscus* (Hibiscus sabdariffa L.)

Hibiscus (*H. sabdariffa* L.) is an annual shrub native to Southeast Asia and widely cultivated in tropic and subtropic regions of the world. China and Thailand are the largest producers and control much of the world supply. All parts of this plant including the stems, leaves, seeds and flower calyces, i.e. the bright-red fleshy fruit surrounding the seedpod, are edible. However, only the mature calyces are used to make hibiscus tea. Around the world, beverages derived from boiled and/or brewed *H. sabdariffa* calyces have several different names, including agua de Jamaica, karkade, roselle, red rorrel, bissap, zobo, lo shen and sour tea. In the United States, dried calyces of this plant are most often found in commercial herbal tea blends.

Table 4.2 Summary of human intervention studies with chamomile (*Chamomilla recutita*)

Reference	Dosage form	Subjects	Dose	Duration	Outcome
Amsterdam et al. (2009)	Capsule	57 adults with generalised anxiety disorder	Escalated dose: 1–5 capsules/day (standardised to 1.2% apigenin) or placebo	8 weeks	Reduced mean total Hamilton Anxiety Rating score compared to placebo
Becker et al. (2006)	Liquid	255 children with acute, non-complicated diarrhoea (0.5–6 years)	Chamomile/apple pectin preparation or placebo (standardised to 2.5 g/100 g chamazulene)	3 days	Reduced stool frequency compared to placebo
Savino et al. (2005)	Beverage	88 healthy breastfed infants (21–60 weeks)	Standardised herbal mixture (including 71.1 mg/kg/day chamomile) or placebo	1 week	Reduced crying time compared to placebo
Nakamura et al. (2002)	Beverage	Young men	1 serving chamomile tea or hot water	Acute	Greater heart rate decrease and ratings of sadness and depression after drinking tea
De la Motte et al. (1997)	Liquid	79 children with acute, non-complicated diarrhoea (0.5–5.5 years)	Chamomile/apple pectin preparation (standardised to 2.5 g/100 g chamazulene) or placebo	3 days	Diarrhoea ended sooner in treatment than in placebo group; duration was attenuated
Weizman et al. (1993)	Beverage	68 healthy infants (2–8 weeks)	150 mL herbal mixture including chamomile or placebo	1 week	Reduced colic compared to placebo
Gould et al. (1973)	Beverage	12 heart disease patients hospitalised for cardiac catheterisation	2 cups chamomile tea	Acute	Decreased mean brachial artery pressure from baseline

Adapted from McKay and Blumberg 2006.

Although not determined in clinical trials, hibiscus tea is used as a folk remedy for loss of appetite, colds, catarrhs of the upper respiratory tract and stomach, to dissolve phlegm, as a gentle laxative, diuretic, and for circulatory disorders (Klein and Riggins 1998c). In traditional medicine, aqueous extracts of *H. sabdariffa* calyces are used for many complaints including high BP, liver diseases and fever (Ali *et al.* 2005).

Studies conducted in animal models suggest that extracts of *H. sabdariffa* calyces have hepatoprotective (Wang *et al.* 2000; Liu *et al.* 2002; Ali *et al.* 2003; Lin *et al.* 2003), chemopreventive (Chewonarin *et al.* 1999), antitumourigenic (Muller and Franz 1992), immunomodulatory (El-Shabrawy *et al.* 1988; Muller and Franz 1992; Dafallah and al-Mustafa 1996), antispasmodic (Ali *et al.* 1991a, 1991b; Obiefuna *et al.* 1994; Owolabi *et al.* 1995; Adegunloye *et al.* 1996; Sarr *et al.* 2009), cathartic (Ali *et al.* 1991a; Haruna 1997; Salah *et al.* 2002), antiatherosclerotic (El-Saadany *et al.* 1991; Tee *et al.* 2002; Chen *et al.* 2003, Hirunpanich *et al.* 2006; Yang *et al.* 2010), diuretic (Caceres *et al.* 1987; Mojiminiyi *et al.* 2000) and hypotensive effects (Jonadet *et al.* 1990; Ali *et al.* 1991b; Adegunloye *et al.* 1996; Onyenekwe *et al.* 1999; Odigie *et al.* 2003; Mojiminiyi *et al.* 2007).

In humans, evidence supporting a cholesterol-lowering effect is limited. In a 4-week trial of 53 type 2 diabetics, mean age 50–55 years, subjects were randomised to receive either two servings per day of hibiscus tea or black tea similarly prepared (Mozaffari-Khosravi *et al.* 2009a). The teas were prepared by brewing 2 g in 240 mL boiled water for 20–30 minutes and adding 5 g sugar. In the group consuming hibiscus tea, HDL increased ($P = 0.002$), while total cholesterol, LDL, triglycerides and Apo-B100 all decreased ($P < 0.05$) compared to baseline values. In the black tea group, only HDL significantly increased from baseline ($P = 0.002$). No comparison of the between-group changes over the course of the intervention were included in the analysis. Therefore, these results merely suggest a possible effect of *H. sabdariffa* on lipid and lipoprotein levels in diabetic patients. In an uncontrolled study by Lin *et al.* (2007), 42 subjects, aged 18–75 years, with cholesterol levels of 175–327 mg/dL who were not taking medications for 1 month prior were randomised to receive 3, 6 or 9 capsules containing 500 mg each of lyophilised *H. sabdariffa* tea daily for 4 weeks. The tea powder was prepared by steeping 150 g of dried calyces in 6 L water for 2 hours before drying. Although reductions in serum cholesterol were observed in the groups receiving three and six capsules per day ($P < 0.05$ compared with baseline), the lack of a placebo group makes the results less than definitive.

Evidence supporting a BP-lowering effect of *H. sabdariffa* in humans is accumulating. To date, studies have been conducted in hypertensive, mildly hypertensive and prehypertensive subjects, as well as in type 2 diabetics with mildly elevated BP. Two studies compared the effects of hibiscus to black tea, one used an inert control beverage and two compared it with antihypertensive drugs. The first human study published study was a 15-day trial of 54 patients with systolic blood pressure (SBP) 160–180 mm Hg and diastolic blood pressure (DBP) 100–114 mm Hg, who were using two or less antihypertensive drugs and had no secondary hypertension or underlying diseases (Haji Faraji and Haji Tarkhani 1999). Subjects were randomised to receive either one-daily serving of hibiscus tea prepared with 'two spoonfuls of blended tea per glass' brewed in boiling water for 20–30 minutes or black tea prepared similarly. Within 12 days, SBP was 17.6 mm Hg lower in the group receiving hibiscus tea compared with a 6.4 mm Hg reduction in the group receiving black tea ($P < 0.001$), and DBP was 10.9 mm Hg lower with hibiscus tea compared to a 3.5 mm Hg reduction with black tea ($P < 0.001$). Three days after both treatments were stopped, SBP increased 7.9 mm Hg and DBP 5.6 mm Hg in the hibiscus group, compared with minimal increases in the black tea group. Black tea was also used as the control beverage in a randomised trial of 53 type 2 diabetic patients with mild hypertension, i.e. SBP/DBP <140/85 mm Hg (Mozaffari-Khosravi *et al.* 2009b). Subjects in this study had diabetes for >5 years, most were taking oral antihyperglycaemic agents, but no cholesterol or BP-lowering drugs, and had no secondary hypertension or other diseases. Subjects consumed either two servings per day of a beverage prepared by brewing 2 g of dried *H. sabdariffa* in 240 mL boiled water for 20–30 minutes or black tea prepared similarly. After 30 days, SBP was 21.7 mm Hg lower in the hibiscus group compared with an 8.7 mm Hg increase in the black tea group ($P < 0.001$), and DBP was 1.1 mm Hg lower with hibiscus compared to a 3.3 mm Hg increase with black tea ($P = 0.04$). Given that subjects were asked to refrain from consuming any other tea beverages over the course of this study, it is plausible that the higher caffeine intake in the black tea group contributed to the increases in SBP and DBP observed in this group. Although chronic black tea consumption has little to no effect on BP (Taubert *et al.* 2007; Grassi *et al.* 2009), it has been shown to raise BP acutely (Hooper *et al.* 2008) and improve brachial artery flow-mediated dilation (Duffy *et al.* 2001), disqualifying it as an appropriate control beverage in hypertension studies.

In a randomised, double-blind, placebo-controlled clinical trial, McKay *et al.* (2010) demonstrated a significant BP-lowering effect of consuming 3 cups per day of hibiscus tea in pre- and mildly hypertensive subjects with SBP/DBP <150/95 mm Hg who were not taking any hypertensive medications and had no major chronic diseases. The control beverage used in this study was an inert, artificially flavoured and coloured beverage devoid of nutrients and polyphenols. Each serving of the *H. sabdariffa* tea was prepared by brewing 1.25 g dried calyces in 240 mL water for 6 minutes. After 6 weeks, SBP was 7.2 mm Hg lower in the hibiscus tea group compared with a 1.3 mm Hg reduction in the placebo group ($P = 0.03$). Although DBP was 3.1 mm Hg lower with hibiscus tea and 0.5 mm Hg lower with placebo, the between-group difference was not statistically significant ($P = 0.16$). They authors also observed a greater reduction in SBP among subjects with higher initial baseline levels (Figure 4.1).

The effects of *H. sabdariffa*-derived beverages were compared directly with two angiotensin converting enzyme (ACE) inhibitors, captopril and lisinopril, in separate randomised trials (Herrera-Arellano *et al.* 2004, 2007). In the first trial, 75 hypertensive patients, aged 30–80 years, who were not taking any antihypertensive medications for at least 1 month prior, received either a daily infusion of *H. sabdariffa*, prepared by brewing 10 g dry calyces in 0.5 L boiled water for 10 minutes, or 25 mg captopril administered twice daily. After 4 weeks, *H. sabdariffa* lowered SBP 15.4 mm Hg from baseline ($P < 0.03$) and DBP 11.6 mm Hg from baseline ($P < 0.06$). No significant differences were detected between the changes observed in the hibiscus and captopril groups ($P > 0.25$), suggesting a comparable effect of both treatments in this cohort. The second study compared the effects of a dried *H. sabdariffa* extract, standardised to 250 mg anthocyanins, with 10 mg per day lisinopril in 171 stage I or II hypertensive subjects, aged 25–61 years, after 4 weeks. Even though the hibiscus extract lowered

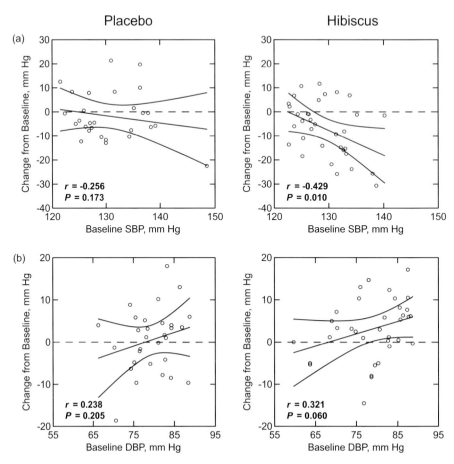

Figure 4.1 Changes in SBP (a) and DBP (b) relative to baseline values in prehypertensive and mildly hypertensive adults following a 6-week intervention with placebo beverage ($n = 30$) or hibiscus tea ($n = 35$). Linear regression lines and 95% CI are depicted. Correlation coefficients r and nominal P-values were calculated with the Pearson test.

SBP 16.6 mm Hg and DBP 11.8 mm Hg ($P < 0.05$ compared with baseline), the reductions achieved with lisinopril were significantly greater ($P < 0.05$). Nevertheless, the therapeutic effectiveness of the hibiscus extract was determined to be 65%, with 100% tolerability and safety, suggesting a potential role as a therapeutic agent or adjuvant for hypertensive patients. A summary of human studies with *H. sabdariffa* is presented in Table 4.3.

No toxicological studies of hibiscus have been reported in humans, but there have been a few reports in animals. Onyenekwe *et al.* (1999) administered a concentrated extract of hibiscus calyces to mice i.p. at 1000, 3000 and 5000 mg/kg. The higher the dose, the less active the mice became, although all recovered within 30 minutes. They estimated the LD50 of hibiscus to be >5000 mg/kg in mice. Akindahunsi and Olaleye (2003) also reported behavioural changes, including prostration, stretching,

Table 4.3 Summary of human intervention studies with hibiscus (*Hibiscus sabdariffa*)

Reference	Dosage form	Subjects	Dose	Duration	Outcome
McKay et al. (2010)	Beverage	65 subjects with mild hypertension (<150/95 mm Hg)	1.25 g/240 mL boiled water, brewed 6 min, or 240 mL artificially flavoured/coloured placebo, three times/day	6 weeks	Reduced SBP and DBP compared to baseline, and SBP compared to placebo
Mozaffari-Khosravi et al. (2009)	Beverage	53 type 2 diabetics (50–55 years)	2 g hibiscus or black tea/240 mL boiled water, brewed 20–30 min, + 5 g sugar, two times/day	4 weeks	Reduced total cholesterol, LDL, triglycerides and increased HDL compared to baseline; no between-group changes
Mozzafari-Khosravi et al. (2009b)	Beverage	53 type 2 diabetics (<140/85 mm Hg; 50–55 years)	2 g hibiscus or black tea/240 mL boiled water, brewed 20–30 min, + 5 g sugar, two times/day	4 weeks	Reduced SBP and DBP compared to black tea
Lin et al. (2007)	Dried tea powder	42 hypercholesterolaemics (18–75 years)	3, 6 or 9 capsules/day, each with 500 mg lyophilised hibiscus tea powder; no placebo	4 weeks	Reduced cholesterol with 3 and 6 capsules/day compared to baseline levels
Herrera-Arellano et al. (2007)	Dried extract	171 hypertensives (25–61 years)	Hibiscus extract (standardised to 250 mg anthocyanins)/250 mL or 10 mg lisinopril, once/day	4 weeks	Reduced SBP and DBP compared to baseline, but not compared to hypertensive drug
Herrera-Arellano et al. (2004)	Beverage	75 hypertensives (30–80 years)	10 g dry calyces/0.5 L boiled water, brewed 10 min, or 25 mg captopril, once/day	4 weeks	No significant differences between treatments
Haji Faraji and Haji Tarkhani (1999)	Beverage	54 hypertensives (160–180/100–114 mm Hg)	2 spoonfuls of hibiscus or black tea/glass brewed in boiling water 20–30 min/day	12 days	Reduced SBP and DBP compared to black tea

sluggishness and slow response to external stimuli, within 5 minutes after administration of 250 mg/kg p.o. to rats. In this experiment, rats were given 250 mg/kg of an aqueous methanolic extract of hibiscus once daily for 1, 3, 5, 10 or 15 days. Compared to the control group receiving saline alone, serum alanine aminotransferase was higher in all but the 1-day group, aspartate aminotransferase was higher in the 10- and 15-day groups, and alkaline phosphatase and lactate dehydrogenase were not significantly affected. Serum albumin also increased in the 5-, 10- and 15-day groups. Elevation of these biochemical indices is indicative of injury to the liver or heart, but no pathological features were observed in these organs for any treatment level. The investigators concluded that prolonged use of hibiscus extract at 250 mg/kg could affect the liver, but the experimental design did not allow the determination of a safe level with regular consumption.

4.3.5 *Rooibos* (Aspalathus linearis)

Rooibos (*A. linearis*) is a shrub-like leguminous bush native to the Cedarberg mountains of South Africa, where it is extensively cultivated for its commercial use as an herbal tea or tisane (McKay and Blumberg 2007). Upon harvesting, the needle-like leaves and stems can be either bruised and fermented prior to drying or dried immediately without fermentation. During fermentation, the colour changes from green to red with oxidation of its constituent polyphenols. The unfermented product is called 'green rooibos', while the fermented product is called 'red tea' or 'red bush tea'. Brewed beverages of this plant may also be referred to as rooibos tea, rooibosch, rooitea or rooitee.

In folk medicine, rooibos tea is used to alleviate infantile colic, indigestion, heartburn and nausea, stimulate the appetite, reduce nervousness and promote sound sleep (Joubert *et al.* 2008). Topical applications of rooibos extract are believed to alleviate dermatological problems, i.e. eczema, acne and nappy rash. The modern use of rooibos tea began in the early twentieth century. It was considered an alternate to 'Oriental' tea and consumed as a strong, hot brew with milk and sugar added. During the summer months in South Africa, it is enjoyed cold, usually with lemon juice and sugar added. More recently, it has become popular to brew and serve rooibos in the style of ordinary espresso, creating new variations of coffee drinks such as red cappuccinos and lattes.

Studies conducted in animal models suggest that rooibos reduces lipid peroxidation (Inanami *et al.* 1995; Shimoi *et al.* 1996), enhances endogenous antioxidant defenses (Marnewick *et al.* 2003; Baba *et al.* 2009), and has immune-modulating (Kunishiro *et al.* 2001), hepatoprotective (Ulicna *et al.* 2003; Kucharska *et al.* 2004), antimutagenic (Sasaki *et al.* 1993; Shimoi *et al.* 1996; Marnewick *et al.* 2004) and antispasmodic effects (Snyckers and Salemi 1974; Gilani *et al.* 2006; Khan and Gilani 2006). Although this evidence suggests that a potential health benefit of rooibos consumption is accumulating, none of these actions has yet been demonstrated in humans.

Only three human studies of rooibos tea have been published in the scientific literature to date: two are related to its effect on iron absorption and one on its potential antiallergenic properties. Hesseling *et al.* (1979) studied the effect of rooibos *vs.* black tea and water on iron absorption in 30 healthy young men aged 21–34 years.

No between-group differences with regard to iron status (haemoglobin, ferritin and transferrin) were detected at baseline. Subjects were administered 1 µCi radiolabelled ^{59}Fe plus 16 mg elemental iron followed by rooibos tea, black tea or water. Mean iron absorption, measured 2 weeks later, was 7.25% with rooibos tea, 1.70% with black tea and 9.34% with water. Therefore, unlike black tea, rooibos does not appear to interfere with iron absorption. More recently, Breet *et al.* (2005) examined the iron status of undernourished African school children following 16 weeks' consumption of either 400 mL per day black tea or rooibos tea taken with a non-haeme iron–containing meal. They found no statistically significant changes in any iron status parameters between the black tea and rooibos groups. A summary of human studies with *A. linearis* is presented in Table 4.4.

Hesseling and Joubert (1982) also tested whether rooibos tea had an antiallergenic effect in subjects diagnosed with asthma or hay fever ($n = 7$). Skin prick tests of 16 commonly inhaled allergens were administered on both forearms 11–14 and 7 days prior to treatment (repeated controls) and then again following rooibos treatment. On the day of treatment, subjects consumed rooibos tea prepared as 25 g/L in boiling water for 5 minutes and provided in three 500-mL doses 3 hours apart. Following the two control tests, 14 of the 16 antigens were of similar size on both arms of each patient, while two antigens produced a significant positive reaction. Following the ingestion of rooibos tea, 12 of the antigens remained unchanged from controls, while reactions to four antigens (house dust, grass pollen, dog epithelia and *Aspergillus fumigatus*)

Table 4.4 Summary of human studies with rooibos (*Aspalathus linearis*)

Reference	Dosage form	Subjects	Dose	Duration	Outcome
Breet *et al.* (2005)	Beverage	150 undernourished children (6–15 years)	400 mL/day rooibos tea or black tea	16 weeks	No change in iron status between groups
Hesseling and Joubert (1982)	Beverage	7 subjects with asthma or hay fever	25 g rooibos/L boiling water, provided in three 500-mL doses given 3 h apart on same day skin prick tests of antigens were administered	1 day	No antihistaminic effects were observed compared to control tests given without rooibos treatment
Hesseling *et al.* (1979)	Beverage	30 healthy men (21–34 years)	Rooibos tea, black tea or water following acute ingestion of radiolabelled iron	2 weeks	Compared to black tea, rooibos had no significant effect on Fe absorption, and was no different than water

Adapted from McKay and Blumberg 2007.

were significantly larger. In another experiment, a rooibos poultice was applied to one forearm of each subject for 15 minutes and followed by skin prick tests that gave similar results. Neither ingestion of rooibos tea nor its topical application exhibited any antihistaminic effects in this study. Therefore, its therapeutic value as an antiallergy treatment has not been substantiated in humans.

No adverse effects were reported by Marnewick *et al.* (2003) who examined a variety of safety indices for green and red rooibos following a 10-week intervention with 2 g/100 mL aqueous extracts in rats. Neither tea affected body weight, liver weight, or liver and kidney parameters, including serum aspartate transaminase, alanine transaminase, alkaline phosphatase, creatinine, total and unconjugated bilirubin, total protein, total cholesterol or iron status. However, potential interactions between herbal products and prescription medications must always be considered. In rats, aqueous extracts of rooibos have been shown to inhibit the activity of the cytochrome P450 isoenzyme, CYP2C11, after a 3-day infusion (Jang *et al.* 2004), and increase the activity of CYP3A in rat intestine, but not liver, after continuous ingestion for 2 weeks (Matsuda *et al.* 2007).

4.3.6 Honeybush (Cycolia intermedia)

Honeybush (*C. intermedia*) is a short, woody shrub grown in South Africa (McKay and Blumberg 2007). Unlike other herbal tea ingredients, honeybush is not widely cultivated, and most of the commercially available product is collected from natural plant populations. Tisanes of *C. intermedia* leaves, stems and flowers may be referred to as honeybush tea, Heuningtee, Bergtee, Boertee, Bossiestee or Bush tea. During harvesting the plant material is cut to disrupt cellular integrity, fermented in either a curing heap or at elevated temperatures in a preheated baking oven, and then allowed to dry. During fermentation, the plant material changes colour from green to dark brown as the phenolic compounds are oxidised.

In folk medicine, *Cyclopia* species other than *C. intermedia* have been used as a restorative, an expectorant in chronic catarrh and pulmonary tuberculosis, an appetite stimulant, digestive aid, and in alleviating heartburn and nausea (Joubert *et al.* 2008).

Anecdotal reports suggest it can stimulate milk production in breastfeeding women and treat colic in babies (Rood 1994). More recently, infusions from the fermented herb have become increasingly popular as a beverage for everyday use due to their characteristic honey-like flavour, low tannin content, absence of caffeine and potential health effects related to their antimutagenic and antioxidant properties (Kokotkiewicz and Luczkiewicz 2009).

No clinical trials or human studies examining the effects of *C. intermedia* or honeybush tea have been reported. However, in animal models, *C. intermedia* has demonstrated antioxidant and chemopreventive actions. Marnewick *et al.* (2003) examined the effects of honeybush tea on antioxidant capacity and hepatic phase II metabolising enzymes in male Fischer rats ($n = 10$ per group) given either water, a 2% solution of green tea, black tea, rooibos tea (unfermented or fermented) or 4% honeybush tea (unfermented or fermented) as their sole source of drinking fluid for 10 weeks. Each tea solution was prepared by adding boiling water to tea leaves and stems, and allowing them to brew at room temperature for 30 minutes prior to filtration. No changes in antioxidant capacity, measured with the oxygen radical absorbance capacity (ORAC) assay, were observed in liver homogenates of rats treated with either type of honeybush or rooibos tea. However, all teas tested lowered hepatic oxidised glutathione (GSSG) when compared to water. GSSG concentrations in both the unfermented (0.51 ± 0.18 µM/mg protein) and fermented (0.70 ± 0.16 µM/mg) honeybush tea groups were comparable to levels found in the green (0.76 ± 0.15 µM/mg) and black (0.87 ± 0.29 µM/mg) tea groups, although levels in the unfermented and fermented rooibos tea groups were significantly lower (0.42 ± 0.13 and 0.40 ± 0.13 µM/mg, respectively). The ratios of reduced glutathione (GSH) to GSSG were significantly higher in the livers of rats given unfermented honeybush and marginally higher in the fermented honeybush group, while no changes were observed in those given green or black tea. Cytosolic glutathione S-transferase α (GST-α) activity was higher only in rats given the honeybush or rooibos teas, and microsomal glucuronosyl transferase (UDP-GT) activity was modestly higher with the unfermented honeybush and rooibos teas only. The results of this study suggest that honeybush tea may protect against oxidative damage *in vivo*.

Marnewick *et al.* (2004) used the same study design to examine the effects of honeybush tea on the *ex vivo* modulation of chemical-induced mutagenesis in rats. Using microsomal and cytosolic liver fractions isolated from the honeybush tea-treated rats, mutagenicity induced with 2-acetylaminofluorene (AAF) and aflatoxin B_1 (AFB_1) in the *Salmonella* assay was suppressed. Using *Salmonella typhimurium* strain TA100 and AFB_1, the number of histidine revertants per plate was reduced with cytosolic fractions (0.25 mg/mL protein) from rats treated with unfermented and fermented honeybush (303.2 and 306.1, respectively) compared to rats given water (385.0, $P < 0.05$). Unfermented honeybush also reduced mutagenesis determined with TA98 and AAF (161.3 *vs.* 220.3 in control animals, $P < 0.05$), but the effect of fermented honeybush was only marginal (191.1, $P < 0.10$). No protective effects were observed when higher cytosolic protein concentrations (1 mg/mL) from honeybush-treated rats were used. The hepatic microsomal fractions from rats consuming unfermented, but not fermented, honeybush teas were able to protect against mutagenicities induced by AFB_1 in strain TA100, but increased the number of revertants obtained with AAF induction of TA100 (79%); in comparison, microsomal fractions from the green and black tea groups more potently increased the mutagenicity of AAF (132% and 128%, respectively). These results suggest that honeybush tea may alter the potency of hepatocarcinogens *in vivo*.

Marnewick *et al.* (2009) also examined the *in vivo* cancer-modulating properties of these teas. Male Fischer rats ($n = 10$ per group) were given a single dose of diethylnitrosamine (DEN; 200 mg/kg body weight) to initiate cancer. Teas were offered as the sole source of drinking fluid starting 1 week after initiation; 3 weeks later a diet containing cancer-promoting fumonisin B (FB_1) mycotoxins (250 mg/kg) was administered for an additional 3 weeks. Black tea and fermented honeybush significantly ($P < 0.05$) and unfermented honeybush marginally ($P < 0.10$) counteracted the increase in serum creatinine, suggesting a protective effect against FB_1-induced nephrotoxicity. Honeybush (both unfermented and fermented) and green tea also protected against FB_1-induced accumulation of serum cholesterol ($P < 0.05$), while honeybush (both fermented and unfermented) and fermented rooibos reduced FB_1-induced lipid peroxidation in the liver ($P < 0.05$). With regard to the induction of glutathione-S-transferase (GST) foci, unfermented rooibos had a significant effect ($P < 0.05$) and honeybush a marginal effect ($P < 0.10$) on reducing the total number of foci (>10 μm). Fermentation seems to reduce the protective effect of both honeybush and rooibos teas, perhaps due to differences in their major polyphenolic components. Further examination is necessary to confirm the cancer-inhibiting effects of honeybush tea consumption *in vivo*.

No adverse effects or toxicities of *C. intermedia* have been reported, although the presence of microbial contaminants during the fermentation and drying of honeybush tea was reported by Du Toit *et al.* (1999). However, processing the plant materials at a temperature $>60°C$ with controlled curing conditions eliminated these microbial contaminants.

4.4 Summary and conclusions

A large body of evidence has accumulated demonstrating the influence of specific dietary patterns and individual foods on promoting health and reducing the risk of chronic disease. Interestingly, this work has largely ignored the potential contribution of beverages to overall health. For example, the Dietary Guidelines for Americans addresses only fortified low-fat fluid milk as an important source of calcium and vitamin D and water for meeting general hydration needs. More recently, several studies have tried to assess the impact of sugar-sweetened beverages on the risk for obesity.

Particularly, during the last decade, an appreciation has emerged that positive beverage choices may contribute importantly to health. Teas, especially, and tisanes have been proposed as being more healthful choices than milk, diet sodas and fruit juices (Popkin *et al.* 2006); when evaluated by criteria of calories and nutrient density, teas and herbal teas were ranked very highly.

Evidence supporting the potential health benefits of tea and tisanes covers a broad range of conditions. Nonetheless, it is critical to note the great variation in the strength of the available evidence for each outcome. For example, evidence from basic research, observational studies and clinical trials for a beneficial effect of tea (*C. sinensis*) on cardiovascular health is compelling, while its impact on other diseases is mixed or more limited. Attention is now warranted for studies that apply '-omics' and systems biology approaches to better predict which conditions, and perhaps even which individuals, are most likely affected by regular tea consumption. As most of the available research on tisanes has been conducted *in vitro* and in animal models, advances towards understanding their contribution to health promotion requires evidence obtained from human studies including intermediary biomarkers of disease, physiological responses and the modification of pathological processes. For new research on teas and, particularly, on tisanes, attention should also be directed to the potential for untoward responses as well as adverse interactions with medications.

Research on tea and tisanes has understandably been directed towards its role in the primary prevention of chronic disease. However, it is worthwhile noting the therapeutic role of these beverages in the practice of Ayurvedic, Chinese and other traditional medicines. Substantiating the value of this long history of such use presents another avenue for future studies.

References

Abebe, W. (2002) Herbal medication: potential for adverse interactions with analgesic drugs. *J. Clin. Pharm. Ther.*, **27**, 391–401.

Achterrath-Tuckermann, U., Kunde, R., Flaskamp, E. *et al.* (1980) Pharmacological investigations with compounds of chamomile; V. Investigations on the spasmolytic effect of compounds of chamomile and Kamillosan on the isolated guinea pig ileum. *Planta Med.*, **9**, 38–50.

Adegunloye, B.J., Omoniyi, J.O., Owolabi, O.A. *et al.* (1996) Mechanisms of the blood pressure lowering effect of the calyx extract of *Hibiscus sabdariffa* in rats. *Afr. J. Med. Med. Sci.*, **25**, 235–238.

Akindahunsi, A.A. and Olaleye, M.T. (2003) Toxicological investigation of aqueous-methanolic extract of the calyces of *Hibiscus sabdariffa* L. *J. Ethnopharmacol.*, **89**, 161–164.

Alexopoulos, N., Vlachopoulos, C., Aznaouridis, K. *et al.* (2008) The acute effect of green tea consumption on endothelial function in healthy individuals. *Eur. J. Cardiovasc. Prev. Rehabil.*, **15**, 300–305.

Al-Hindawi, M.K., Al-Deen, I.H., Nabi, M.H. *et al.* (1989) Antiinflammatory activity of some Iraqi plants using intact rats. *J. Ethnopharmacol.*, **26**, 163–168.

Ali, M.B., Mohamed, A.H., Salih, W.M. *et al.* (1991a) Effect of an aqueous extract of *Hibiscus sabdariffa* calyces on the gastrointestinal tract. *Fitoterapia*, **62**, 475–479.

Ali, M.B., Salih, W.M., Mohamed, A.H. *et al.* (1991b) Investigation of the antispasmodic potential of *Hibiscus sabdariffa* calyces. *J. Ethnopharmacol.*, **31**, 249–257.

Ali, B.H., Mousa, H.M. and El-Mougy, S. (2003) The effect of a water extract and anthocyanins of *Hibiscus sabdariffa* L. on paracetamol-induced hepatoxicity in rats. *Phytother. Res.*, **17**, 56–59.

Ali, B.H., Al Wabel, N. and Blunden, G. (2005) Phytochemical, pharmacological and toxicological aspects of *Hibiscus sabdariffa* L.: a review. *Phytother. Res.*, **19**, 369–375.

Al-Jubouri, H.H.F., Al-Jalil, B.H., Farid, I. *et al.* (1990) The effect of chamomile on hyperlipidemias in rats. *J. Fac. Med. Baghdad*, **32**, 5–11.

Amsterdam, J.D., Li, Y., Soeller, I. *et al.* (2009) A randomized, double-blind, placebo-controlled trial of oral *Matricaria recutita* (chamomile) extract therapy for generalized anxiety disorder. *J. Clin. Psychopharmacol.*, **29**, 378–382.

Andersen, T. and Fogh, J. (2001) Weight loss and delayed gastric emptying following a South American herbal preparation in overweight patients. *J. Hum. Nutr. Diet.*, **14**, 243–250.

Arab, L. and Il'Yasova, D. (2003) The epidemiology of tea consumption and colorectal cancer incidence. *J. Nutr.*, **133**, 3310S–3318S.

Auvichayapat, P., Prapochanung, M., Tunkamnerdthai, O. *et al.* (2008) Effectiveness of green tea on weight reduction in obese Thais: a randomized, controlled trial. *Physiol. Behav.*, **93**, 486–491.

Baba, H., Ohtsuka, Y., Haruna, H., *et al.* (2009) Studies of anti-inflammatory effects of Rooibos tea in rats. *Pediatr. Int.*, **51**, 700–704.

Baker, J.A., Beehler, G.P., Sawant, A.C., *et al.* (2006) Consumption of coffee, but not black tea, is associated with a decreased risk of premenopausal breast cancer. *J. Nutr.*, **136**, 166–171.

Baker, J.A., Boakye, K., McCann, S.E. *et al.* (2007) Consumption of black tea or coffee and risk of ovarian cancer. *Int. J. Gynecol. Cancer*, **17**, 50–54.

Bates, M.N., Hopenhayn, C., Rey, O.A. *et al.* (2007) Bladder cancer and Mate consumption in Argentina: a case-control study. *Cancer Lett.*, **46**, 268–273.

Becker, B., Kuhn, U. and Hardewig-Budny, B. (2006) Double-blind, randomized evaluation of clinical efficacy and tolerability of an apple pectin-chamomile extract in children with unspecific diarrhea. *Arzneimittelforschung*, **56**, 387–393.

Belza, A., Toubro, S. and Astrup, A. (2009) The effect of caffeine, green tea and tyrosine on thermogenesis and energy intake. *Eur. J. Clin. Nutr.*, **63**, 57–64.

Bianco, M.I., Luquez, C., de Jong, L.I. *et al.* (2008) Presence of *Clostridium botulinum* spores in *Matricaria chamomilla* (chamomile) and its relationship with infant botulism. *Int. J. Food Microbiol.*, **121**, 357–360.

Bixby, M., Spieler, L., Menini, T. *et al.* (2005) *Ilex paraguariensis* extracts are potent inhibitors of nitrosative stress: a comparative study with green tea and wines using a protein nitration model and mammalian cell cytotoxicity. *Life Sci.*, **77**, 345–358.

Bracesco, N., Dell, M., Rocha, A. *et al.* (2003) Antioxidant activity of a botanical extract preparation of *Ilex paraguariensis*: prevention of DNA double-strand breaks in *Saccharomyces cerevisiae* and human low-density lipoprotein oxidation. *J. Altern. Complement. Med.*, **9**, 379–387.

Breet, P., Kruger, H.S., Jerling, J.C. *et al.* (2005) Actions of black tea and rooibos on the iron status of primary school children. *Nutr. Res.*, **25**, 983–994.

Bryans, J.A., Judd, P.A. and Ellis, P.R. (2007) The effect of consuming instant black tea on postprandial plasma glucose and insulin concentrations in healthy humans. *J. Am. Coll. Nutr.*, **26**, 471–477.

Burns, J., Yokota, T., Ashihara, H. *et al.* (2002) Plant foods and herbal sources of resveratrol. *J. Agric. Food Chem.*, **50**, 3337–3340.

Cabrera, C., Artacho, R., Giménez, R. (2006) Beneficial effects of green tea – a review. *J. Am. Col. Nutr.*, **25**, 79–99.

Caceres, A., Giron, L.M. and Martinez, A.M. (1987) Diuretic activity of plants used for the treatment of urinary ailments in Guatemala. *J. Ethnopharmacol.*, **19**, 233–245.

Cassidy, A., Hanley, B. and Lamuela-Raventos, R.M. (2000) Isoflavones, lignans and stilbenes: origins, metabolism and potential importance to human health. *J. Sci. Food Agric.*, **80**, 1044–1062.

Cemek, M., Kaga, S., Simsek, N., *et al.* (2008) Antihyperglycemic and antioxidative potential of *Matricaria chamomilla* L. in streptozotocin-induced diabetic rats. *J. Nat. Med.*, **62**, 284–293.

Chan, C.C., Koo, M.W., Ng, E.H., *et al.* (2006) Effects of Chinese green tea on weight, and hormonal and biochemical profiles in obese patients with polycystic ovary syndrome – a randomized placebo-controlled trial. *J. Soc. Gynecol.Investig*, **13**, 63–68.

Chen, C.C., Hsu, J.D., Wang, S.F. *et al.* (2003) *Hibiscus sabdariffa* extract inhibits the development of atherosclerosis in cholesterol-fed rabbits. *J. Agric. Food Chem.*, **51**, 5472–5477.

Chen, Y.K., Lee, C.H., Wu, I.C. *et al.* (2009) Food intake and the occurrence of squamous cell carcinoma in different sections of the esophagus in Taiwanese men. *Nutrition*, **25**, 753–761.

Cheng, Y., Zhang, H.T., Sun, L. *et al.* (2006) Involvement of cell adhesion molecules in polydatin protection of brain tissues from ischemia-reperfusion injury. *Brain Res.*, **1110**, 193–200.

Chewonarin, T., Kinouchi, T., Kataoka, K. *et al.* (1999) Effects of roselle (*Hibiscus sabdariffa* Linn.), a Thai medicinal plant, on the mutagenicity of various known mutagens in Salmonella typhimurium and on formation of aberrant crypt foci induced by the colon carcinogens azoxymethane and 2-amino-1-methyl-6-phenylimidazo[4,5-b]pyridine in F344 rats. *Food Chem. Toxicol.*, **37**, 591–601.

Choan, E., Segal, R., Jonker, D. *et al.* (2005) A prospective clinical trial of green tea for hormone refractory prostate cancer: an evaluation of the complementary/alternative therapy approach. *Urol. Oncol.*, **23**, 108–113.

Counet, C., Callemien, D. and Collin, S. (2006) Chocolate and cocoa: new sources of trans-resveratrol and trans-piceid. *Food Chem.*, **98**, 649–657.

Crozier, A., Jaganath, I.B., Clifford, M.N. (2009) Dietary phenolics: chemistry, bioavailability and effects on health. *Nat Prod Rep.*, **26**, 1001–1043.

Dafallah, A.A. and al-Mustafa, Z. (1996) Investigation of the anti-inflammatory activity of *Acacia nilotica* and *Hibiscus sabdariffa*. *Am. J. Chin. Med.*, **24**, 263–269.

Dai, Q., Shu, X.O., Li, H. *et al.* (2010) Is green tea drinking associated with a later onset of breast cancer? *Ann. Epidemiol.*, **20**, 74–81.

De la Motte, S., Bose-O'Reilly, S., Heinisch, M. *et al.* (1997) Double-blind comparison of a preparation of pectin/chamomile extract and placebo in children with diarrhoea. *Arzneimittelforschung*, **47**, 1247–1249.

De la Torre Morin, F., Sanchez Machin, I., Garcia Robaina, J.C. *et al.* (2001) Clinical cross-reactivity between *Artemisia vulgaris* and *Matricaria chamomilla* (chamomile). *J. Investig. Allergol. Clin. Immunol.*, **11**, 118–122.

De Mejia, E.G., Song, Y.S., Ramirez-Mares, M.V. *et al.* (2005) Effect of yerba mate (*Ilex paraguariensis*) tea on topoisomerase inhibition and oral carcinoma cell proliferation. *J. Agric. Food Chem.*, **53**, 1966–1973.

De Morais, E.C., Stefanuto, A., Klein, G.A. *et al.* (2009) Consumption of yerba mate (*Ilex paraguariensis*) improves serum lipid parameters in healthy dyslipidemic subjects and provides an additional LDL-cholesterol reduction in individuals on statin therapy. *J. Agric. Food Chem.*, **57**, 8316–8324.

De Stefani, E., Correa, P., Fierro, L. *et al.* (1991) Black tobacco, maté, and bladder cancer. A case-control study from Uruguay. *Cancer*, **67**, 536–540.

De Stefani, E., Fierro, L., Correa, P. *et al.* (1996) Mate drinking and risk of lung cancer in males: a case-control study from Uruguay. *Cancer Epidemiol. Biomarkers Prev.*, **5**, 515–519.

De Stefani, E., Fierro, L., Mendilaharsu, M. *et al.* (1998) Meat intake, 'mate' drinking and renal cell cancer in Uruguay: a case-control study. *Br. J. Cancer.*, **78**, 1239–1243.

De Stefani, E., Boffetta, P., Deneo-Pellegrini, H. *et al.* (2007) Non-alcoholic beverages and risk of bladder cancer in Uruguay. *BMC Cancer*, **7**, 57.

Della Loggia, R., Traversa, U., Scarcia, V. *et al.* (1982) Depressive effects of *Chamomilla recutita* (L.) Raush, tubular flowers, on central nervous system in mice. *Pharmacol. Res. Commun.*, **14**, 153–162.

Devine, A., Hodgson, J.M., Dick, I.M. *et al.* (2007) Tea drinking is associated with benefits on bone density in older women. *Am. J. Clin. Nutr.*, **86**, 1243–1247.

Diepvens, K., Kovacs, E.M.R., Nijs, I.M.T. *et al.* (2005) Effects of green tea on resting energy expenditure and substrate oxidation during weight loss in overweight females. *Br. J. Nutr.*, **94**, 1026–1034.

Diepvens, K., Kovacs, E.M.R., Vogels, N. *et al.* (2006) Metabolic effects of green tea and phases of weight loss. *Physiol. Behav.*, **87**, 185–191.

Du Toit, J., Joubert, E. and Britz, T.J. (1999) Identification of microbial contaminants present during the curing of honeybush tea (*Cyclopia*). *J. Sci. Food Agric.*, **79**, 2040–2044.

Duffy, S.J., Keaney, J.F. Jr., Holbrook, M. *et al.* (2001) Short- and long-term black tea consumption reverses endothelial dysfunction in patients with coronary artery disease. *Circulation*, **104**, 151–156.

El-Saadany, S.S., Sitohy, M.Z., Labib, S.M. *et al.* (1991) Biochemical dynamics and hypocholesterolemic action of *Hibiscus sabdariffa* (Karkade). *Nahrung*, **35**, 567–576.

Fagundes, R.B., Abnet, C.C., Strickland, P.T. *et al.* (2006) Higher urine 1-hydroxy pyrene glucuronide (1-OHPG) is associated with tobacco smoke exposure and drinking maté in healthy subjects from Rio Grande do Sul, Brazil. *BMC Cancer*, **6**, 139–145.

Forster, H.B., Niklas, H. and Lutz, S. (1980) Antispasmodic effects of some medicinal plants. *Planta Med.*, **140**, 309–319.

Fujita, H. and Yamagami, T. (2008) Antihypercholesterolemic effect of Chinese black tea extract in human subjects with borderline hypercholesterolemia. *Nutr. Res.*, **28**, 450–546.

Fukino, Y., Shimbo, M., Aoki, N. *et al.* (2005) Randomized controlled trial for an effect of green tea consumption on insulin resistance and inflammation markers. *J. Nutr. Sci. Vitaminol. (Tokyo)*, **51I**, 335–342.

Fukino, Y., Ikeda, A., Maruyama, K. *et al.* (2008) Randomized controlled trial for an effect of green tea-extract powder supplementation on glucose abnormalities. *Eur. J. Clin. Nutr.*, **62**, 953–960.

Gardner, E.J., Ruxton, C.H.S. and Leeds, A.R. (2003) Black tea – Helpful or harmful? A review of the evidence. *Eur. J. Clin. Nutr.*, **61**, 3–18.

Gilani, A.H., Khan, A.U., Ghayur, M.N. *et al.* (2006) Antispasmodic effects of Rooibos tea (*Aspalathus linearis*) is mediated predominantly through K^+-channel activation. *Basic Clin. Pharmacol. Toxicol.*, **99**, 365–373.

Goldenberg, D. (2002) Maté: a risk factor for oral and oropharyngeal cancer. *Oral Oncol.*, **38**, 646–649.

Goldenberg, D., Golz, A. and Joachims, H.Z. (2003) The beverage mate: a risk factor for cancer of the head and neck. *Head Neck*, **25**, 595–601.

Gosmann, G. and Schenkel, E.P. (1989) A new saponin from mate, Ilex paraguarensis. *J. Nat. Prod.*, **52**, 1367–1370.

Gould, L., Reddy, C.V. and Gomprecht, R.F. (1973) Cardiac effects of chamomile tea. *J. Clin. Pharmacol. New Drugs*, **13**, 475–479.

Grassi, D., Mulder, T.P., Draijer, R. *et al.* (2009) Black tea consumption dose-dependently improves flow-mediated dilation in healthy males. *J. Hypertens.*, **27**, 774–781.

Gugliucci, A. (1996) Antioxidant effects of *Ilex paraguariensis*: induction of decreased oxidability of human LDL *in vivo. Biochem. Biophys. Res. Commun.*, **224**, 338–344.

Gugliucci, A. and Stahl, A.J.C. (1995) Low density lipoprotein oxidation is inhibited by extracts of *Ilex paraguariensis*. *Biochem. Mol. Biol. Int.*, **35**, 47–56.

Gugliucci, A., Bastos, D.H., Schulze, J. *et al.* (2009) Caffeic and chlorogenic acids in *Ilex paraguariensis* extracts are the main inhibitors of AGE generation by methylglyoxal in model proteins. *Fitoterapia*, **80**, 339–434.

Haji Faraji, M. and Haji Tarkhani, A. (1999) The effect of sour tea (*Hibiscus sabdariffa*) on essential hypertension. *J. Ethnopharmacol.*, **65**, 231–236.

Haruna, A.K. (1997) Cathartic activity of soborodo: the aqueous extract of calyx of *Hibiscus sabdariffa* L. *Phytother. Res.*, **11**, 307–308.

Heck, C.I. and De Mejia, E.G. (2007) Yerba maté tea (*Ilex paraguariensis*): a comprehensive review on chemistry, health implications, and technological considerations. *J. Food Sci.*, **72**, R138–R151.

Heck, AM., DeWitt, B.A. and Lukes, A.L. (2000) Potential interactions between alternative therapies and warfarin. *Am. J. Health Syst. Pharm.*, **57**, 1221–1227.

Heidari, M.R., Dadollahi, Z., Mehrabani, M. *et al.* (2009) Study of antiseizure effects of *Matricaria recutita* extract in mice. *Ann. N. Y. Acad. Sci.*, **1171**, 300–304.

Henning, S.M., Aronson, W., Niu, Y. *et al.* (2006) Tea polyphenols and theaflavins are present in prostate tissue of humans and mice after green and black tea consumption. *J. Nutr.*, **136**, 1839–1843.

Herrera-Arellano, A., Flores-Romero, S., Chavez-Soto, M.A. *et al.* (2004) Effectiveness and tolerability of a standardized extract from Hibiscus sabdariffa in patients with mild to moderate hypertension: a controlled and randomized clinical trial. *Phytomedicine*, **11**, 375–382.

Herrera-Arellano, A., Miranda-Sanchez, J., Avila-Castro, P. *et al.* (2007) Clinical effects produced by a standardized herbal medicinal product of Hibiscus sabdariffa on patients with hypertension. A randomized, double-blind, lisinopril-controlled clinical trial. *Planta Med*, **73**, 6–12.

Hertog, M.G.L., Kromhout, D., Aravanis, C. *et al.* (1995) Flavonoid intake and long-term risk of coronary heart disease and cancer in the Seven Countries Study. *Arch. Intern. Med.*, **155**, 381–386.

Hesseling, P.B. and Joubert, J.R. (1982) The effect of rooibos tea on the type I allergic reaction. *S. Afr. Med. J.*, **62**, 1037–1038.

Hesseling, P.B., Klopper, J.F. and van Heerden, P.D. (1979) The effect of rooibos tea on iron absorption. *S. Afr. Med. J.*, **55**, 631–632.

Hintikka, J., Tolmunen, T., Honkalampi, K. *et al.* (2005) Daily tea drinking is associated with a low level of depressive symptoms in the Finnish General population. *Eur. J. Epidemiol.*, **20**, 359–363.

Hirata, K., Shimada, K., Watanabe, H. *et al.* (2004) Black tea increases coronary flow velocity reserve in healthy male subjects. *Am. J. Cardiol.*, **93**, 1384–1388.

Hirunpanich, V., Utaipat, A., Morales, N.P. *et al.* (2006) Hypocholesterolemic and antioxidant effects of aqueous extracts from the dried calyx of *Hibiscus sabdariffa* L. in hypercholesterolemic rats. *J. Ethnopharmacol.*, **103**, 252–260.

Hooper, L., Kroon, P.A., Rimm, E.B., *et al.* (2008) Flavonoids, flavonoid-rich foods, and cardiovascular risk: a meta-analysis of randomized controlled trials. *Am. J. Clin. Nutr.*, **88**, 38–50.

Hoshiyama, Y., Kawaguchi, T., Miura, Y. *et al.* (2004) A nested case-control study of stomach cancer in relation to green tea consumption in Japan. *Br. J. Cancer.*, **90**, 135–138.

Hosoda, K., Wang, M.F., Liao, M.L. *et al.* (2003) Antihyperglycemic effect of oolong tea in type 2 diabetes. *Diabetes Care*, **26**, 1714–1718.

Hozawa, A., Kuriyama, S., Nakaya, N. *et al.* (2009) Green tea consumption is associated with lower psychological distress in a general population: the Ohsaki Cohort 2006 Study. *Am. J. Clin. Nutr.*, **90**, 1390–1396.

Hsu, C., Tsai, T., Kao, Y. *et al.* (2008) Effect of green tea extract on obese women: a randomized, double-blind, placebo-controlled clinical trial. *Clin. Nutr.*, **27**, 363–370.

Huffman, M.A. (2003) Animal self-medication and ethnomedicine: exploration and exploitation of the medicinal properties of plants. *Proc. Nutr. Soc.*, **62**, 371–381.

Ide, R., Fujino, Y., Hoshiyama, Y. *et al.* (2007) A prospective study of green tea consumption and oral cancer incidence in Japan. *Ann. Epidemiol.*, **17**, 821–826.

Il'Yasova, D., Arab, L., Martinchik, A. *et al.* (2003) Black tea consumption and risk of rectal cancer in Moscow population. *Ann. Epidemiol.*, **13**, 405–411.

Inanami, O., Asanuma, T., Inukai, N. *et al.* (1995) The suppression of age-related accumulation of lipid peroxides in rat brain by administration of Rooibos tea (*Aspalathus linearis*). *Neurosci. Lett.*, **196**, 85–88.

Inoue, M., Robien, K., Wang, R. *et al.* (2008) Green tea intake, MTHFR/TYMS genotype and breast cancer risk: the Singapore Chinese Health Study. *Carcinogenesis*, **29**, 1967–1972.

Islami, F., Boffetta, P., Ren, J. *et al.* (2009) High-temperature beverages and foods and esophageal cancer risk-A systematic review. *Int. J. Cancer*, **125**, 491–524.

Iso, H., Date, C., Wakai, K. *et al.* (2006) The relationship between green tea and total caffeine intake and risk for self-reported type 2 diabetes among Japanese adults. *Ann. Intern. Med.*, **144**, 554–562.

Jang, E.H., Park, Y.C. and Chung, W.G. (2004) Effects of dietary supplements on induction and inhibition of cytochrome P450s protein expression in rats. *Food Chem. Toxicol.*, **42**, 1749–1756.

Janjua, R., Munoz, C., Gorell, E. *et al.* (2009) A two-year, double-blind, randomized placebo-controlled trial of oral green tea polyphenols on the long-term clinical and histological appearance of photoaging skin. *Dermatol. Surg.*, **35**, 1057–1065.

Jatoi, A., Ellison, N., Burch, P.A., *et al.* (2003) A phase II trial of green tea in the treatment of patients with androgen independent metastatic prostate cancer. *Cancer*, **97**, 1442–1446.

Jian, L., Xie, L.P., Lee, A.H. *et al.* (2004) Protective effect of green tea against prostate cancer: a case-control study in southeast china. *Int. J. Cancer*, **108**, 130–135.

Jonadet, M., Bastide, J., Bastide, P. *et al.* (1990) *In vitro* enzyme inhibitory and *in vivo* cardio-protective activities of hibiscus (*Hibiscus sabdariffa* L.). *J. Pharm. Belg.*, **45**, 120–124.

Jordan, S.J., Purdie, D.M., Green, A.C. *et al.* (2004) Coffee, tea and caffeine and risk of epithelial ovarian cancer. *Cancer Causes Control*, **15**, 359–365.

Joubert, E., Gelderblom, W.C.A., Louwd, A. *et al.* (2008) South African herbal teas: *Aspalathus linearis, Cyclopia spp.* and *Athrixia phylicoides* – a review. *J. Ethnopharmacol.*, **119**, 376–412.

Kakuta, Y., Nakaya, N., Nagase, S., *et al.* (2009) Case-control study of green tea consumption and the risk of endometrial endometrioid adenocarcinoma. *Cancer Causes Control*, **20**, 617–24.

Kato, A., Minoshima, Y., Yamamoto, J. *et al.* (2008) Protective effects of dietary chamomile tea on diabetic complications. *J. Agric. Food Chem.*, **56**, 8206–8211.

Khan, A.U. and Gilani, A.H. (2006) Selective bronchodilatory effect of Rooibos tea (*Aspalathus linearis*) and its flavonoid, chrysoeriol. *Eur. J. Nutr.*, **45**, 463–469.

Kikuchi, N., Ohmori, K., Shimazu, T. *et al.* (2006) No association between green tea and prostate cancer risk in Japanese men: the Ohsaki Cohort Study. *Brit. J. Cancer*, **95**, 371–373.

Kim, K.W., Ha, K.T., Park, C.S. *et al.* (2007) *Polygonum cuspidatum*, compared with baicalin and berberine, inhibits inducible nitric oxide synthase and cyclooxygenase-2 gene expressions in RAW 264.7 macrophages. *Vascul. Pharmacol.*, **47**, 99–107.

Kimura, Y., Ohminami, H., Okuda, H. *et al.* (1983) Effects of stilbene components of roots of *Polygonum* ssp. on liver injury in peroxidized oil-fed rats. *Planta Med.*, **49**, 51–54.

Kimura, Y., Okuda, H. and Arichi, S. (1985) Effects of stilbenes on arachidonate metabolism in leukocytes. *Biochim. Biophys. Acta*, **834**, 275–278.

Klein, S. and Riggins, C. (1998a) Maté. In M. Blumenthal and A. Goldberg (eds), *The Complete German Commission E Monographs: Therapeutic Guide to Herbal Medicines*. American Botanical Council, Austin, TX; Integrative Medicine Communications, Boston, MA, pp. 167–168.

Klein, S. and Riggins, C. (1998b) Chamomile flower, German. In M. Blumenthal and A. Goldberg (eds), *The Complete German Commission E Monographs: Therapeutic Guide to Herbal Medicines*. American Botanical Council, Austin, TX; Integrative Medicine Communications, Boston, MA, p. 107.

Klein, S. and Riggins, C. (1998c) Hibiscus. In M. Blumenthal and A. Goldberg (eds), *The Complete German Commission E Monographs: Therapeutic Guide to Herbal Medicines*. American Botanical Council, Austin, TX; Integrative Medicine Communications, Boston, MA, p. 336.

Kobayashi, Y., Nakano, Y., Inayama, K. *et al.* (2003) Dietary intake of the flower extracts of German chamomile (*Matricaria recutita* L.) inhibited compound 48/80-induced itch-scratch responses in mice. *Phytomedicine*, **10**, 657–664.

Kobayashi, Y., Takahashi, R. and Ogino, F. (2005) Antipruritic effect of the single oral administration of German chamomile flower extract and its combined effect with antiallergic agents in ddY mice. *J. Ethnopharmacol.*, **101**, 308–312.

Kokotkiewicz, A. and Luczkiewicz, M. (2009) Honeybush (*Cyclopia sp.*) – a rich source of compounds with high antimutagenic properties. *Fitoterapia*, **80**, 3–11.

Komatsu, T., Nakamori, M., Komatsu, K. *et al.* (2003) Oolong tea increases energy metabolism in Japanese females. *J. Med.Invest.*, **50**, 170–175.

Kong, L.D., Cai, Y., Huang, W.W. *et al.* (2000) Inhibition of xanthine oxidase by some Chinese medicinal plants used to treat gout. *J. Ethnopharmacol.*, **73**, 199–207.

Kovacs, E.M., Lejeune, M.P., Nijs, I. *et al.* (2004) Effects of green tea on weight maintenance after body-weight loss. *Brit. J. Nutr.*, **91**, 431–437.

Koyama, Y., Kuriyama, S., Aida, J. *et al.* (2010) Association between green tea consumption and tooth loss: cross-sectional results from the Ohsaki Cohort 2006 study. *Prev. Med.*, **50**, 173–179.

Kromhout, D., Bloemberg, B.P., Feskens, E.J. et al. (1996) Alcohol, fish, fibre and antioxidant vitamins intake do not explain population differences in coronary heart disease mortality. *Int. J. Epidemiol.*, **25**, 753–759.

Kucharska, J., Ulicna, O., Gvozdjakova, A. *et al.* (2004) Regeneration of coenzyme Q9 redox state and inhibition of oxidative stress by Rooibos tea (*Aspalathus linearis*) administration in carbon tetrachloride liver damage. *Physiol. Res.*, **53**, 515–521.

Kunishiro, K., Tai, A. and Yamamoto, I. (2001) Effects of rooibos tea extract on antigen-specific antibody production and cytokine generation *in vitro* and *in vivo*. *Biosci. Biotechnol. Biochem.*, **65**, 2137–2145.

Kuo, Y., Yu, C.L., Liu, C.Y. *et al.* (2009) A population-based, case-control study of green tea consumption and leukemia risk in southwestern Taiwan. *Cancer Causes Control*, **20**, 57–65.

Kurahashi, N., Sasazuki, S., Iwasaki, M. *et al.* (2008a) Green tea consumption and prostate cancer risk in Japanese men: a prospective study. *Am. J. Epidemiol.*, **167**, 71–77.

Kurahashi, N., Inoue, M., Iwasaki, M. *et al.* (2008b) Coffee, green tea, and caffeine consumption and subsequent risk of bladder cancer in relation to smoking status: a prospective study in Japan. *Cancer Sci.*, **100**, 284–291.

Kuriyama, S., Shimazu, T., Ohmori, K., *et al.* (2006a) Green tea consumption and mortality due to cardiovascular disease, cancer, and all causes in Japan: the Ohsaki Study. *JAMA*, **296**, 1255–1265.

Kuriyama, S., Hozawa, A., Ohmori, K. *et al.* (2006b) Green tea consumption and cognitive function: a cross-sectional study from the Tsurugaya Project. *Am. J. Clin. Nutr.*, **83**, 355–361.

Lagiou, P., Talamini, R., Samoli, E. *et al.* (2009) Diet and upper-aerodigestive tract cancer in Europe: the ARCAGE study. *Int. J. Cancer*, **214**, 2671–2676.

Lanzetti, M., Bezerra, F.S., Romana-Souza, B, *et al.* (2008) Maté tea reduced acute lung inflammation in mice exposed to cigarette smoke. *Nutrition*, **24**, 375–381.

Larson, S.C. and Wolk, A. (2005) Tea consumption and ovarian cancer risk in a population-based cohort. *Arch. Intern. Med.*, **165**, 2683–2686.

Larzelere, M.M. and Wiseman P. (2002) Anxiety, depression, and insomnia. *Prim. Care*, **29**, 339–360.

Laurie, S.A., Miller, V.A., Grant, S.C. *et al.* (2005) Phase I study of green tea extract in patients with advanced lung cancer. *Cancer Chemother. Pharmacol.*, **55**, 33–38.

Li, Q., Kakizaki, M., Kuriyama, S. *et al.* (2008) Green tea consumption and lung cancer risk: the Ohsaki Study. *Brit. J. Cancer*, **99**, 1179–1184.

Liang, W., Lee, A.H., Binns, C.W. *et al.* (2009) Tea consumption and ischemic stroke risk: a case-Control study in Southern China. *Stroke*, **40**, 2480–2485.

Lin, W.L., Hsieh, Y.J., Chou, F.P. *et al.* (2003) Hibiscus protocatechuic acid inhibits lipopolysaccharide-induced rat hepatic damage. *Arch. Toxicol.*, **77**, 42–47.

Lin, T.L., Lin, H.H., Chen, C.C. *et al.* (2007) *Hibiscus sabdariffa* extract reduces serum cholesterol in men and women. *Nutr. Res.*, **27**, 140–145.

Lin, Y., Kikuch, S., Tamakoshi, A. *et al.* (2008) Green tea consumption and the risk of pancreatic cancer in Japanese adults. *Pancreas*, **237**, 25–30.

Liu, C., Wang, J.M., Chu, C. *et al.* (2002) *In vivo* protective effect of protocatechuic acid on tert-butyl hydroperoxide-induced rat hepatotoxicity. *Food Chem. Toxicol.*, **40**, 635–641.

Lu, C., Lan, S., Lee, Y. et al. (1999) Tea consumption: fluid intake and bladder cancer risk in Southern Taiwan. *Urology*, **54**, 823–828.

Lunceford, N. and Gugliucci, A. (2005) *Ilex paraguariensis* extracts inhibit AGE formation more efficiently than green tea. *Fitoterapia*, **76**, 419–427.

Luo, J., Inoue, M., Iwasaki, M. et al. (2007) Green tea and coffee intake and risk of pancreatic cancer in a large-scale, population-based cohort study in Japan (JPHC Study). *Eur. J. Cancer Prev.*, **6**, 542–548.

Maki, K.C., Reeves, M.S., Farmer, M. et al. (2009) Green tea catechin consumption enhances exercise-induced abdominal fat loss in overweight and obese adults. *J. Nutr.*, **139**, 264–270.

Mann, C. and Staba, E.J. (1986) The chemistry, pharmacology, and commercial formulations of Chamomile. In: L.E. Craker and J.E. Simon (eds), Herbs, Spices, and Medicinal Plants: Recent Advances in Botany, Horticulture, and Pharmacology Volume 1. Oryx Press, Phoenix, AZ, p. 235.

Marnewick, J.L., Joubert, E., Swart, P. et al. (2003) Modulation of hepatic drug metabolizing enzymes and oxidative status by rooibos (*Aspalathus linearis*) and honeybush (*Cyclopia intermedia*), green and black (*Camellia sinensis*) teas in rats. *J. Agric. Food Chem.*, **51**, 8113–8119.

Marnewick, J.L., Batenburg, W., Swart, P. et al. (2004) *Ex vivo* modulation of chemical-induced mutagenesis by subcellular liver fractions of rats treated with rooibos (*Aspalathus linearis*) tea, honeybush (*Cyclopia intermedia*) tea, as well as green and black (*Camellia sinensis*) teas. *Mutat. Res.*, **558**, 145–154.

Marnewick, J.L., Van Der Westhuizen, F.H., Joubert, E. et al. (2009) Chemoprotective properties of rooibos (*Aspalathus linearis*), honeybush (*Cyclopia intermedia*) herbal and green and black (*Camellia sinensis*) teas against cancer promotion induced by fumonisin B1 in rat liver. *Food Chem. Toxicol.*, **47**, 220–229.

Matsuda, K., Nishimura, Y., Kurata, N., et al. (2007) Effects of continuous ingestion of herbal teas on intestinal CYP3A in the rat. *J. Pharmacol. Sci.*, **103**, 214–221.

Matsuyama, T., Tanaka, Y., Kamimaki, I. et al. (2008) Catechin safely improved higher levels of fatness, blood pressure, and cholesterol in children. *Obesity*, **16**, 1338–1348.

Matsumoto, R.L., Bastos, D.H., Mendonça, S. et al. (2009) Effects of mate tea (*Ilex paraguariensis*) ingestion on mRNA expression of antioxidant enzymes, lipid peroxidation, and total antioxidant status in healthy young women. *J. Agric. Food Chem.*, **57**, 1775–1780.

McKay, D.L. and Blumberg, J.B. (2002) The role of tea in human health: an update. *J. Am. Coll. Nutr.*, **21**, 1–13.

McKay, D.L. and Blumberg, J.B. (2006) A review of the bioactivity and potential health benefits of chamomile tea (*Matricaria recutita* L.). *Phytother Res.*, **20**, 519–530.

McKay, D.L. and Blumberg, J.B. (2007) A review of the bioactivity of South African herbal teas: rooibos (*Aspalathus linearis*) and honeybush (*Cyclopia intermedia*). *Phytother Res.*, **21**, 1–16.

McKay, D.L., Chen, C.Y., Saltzman, E. et al. (2010) *Hibiscus sabdariffa* L. tea (tisane) lowers blood pressure in prehypertensive and mildly hypertensive adults. *J. Nutr.*, **140**, 298–303.

Miller, L.G. (1998) Herbal medicinals: selected clinical considerations focusing on known or potential drug-herb interactions. *Arch. Intern. Med.*, **158**, 2200–2211.

Miranda, D.D.C., Arcar, D.P., Pedrazzoli J. et al. (2008) Protective effects of maté tea (*Ilex paraguariensis*) on H_2O_2-induced DNA damage and DNA repair in mice. *Mutagenesis.*, **23**, 261–265.

Mojiminiyi, F.B.O., Adegunloye, B.J., Egbeniyi, Y.A. et al. (2000) An investigation of the diuretic effect of an aqueous extract of the petals of *Hibiscus sabdariffa*. *J. Med. Medical Sci.*, **2**, 77–80.

Mojiminiyi, F.B., Dikko, M., Muhammad, B.Y. et al. (2007) Antihypertensive effect of an aqueous extract of the calyx of *Hibiscus sabdariffa*. *Fitoterapia*, **78**, 292–297.

Montella, M., Polesel, J., La Vecchia, C. *et al.* (2007) Coffee and tea consumption and risk of hepatocellular carcinoma in Italy. *Int. J. Cancer*, **120**, 1555–1559.

Mosimann, A.L., Wilhelm-Filho, D. and da Silva, E.L. (2006) Aqueous extract of *Ilex paraguariensis* attenuates the progression of atherosclerosis in cholesterol-fed rabbits. *Biofactors*, **26**, 59–70.

Mozaffari-Khosravi, H., Jalali-Khanabadi, B.A., Afkhami-Ardekani, M. *et al.* (2009a) Effects of sour tea (*Hibiscus sabdariffa*) on lipid profile and lipoproteins in patients with type II diabetes. *J. Altern. Complem. Med.*, **15**, 899–903.

Mozaffari-Khosravi, H., Jalali-Khanabadi, B.A., Afkhami-Ardekani, M. *et al.* (2009b) The effects of sour tea (*Hibiscus sabdariffa*) on hypertension in patients with type II diabetes. *J. Hum. Hyptertens.*, **23**, 48– 54.

Mu, L.N., Lu, Q.Y., Yu, S.Z. *et al.* (2005) Green tea drinking and multigenetic index on the risk of stomach cancer in a Chinese population. *Int. J. Cancer*, **116**, 972–983.

Mukamal, K.J., Maclure, M., Muller, J.E. *et al.* (2002) Tea consumption and mortality after acute myocardial infarction. *Circulation*, **105**, 2476–2481.

Mukamal, K.J., MacDermott, K., Vinson, J.A. *et al.* (2007) A 6-month randomized pilot study of black tea and cardiovascular risk factors. *Am. Heart J.*, **154**, 724.e1–724.e6.

Muller, B.M. and Franz, G. (1992) Chemical structure and biological activity of polysaccharides from *Hibiscus sabdariffa*. *Planta Med.*, **158**, 60–67.

Naganuma, T., Kuriyama, S., Kakizaki, M. *et al.* (2009) Green tea consumption and hematologic malignancies in Japan. *Am. J. Epidemiol.*, **170**, 730–738.

Nagao, T., Hase, T. and Tokimitsu, I. (2007) A green tea extract high in catechins reduces body fat and cardiovascular risks in humans. *Obesity*, **15**, 1473–1483.

Nagaya, N., Yamamoto, H., Uematsu, M. *et al.* (2004) Green tea reverses endothelial dysfunction in healthy smokers. *Heart*, **90**, 1485–1486.

Nakamura, H., Moriya, K., Oda, S. *et al.* (2002) Changes in the parameters of autonomic nervous system and emotion spectrum calculated from encephalogram after drinking chamomile tea [Japanese]. *Aroma Res.*, **3**, 251–255.

Nijveldt, R.J., van Nood, E., van Hoorn, D.E.C. *et al.* (2001) Flavonoids: a review of probable mechanisms of action and potential applications. *Am. J. Clin. Nutr.*, **74**, 418–425.

Niu, K., Hozawa, A., Kuriyama, S. *et al.* (2009) Green tea consumption is associated with depressive symptoms in the elderly. *Am. J. Clin. Nutr.*, **90**, 1615–1622.

O'Hara, M., Kiefer, D., Farrell, K. *et al.* (1998) A review of 12 commonly used medicinal herbs. *Arch. Fam. Med.*, **17**, 523–536.

Obiefuna, P.C.M., Owolabi, O.A., Adegunloye, B.J. *et al.* (1994) The petal extract of *Hibiscus sabdariffa* produces relaxation of isolated rat aorta. *Int. J. Pharmacogn.*, **32**, 69–74.

Odegaard, A.O., Pereira, M.A., Koh, W. et al. (2009) Coffee, tea, and incident type 2 diabetes: the Singapore Chinese Health Study. *Am. J. Clin. Nutr.*, **88**, 979–985.

Odigie, I.P., Ettarh, R.R. and Adigun, S.A. (2003) Chronic administration of aqueous extract of *Hibiscus sabariffa* attenuates hypertension and reverses cardiac hypertrophy in 2K-1C hypertensive rats. *J. Ethnopharmacol.*, **86**, 181–185.

Okomoto, K. (2006) Habitual green tea consumption and risk of an anuerysmal rupture subarachnoid hemorrhage: a case-control study in Nagoya, Japan. *Eur. J. Epidemiol.*, **21**, 367–371.

Onyenekwe, P.C., Ajan, E.O., Ameh, D.A. *et al.* (1999) Antihypertensive effect of roselle (*Hibiscus sabdariffa*) calyx infusion in spontaneously hypertensive rats and a comparison of its toxicity with that in Wistar rats. *Cell Biochem. Funct.*, **17**, 199–206.

Opala, T., Rzymskip, P., Pischel, I. *et al.* (2006) Efficacy of 12 weeks supplementation of a botanical extract-based weight loss formula on bodyweight, body composition and blood chemistry in healthy, overweight subjects – a randomized double-blind placebo-controlled clinical trial. *Eur. J. Med. Res.*, **11**, 343–350.

Owolabi, O.A., Adegunloye, B.J., Ajagbona, O.P. et al. (1995) Mechanism of relaxant effect mediated by an aqueous extract of *Hibiscus sabdariffa* petals in isolated rat aorta. *Int. J. Pharmacogn.*, **33**, 210–214.

Park, C.S., Lee, Y.C., Kim, J.D. et al. (2004) Inhibitory effects of *Polygonum cuspidatum* water extract (PCWE) and its component resveratrol on acyl-coenzyme A-cholesterol acyltransferase activity for cholesteryl ester synthesis in HepG2 cells. *Vascul. Pharmacol.*, **40**, 279–284.

Pereira, F., Santos, R., Pereira, A. (1997) Contact dermatitis from chamomile tea. *Contact Dermatitis*, **136**, 307.

Pintos, J., Franco, E.L., Oliveira, B.V. et al. (1994) Maté, coffee, and tea consumption and risk of cancers of the upper aerodigestive tract in southern Brazil. *Epidemiology*, **5**, 583–590.

Pisters, K.M., Newman, R.A., Coldman, B. et al. (2001) Phase I trial of oral green tea extract in adult patients with solid tumors. *J. Clin. Oncol.*, **19**, 1830–1838.

Popkin, B.M., Armstrong, L.E., Bray, G.M. et al. (2006) A new proposed guidance system for beverage consumption in the United States. *Am. J. Clin. Nutr.*, **83**, 529–542.

Ragab, A.R., Van Fleet, J., Jankowski, B. et al. (2006) Detection and quantitation of resveratrol in tomato fruit (*Lycopersicon esculentum* Mill.). *J. Agric. Food Chem.*, **54**, 7175–7179.

Ramirez-Mares, M.V., Chandra, S. and de Mejia, E.G. (2004) *In vitro* chemopreventive activity of *Camellia sinensis*, *Ilex paraguariensis* and *Ardisia compressa* tea extracts and selected polyphenols. *Mutat. Res.*, **554**, 53–65.

Rees, J.R., Stukel, T.A., Perry, A.E. et al. (2007) Tea consumption and basal cell and squamous cell skin cancer: results of a case control study. *J. Am. Acad. Dermatol.*, **56**, 781–785.

Reider, N., Sepp, N., Fritsch, P. et al. (2000) Anaphylaxis to camomile: clinical features and allergen cross-reactivity. *Clin. Exp. Allergy*, **30**, 1436–1443.

Reinback, H.C., Smeets, A., Martinussen, T. et al. (2009) Effects of capsaicin, green tea, and CH-19 sweet pepper on appetite and energy intake in humans in negative and positive energy balance. *Clin. Nutr.*, **28**, 260–265.

Rodriguez-Serna, M., Sanchez-Motilla, J.M., Ramon, R. et al. (1998) Allergic and systemic contact dermatitis from *Matricaria chamomilla* tea. *Contact Dermatitis*, **39**, 192–193.

Rood, B. (1994) *Uit die Veldapteek*. Tafelberg-Uitgewers Bpk, Cape Town, South Africa, p. 51.

Rumpler, W., Seale, J., Clevidence, B. et al. (2001) Oolong tea increases metabolic rate and fat oxidation in men. *J. Nutr.*, **131**, 2848–2852.

Rycroft, R.J. (2003) Recurrent facial dermatitis from chamomile tea. *Contact Dermatitis*, **48**, 229.

Salah, A.M., Gathumbi, J. and Vierling, W. (2002) Inhibition of intestinal motility by methanol extracts of *Hibiscus sabdariffa* L. (Malvaceae) in rats. *Phytother. Res.*, **16**, 283–285.

Sano, J., Inami, S., Seimiya, K. et al. (2004) Effects of green tea intake on the development of coronary artery disease. *Circ. J.*, **68**, 665–670.

Sarr, M., Ngom, S., Kane, M.O. et al. (2009) *In vitro* vasorelaxation mechanisms of bioactive compounds extracted from *Hibiscus sabdariffa* on rat thoracic aorta. *Nutr. Metab.*, **26**, 45.

Sasaki, Y.F., Yamada, H., Shimoi, K., et al. (1993) The clastogen-suppressing effects of green tea, Po-lei tea and Rooibos tea in CHO cells and mice. *Mutat. Res.*, **286**, 221–232.

Sasazuki, S., Inoue, M., Hanaoka, T. et al. (2004) Green tea consumption and subsequent risk of gastric cancer by subsite: the JPHC Study. *Cancer Causes Control*, **15**, 483–491.

Sato, M., Maulik, G., Bagchi, D. et al. (2000) Myocardial protection by protykin, a novel extract of trans-resveratrol and emodin. *Free Radic. Res.*, **32**, 135–144.

Savino, F., Cresi, F., Castagno, E. et al. (2005) A randomized double-blind placebo-controlled trial of a standardized extract of *Matricariae recutita*, *Foeniculum vulgare* and *Melissa officinalis* (ColiMil) in the treatment of breastfed colicky infants. *Phytother. Res.*, **9**, 335–340.

Sewram, V., De Stefani, E., Brennan, P. *et al.* (2003) Maté consumption and the risk of squamous cell esophageal cancer in Uruguay. *Cancer Epidemiol. Biomarkers Prev.*, **12**, 508–513.

Shimada, K., Kawarabayashi, T., Tanaka, A. *et al.* (2004) Oolong tea increases plasma adiponectin levels and low-density lipoprotein particle size in patients with coronary artery disease. *Diabetes Res. Clin. Pract.*, **65**, 227–234.

Shimazu, T., Inoue, M., Sasazuki, S. *et al.* (2008) Coffee consumption and risk of endometrial cancer: a prospective study in Japan. *Int. J. Cancer*, **123**, 2406–2410.

Shimbo, M., Nakamura, K., Jing Shi, H. *et al.* (2005) Green tea consumption in everyday life and mental health. *Public Health Nutr.*, **8**, 1300–1306.

Shimizu, M. Fukutomi, Y., Ninomiya, M. *et al.* (2008) Green tea extracts for the prevention of metachronous colorectal adenomas: a pilot study. *Cancer Epidemiol. Biomarkers Prev.*, **17**, 3020–3025.

Shimoi, K., Masuda, S., Shen, B. *et al.* (1996) Radioprotective effects of antioxidative plant flavonoids in mice. *Mutat. Res.*, **350**, 153–161.

Shipochliev, T., Dimitrov, A and Aleksandrova, E. (1981) Antiinflammatory action of a group of plant extracts. *Vet. Med. Nauki.*, **18**, 87–94.

Shrubsole, M.J., Lu, W., Chen, Z. *et al.* (2009) Drinking green tea modestly reduces breast cancer risk. *J. Nutr.*, **139**, 310–316.

Snyckers, F.O. and Salemi, G. (1974) Studies of South African medicinal plants. Part 1: quercetin as the major in vitro active component of rooibos tea. *J. South African Chem. Inst.*, **27**, 5–7.

Steevens, J., Schouten, L.J., Verhage, B.A.J. *et al.* (2007) Tea and coffee drinking and ovarian cancer risk: results from the Netherland Cohort Study and a meta-analysis. *Brit. J. Cancer*, **97**, 1291–1294.

Steptoe, A., Gibson, E.L., Vuononvirta, R. *et al.* (2007) The effects of tea on psychophysiological stress responsivity and post-stress recovery: a randomized double-blind trial. *Psychopharmacol.*, **190**, 81–89.

Subiza, J., Subiza, J.L., Hinojosa, M. *et al.* (1989) Anaphylactic reaction after the ingestion of chamomile tea: a study of cross-reactivity with other composite pollens. *J. Allergy Clin. Immunol.*, **84**, 353–358.

Sun, C., Yuan, J., Koh, W.P. *et al.* (2007) Green tea and black tea consumption in relation to colorectal cancer risk: the Singapore Chinese Health Study. *Carcinogenesis*, **28**, 2143–2148.

Tan, L.C., Koh, W.P., Yuan, J.M. *et al.* (2008) Differential effects of black versus green tea on risk of Parkinson's Disease in the Singapore Chinese Health Study. *Am. J. Epidemiol.*, **167**, 553–560.

Tanabe, N. Suzuki, H., Aizawa, Y. *et al.* (2008) Consumption of green and roasted teas and the risk of stroke incidence: results from the Tokamachi-Nakasato Cohort Study in Japan. *Int. J. Epidemiol.*, **37**, 1030–1040.

Taubert, D., Roesen, R. and Schömig, E. (2007) Effect of cocoa and tea intake on blood pressure: a meta-analysis. *Arch. Intern. Med.*, **167**, 626–634.

Tee, P., Yusof, S., Suhaila, M. *et al.* (2002) Effect of roselle (*Hibiscus sabdariffa* L.) on serum lipids of Sprague Dawley rats. *Nutr. Food Sci.*, **32**, 190–196.

Ulicna, A.O., Greksak, M., Vancov, A.O. *et al.* (2003) Hepatoprotective effect of rooibos tea (*Aspalathus linearis*) on CCl4-induced liver damage in rats. *Physiol. Res.*, **52**, 461–466.

Vastano, B.C., Chen, Y., Zhu, N. *et al.* (2000) Isolation and identification of stilbenes in two varieties of *Polygonum cuspidatum*. *J. Agric. Food Chem.*, **48**, 253–256.

Venables, M.C., Hulston, C.J., Cox, H.R. *et al.* (2008) Green tea extract ingestion, fat oxidation, and glucose tolerance in healthy humans. *Am. J. Clin. Nutr.*, **87**, 778–784.

Vlachopoulos, C., Alexopoulos, N., Dima, I. *et al.* (2006) Acute effect of black and green tea on aortic stiffness and wave reflections. *J. Am. Coll. Nutr.*, **25**, 216–223.

Wang, C.J., Wang, J.M., Lin, W.L. *et al.* (2000) Protective effect of hibiscus anthocyanins against tert-butyl hydroperoxide-induced hepatic toxicity in rats. *Food Chem. Toxicol.*, **38**, 411–416.

Wang, J.M., Tang, L., Sun, G. *et al.* (2006) Etiological study of esophageal squamous cell carcinoma in an endemic region: a population-based case control study in Huaian, China. *BMC Cancer*, **6**, 287.

Wang, J.M., Xu, B., Rao, J.Y. *et al.* (2007a) Diet habits, alcohol drinking, tobacco smoking, green tea drinking, and the risk of esophageal squamous cell carcinoma in the Chinese population. *Eur. J. Gastroenterol. Hepatol.*, **19**, 171–176.

Wang, C., Zhang, D., Ma, H. *et al.* (2007b) Neuroprotective effects of emodin-8-O-β-D-glucoside *in vivo* and *in vitro*. *Eur. J. Pharmacol.*, **577**, 58–63.

Weizman, Z., Alkrinawi, S., Goldfarb, D. *et al.* (1993) Efficacy of herbal tea preparation in infantile colic. *J. Pediatr.*, **122**, 650–652.

Wichtl, M. and Bisset N.G. (eds) (1994) *Herbal Drugs and Phytopharmaceuticals.* Medpharm Scientific Publishers, Stuttgart.

Xing, W.W., Wu, J.Z., Jia, M. *et al.* (2009) Effects of polydatin from *Polygonum cuspidatum* on lipid profile in hyperlipidemic rabbits. *Biomed. Pharmacother.*, **63**, 457–462.

Yang, C.S. and Landau, J.M. (2000) Effects of tea consumption on nutrition and health. *J. Nutr.*, **130**, 2409–2412.

Yang, Y.C., Lu, F.H., Wu, J.S. *et al.* (2004) The protective effect of habitual tea consumption on hypertension. *Arch. Intern. Med.*, **164**, 1534–1540.

Yang, G., Shu, X.O., Li, H. *et al.* (2007) Prospective cohort study of green tea consumption and colorectal cancer risk in women. *Cancer Epidemiol. Biomarkers Prev.*, **16**, 1219–1223.

Yang, M.Y., Peng, C.H., Chan, K.C. *et al.* (2010) The hypolipidemic effect of *Hibiscus sabdariffa* polyphenols via inhibiting lipogenesis and promoting hepatic lipid clearance. *J. Agric. Food Chem.*, **58**, 850–859.

Zhang, M., Lee, A.H., Bins, C.W. *et al.* (2004) Green tea consumption enhances survival of epithelial ovarian cancer. *Int. J. Cancer*, **112**, 465–469.

Zhang, C.Z., Wang, S.X., Zhang, Y. *et al.* (2005) *In vitro* estrogenic activities of Chinese medicinal plants traditionally used for the management of menopausal symptoms. *J. Ethnopharmacol.*, **98**, 295–300.

Zhang, M., Holman, C.D.J., Huang, J. *et al.* (2007) Green tea and the prevention of breast cancer: a case-control study in Southeast China. *Carcinogenesis*, **28**, 1074–1078.

Zhang, M., Zhao, X., Zhang, X. *et al.* (2008) Possible protective effect of green tea intake on risk of adult leukaemia. *Brit. J. Cancer*, **98**, 168–170.

Chapter 5
Phytochemicals in Coffee and the Bioavailability of Chlorogenic Acids

Angelique Stalmach[1], Michael N. Clifford[2], Gary Williamson[3] and Alan Crozier[1]

[1] School of Medicine, College of Medical, Veterinary and Life Sciences, University of Glasgow, Glasgow G12 8QQ, UK
[2] Food Safety Research Group, Centre for Nutrition and Food Safety, Faculty of Health and Medical Sciences, University of Surrey, Guildford, Surrey GU2 7XH, UK
[3] School of Food Science and Nutrition, University of Leeds, Leeds LS2 9JT, UK

5.1 Introduction

In economic terms, coffee is the most valuable agricultural product with exports by Third-World and developing counties, amounting to ~7.2 million metric tonnes in 2009 (International Coffee Organisation 2010). The green coffee bean is the processed, generally non-viable, seed of the coffee cherry. Commercial production exploits the seeds of *Coffea arabica* (so-called arabica coffees), which represent ~70% of the world market, and *Coffea canephora* (so-called robusta coffees), which account for ~30%. There are many wild species some of which are virtually caffeine free (e.g. *Coffea pseudozanguebariae*) (Clifford *et al.* 1989a), but not suitable for commercial exploitation (Crozier and Ashihara 2006). Arabicas originated from the highlands of Ethiopia, whereas the robustas originated at lower altitudes across the Ivory Coast, Congo and Uganda. There are many varieties of arabicas (*C. arabica* var. *arabica* accounting for the majority of commercial production) and robustas, and there has been limited commercial exploitation of various arabica × robusta hybrids, and local use of *Coffea liberica*, *Coffea racemosa* and *Coffea dewevrei* in some African countries (Willson 1999). The establishment of coffee plantations in different parts of the world and the adoption of the beverage are discussed in Chapter 1 and several books including Smith (1985).

Of the ~50 recognised producing countries, the major producers in 2009 were Brazil, Vietnam, Colombia, Indonesia and Uganda (International Coffee Organisation 2010). Most coffee is exported from these countries as green beans, but there are also significant sales of instant coffee powder produced in the coffee-growing countries.

Teas, Cocoa and Coffee: Plant Secondary Metabolites and Health, First Edition.
Edited by Professor Alan Crozier, Professor Hiroshi Ashihara and Professor F. Tomás Barberán.
© 2012 Blackwell Publishing Ltd. Published 2012 by Blackwell Publishing Ltd.

5.2 Harvesting coffee beans, roasting and blending

Increasingly, mechanised harvesting is used. There are basically two methods for releasing the seeds from the harvested fruit, the wet process that requires copious supplies of good-quality water, used principally for arabicas, and the dry process used mainly for robustas. In the dry process, the freshly picked or mechanically harvested cherries are sun dried for 2–3 weeks, followed by mechanical removal of the dried husk. In the wet process, the cherries are soaked and fermented in water to remove the pulp prior to drying. After roasting, most robustas are blended as they are generally considered inferior to arabicas. Robustas are, however, preferred for instant coffee production as they give a higher yield of extractables and a less 'thin' liquor. With improved quality assurance, robustas are now generally of very good quality, but nevertheless subtly different from arabicas (Clarke 1985a; Willson 1999).

The commercial beans are roasted at air temperatures as high as 230°C for a few minutes, or at 180°C for up to ~20 minutes. There is a substantial and exothermic pyrolysis, and a myriad of chemical reactions occur as a consequence of the internal temperature and pressure (5–7 atmospheres) achieved. Pyrolysis loss ranges from 3% to 5% of dry weight for a light roast to 5–8% for a medium roast and 8–14% for a dark roast. The pressure is largely due to entrapped carbon dioxide that effectively ensures an inert atmosphere within roasted whole beans. Grinding releases this, and subsequent extraction produces the beverage ready to drink, or on a commercial scale, a concentrated liquor that is converted to powder, either by spray drying or by freeze drying. Domestic brewing extracts some 25–32% of solids, being highest in espresso, but varying with equipment, coffee particle size and the charge of coffee relative to water. Commercial instantisation with water at up to 200 psi extracts some 50% of solids from the roasted bean. Within the European Economic Community, the yield of solubles is controlled by legislation at not more than 1.0 kg from 2.3 kg of green beans (Clarke 1985b). A cup of coffee contains some 1–2% solids by weight. Green beans may be decaffeinated either by the use of supercritical carbon dioxide or organic solvent (Ashihara and Crozier 1999). European Economic Community legislation requires less than 0.1% caffeine in the decaffeinated green bean that corresponds to less than 0.3% in instant powder (Clarke 1985b).

5.3 Phytochemicals in coffee

Phytochemicals are generally associated with fresh or minimally processed plant material, but with coffee, the unique products associated with roasting should not be overlooked. However, only those compounds that significantly enter the brew are dealt with here. For information about those that remain in the grounds, see the review by Speer and Kölling-Speer (2006).

Green coffee beans are one of the richest dietary sources of chlorogenic acids, which comprise 6–10% on a dry weight basis. At present, they also hold the record for the greatest number of individual chlorogenic acids – at least 72 – plus at least three free cinnamic acids. The term 'chlorogenic acid' was first applied to a fraction isolated from coffee beans in 1846 (Payen 1846a, 1846b), although the correct structure of the dominant constituent was not elucidated until almost a century later by Fischer and Dangschat (1932). This major component, known now in the IUPAC numbering system (IUPAC 1976) as 5-O-caffeoylquinic acid (5-CQA), accounts for some 50% of the total chlorogenic acids. This is accompanied by significant amounts of 3-O- and 4-O-caffeoylquinic acid (3-CQA, 4-CQA), the three analogous feruloylquinic acids (FQA), and 3,4-O-, 3,5-O- and 4,5-O-dicaffeoylquinic acids (diCQA), with smaller amounts of p-coumaroylquinic acids (Figure 5.1) and lesser quantities of the three isomeric monoacyl p-coumaroylquinic acids (Clifford 1999). With the development of HPLC–ion trap MS and HPLC–tandem MS methods (Clifford et al. 2003, 2005),

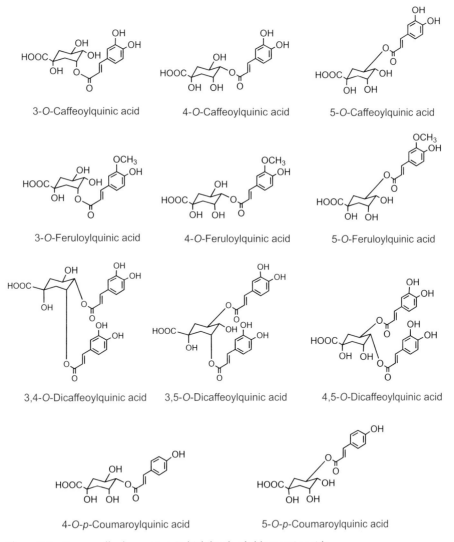

Figure 5.1 Green coffee beans contain high levels of chlorogenic acids.

Table 5.1 Listing of cinnamoyl–amino acid conjugates in green and roasted coffee beans

Cinnamoyl–amino acid conjugates	Reference
p-Coumaroyl-N-tyrosine	(1)
Caffeoyl-N-tyrosine	(1)
Feruloyl-N-tyrosine	(1)
p-Coumaroyl-N-tryptophan	(1)
Caffeoyl-N-tryptophan	(1)
Feruloyl-N-tryptophan	(1)
Caffeoyl-N-phenylalanine	(1)
p-Coumaroyl-N-aspartic acid	(2)
Caffeoyl-N-aspartic acid	(2)

(1) Clifford and Knight (2004); (2) Stark et al. (2006).

numerous minor chlorogenic acids (Clifford et al. 2006a, 2006b; Jaiswal et al. 2010; Jaiswal and Kuhnert 2010) and cinnamoyl–amino acid conjugates have been characterised (Clifford et al. 1989b; Clifford and Knight 2004; Stark et al. 2006; Alonso-Salces et al. 2009a). All the chlorogenic acids so far characterised in coffee have the cinnamate moiety in the *trans*-configuration.

Three components have been tentatively identified as cinnamic acid–hexose conjugates (Alonso-Salces et al. 2009a), and while cinnamoyl-glycosides and cinnamoyl-hexose esters are well known (Clifford et al. 2007) and eminently plausible as components in green coffee beans, the mass spectral and retention time data provided are not entirely convincing. For example, the putative dimethoxycinnamoyl conjugate could occur only as an ester and would then show characteristic 30 amu losses. Tables 5.1 and 5.2 provide a comprehensive listing of cinnamoyl–amino acid conjugates and chlorogenic acids, respectively. The previously noted IUPAC numbering is used throughout this review.

The quantitatively minor chlorogenic acids containing sinapic acid ($3',5'$-dimethoxy-$4'$-hydroxycinnamic acid) are of restricted occurrence. Those containing $3',4'$-dimethoxycinnamic acid or $3',4',5'$-trimethoxycinnamic acid are scarce, and the chlorogenic acid containing $3',5'$-dihydroxy-4-methoxycinnamic acid appears to be unique to coffee (Figure 5.2). The triacyl chlorogenic acids have so far been found only in robusta coffee beans (Jaiswal and Kuhnert 2010). Robustas usually have a greater content of any isomer, but there are exceptions, the best known being a tendency for arabicas to contain more 5-p-coumaroylquinic acid than robustas.

The chlorogenic acids and the cinnamoyl–amino acid conjugates have been used as criteria in chemotaxonomic studies (Clifford et al. 1989a; Anthony et al. 1993) for distinguishing robustas from arabicas (Alonso-Salces et al. 2009b), and to some extent, the profile can be used to define the geographic origin of green coffee beans, especially robutas from Angola (Clifford and Jarvis 1988; Clifford et al. 1989b), Uganda, Vietnam, Cameroun and Indonesia, and for distinguishing South American from African arabicas (Alonso-Salces et al. 2009b).

Beverage astringency has been attributed to unusually high levels of diCQA relative to CQA (Clifford and Ohiokpehai 1983), possibly associated with the harvesting of

Table 5.2 A listing of the chlorogenic acids and other cinnamic acid derivatives characterised in green coffee beans

Caffeic acid *(1)*	Caffeoyl–hexose conjugate[a] *(1)*
Ferulic acid *(1)*	Dicaffeoyl–hexose conjugate[a] *(1)*
3,4-Dimethoxycinnamic acid *(1)*	Dimethoxycinnamoyl–hexose conjugate[a] *(1)*
3-*O*-*p*-Coumaroylquinic acid *(2)*	3-*O*-Caffeoylquinic acid *(2)*
4-*O*-*p*-Coumaroylquinic acid *(2)*	4-*O*-Caffeoylquinic acid *(2)*
5-*O*-*p*-Coumaroylquinic acid *(2)*	5-*O*-Caffeoylquinic acid *(2)*
3-*O*-Feruloylquinic acid *(2)*	3-*O*-Dimethoxycinnamoylquinic acid *(3)*
4-*O*-Feruloylquinic acid *(2)*	4-*O*-Dimethoxycinnamoylquinic acid *(3)*
5-*O*-Feruloylquinic acid *(2)*	5-*O*-Dimethoxycinnamoylquinic acid *(3)*
3-*O*-Sinapoylquinic acid *(4)*	3,4-*O*-Di-*p*-coumaroylquinic acid *(5)*
4-*O*-Sinapoylquinic acid *(4)*	3,5-*O*-Di-*p*-coumaroylquinic acid *(5)*
5-*O*-Sinapoylquinic acid *(4)*	4,5-*O*-Di-*p*-coumaroylquinic acid *(5)*
3,4-*O*-Dicaffeoylquinic acid *(2)*	3,4-*O*-Diferuloylquinic acid *(3)*
3,5-*O*-Dicaffeoylquinic acid *(2)*	3,5-*O*-Diferuloylquinic acid *(3)*
4,5-*O*-Dicaffeoylquinic acid *(2)*	4,5-*O*-Diferuloylquinic acid *(3)*
3-*O*-Caffeoyl-4-*O*-feruloylquinic acid *(2)*	3-*O*-Caffeoyl-4-*O*-*p*-coumaroylquinic acid *(5)*
3-*O*-Feruloyl-4-*O*-caffeoylquinic acid *(2)*	3-*O*-*p*-Coumaroyl-4-*O*-caffeoylquinic acid *(5)*
3-*O*-Caffeoyl-5-*O*-feruloylquinic acid *(2)*	3-*O*-Caffeoyl-5-*O*-*p*-coumaroylquinic acid *(5)*
3-*O*-Feruloyl-5-*O*-caffeoylquinic acid *(2)*	3-*O*-*p*-Coumaroyl-5-*O*-caffeoylquinic acid *(5)*
4-*O*-Caffeoyl-5-*O*-feruloylquinic acid *(2)*	4-*O*-Caffeoyl-5-*O*-*p*-coumaroylquinic acid *(5)*
4-*O*-Feruloyl-5-*O*-caffeoylquinic acid *(2)*	4-*O*-*p*-Coumaroyl-5-*O*-caffeoylquinic acid *(5)*
3-*O*-Dimethoxycinnamoyl-4-*O*-caffeoylquinic acid *(3)*	3-*O*-Dimethoxycinnamoyl-4-*O*-feruloylquinic acid *(3)*
3-*O*-Caffeoyl-4-*O*-dimethoxycinnamoylquinic acid *(1)*	3-*O*-Dimethoxycinnamoyl-5-*O*-feruloylquinic acid *(3)*
3-*O*-Dimethoxycinnamoyl-5-*O*-caffeoylquinic acid *(3)*	4-*O*-Dimethoxycinnamoyl-5-*O*-feruloylquinic acid *(3)*
3-*O*-Caffeoyl-5-*O*-dimethoxycinnamoylquinic acid *(1)*	
4-*O*-Dimethoxycinnamoyl-5-*O*-caffeoylquinic acid *(3)*	
4-*O*-Caffeoyl-5-*O*-dimethoxycinnamoylquinic acid *(1)*	
3-*O*-*p*-Coumaroyl-4-*O*-feruloylquinic acid *(5)*	3-*O*-Caffeoyl-4-*O*-sinapoylquinic acid *(4)*
3-*O*-*p*-Coumaroyl-5-*O*-feruloylquinic acid *(5)*	3-*O*-Sinapoyl-5-*O*-caffeoylquinic acid *(4)*
4-*O*-*p*-Coumaroyl-5-*O*-feruloylquinic acid *(5)*	3-*O*-Sinapoyl-4-*O*-caffeoylquinic acid *(4)*
3-*O*-*p*-Coumaroyl-4-*O*-dimethoxycinnamoylquinic acid *(6)*	3-*O*-Sinapoyl-5-*O*-feruloylquinic acid *(4)*

(continued)

Table 5.2 (Continued)

3-O-p-Coumaroyl-5-O-dimethoxycinnamoylquinic acid (6)	3-O-Feruloyl-4-O-sinapoylquinic acid[b] (4)
4-O-Dimethoxycinnamoyl-5-O-p-coumaroylquinic acid (5)	4-O-Sinapoyl-5-O-feruloylquinic acid (4)
3-O-Trimethoxycinnamoyl-5-O-caffeoylquinic acid (4)	3-O-Trimethoxycinnamoyl-4-O-feruloylquinic acid (4)
4-O-Trimethoxycinnamoyl-5-O-caffeoylquinic acid (4)	3-O-Trimethoxycinnamoyl-5-O-feruloyl-quinic acid (4)
	4-O-Trimethoxycinnamoyl-5-O-feruloylquinic acid (4)
3-O-(3′,5′-Dihydroxy-4′-methoxy)cinnamoyl-4-O-feruloylquinic acid (4)	3,4,5-O-Tricaffeoylquinic acid (6)
3,4-O-Dicaffeoyl-5-O-feruloylquinic acid	3,4-O-Dicaffeoyl-5-O-sinapoylquinic acid (6)
3,5-O-Dicaffeoyl-4-O-feruloylquinic acid	3-O-Sinapoyl-4,5-O-dicaffeoylquinic acid (6)
3-O-Feruloyl-4,5-O-dicaffeoylquinic acid	
3-O-Dimethoxycinnamoyl-4-O-feruloyl-5-O-caffeoylquinic acid[b] (4)	3,4-O-Diferuloyl-5-O-caffeoylquinic acid (6)
	3-O-Caffeoyl-4,5-O-diferuloylquinic acid (6)

(1) Alonso-Salces et al. (2009a); (2) Clifford et al. (2003); (3) Clifford et al. (2006a); (4) Jaiswal et al. (2010); (5) Clifford et al. (2006b); (6) Jaiswal and Kuhnert (2010).
[a] Tentative identification.
[b] Identification correct, but structure as illustrated in original paper is incorrect.

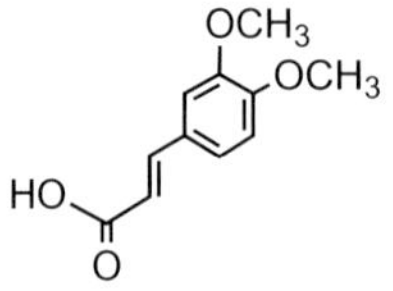

3′,4′-Dimethyoxycinnamic acid

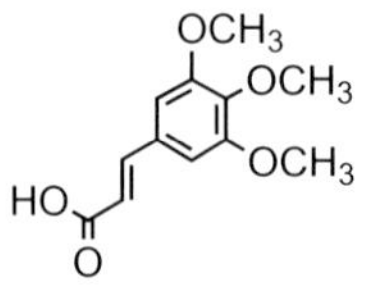

3′,4′,5′-Trimethoxycinnamic acid

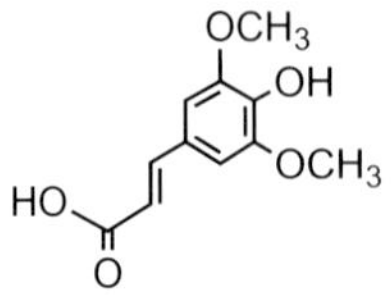

3′,5′-Dimethoxy-4′-hydroxycinnamic acid (sinapic acid)

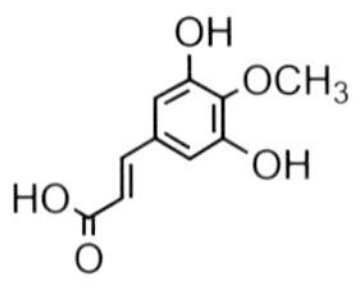

3′,5′-Dihydroxy-4′-methoxycinnamic acid

Figure 5.2 Minor chlorogenic acids of restricted occurrence, based on methoxycinnamic acids, are found in coffee beans. The chlorogenic acid containing 3′,5′-dihydroxy-4′-methoxycinnamic acid appears to be unique to coffee.

immature beans (Clifford and Kazi 1987). Cinnamoyl–amino acid conjugates have been associated with bitterness in cocoa and might also play a similar role in coffee (Stark and Hofmann 2005).

5.3.1 Effects of roasting on the phytochemical content of coffee beans

It has long been known that, during roasting, there is a progressive destruction and transformation of chlorogenic acids, with some 8–10% being lost for every 1% loss of dry matter (Clifford 1985a, 1985b), and this has been confirmed in a recent publication (Moon *et al.* 2009). Nonetheless, substantial amounts of chlorogenic acids survive to be extracted into domestic brews and commercial soluble coffee powders (see Section 5.3.2). While a portion of the chlorogenic acids in green beans is completely destroyed, some is transformed during roasting. Early in roasting when there is still adequate water content, isomerisation (acyl migration) occurs, accompanied by some hydrolysis releasing the cinnamic acids and quinic acid. Later, in roasting the free quinic acid epimerises and lactonises, and several chlorogenic lactones (caffeoylquinides) also form (Figure 5.3; Scholz and Maier 1990; Bennat *et al.* 1994; Schrader *et al.* 1996; Farah *et al.* 2005). The cinnamic acids may be decarboxylated and transformed to a number of simple phenols and range of phenylindans probably via decarboxylation and cyclisation of the vinylcatechol intermediate (Stadler *et al.* 1996). Two of these rather unstable compounds (Figure 5.3) have been found in roasted and instant coffee at 10–15 mg/kg.

Although some chlorogenic lactones have been reported to occur in green coffee beans (Farah *et al.* 2006), they are more commonly found in roasted products (Bennat *et al.* 1994; Schrader *et al.* 1996; Ginz 2001; Farah *et al.* 2005) where they are associated with beverage bitterness (Ginz and Engelhardt, 2000; Ginz 2001; Frank *et al.* 2006).

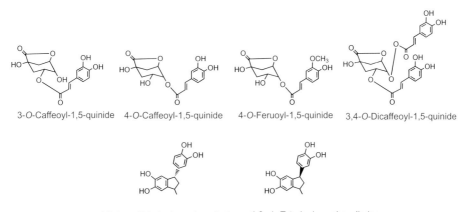

Figure 5.3 During roasting of coffee beans, chlorogenic acids are transformed, resulting in the appearance of lactones (caffeoylquinides) and phenylindans.

At least five caffeoylquinic lactones (CQL), five feruloylquinic lactones (FQL) and two dicaffeoylquinic lactones (diCQL) have been reported (Clifford 2006). They have not been fully characterised, but include 3-O-caffeoylquinic lactone (3-O-caffeoyl-1,5-quinide (3-CQL)), 4-CQL, 4-FQL and 3,4-diCQL (Figure 5.3) (Bennat et al. 1994; Schrader et al. 1996; Farah et al. 2005). It is possible that green coffee beans contain shikimic acid derivatives and these have been confused with the isobaric quinides.

The coffee bean also contains numerous other phytochemicals, many of which enter the beverage and some that are not found elsewhere in the diet. Although not restricted to coffee, the best known is caffeine present in green arabicas at ∼1% and robustas at ∼2% (see Chapter 2). There is some loss by sublimation during roasting but then near quantitative transfer to the brew or instant powder. A survey of 12 instant coffees on the UK market in 1983 reported 2.8–4.8% caffeine (Clifford 1985a).

Coffee contains fatty acyl serotonin (C_{18}; C_{20}; C_{22}; C_{24}) derivatives originating from the wax. They are thought by many to be undesirable irritants, and steaming processes have been developed for their removal (Clifford 1985a; Speer and Kölling-Speer 2006).

Trigonelline, the N-methyl betaine of nicotinic acid, is present in green beans at about 1% and during roasting is partially converted to nicotinic acid (niacin) (Figure 5.4), making coffee beverage a potentially important source of vitamin B_3 (Clifford 1985a). Trigonelline also yields 1-methylpyridinium (up to 0.25%) and 1,2-dimethylpyridinium (Figure 5.4) (up to 25 mg/kg) in proportion to the severity of roasting (Stadler et al. 2002).

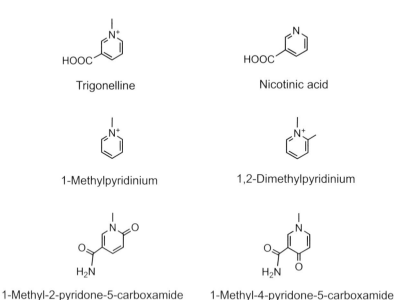

Figure 5.4 During roasting of coffee beans, trigonelline is converted to nicotinic acid (vitamin B_3), 1-methylpyridinium and 1,2-dimethylpyridinium. Following ingestion of trigonelline, 1-methylpyridinium is the main metabolite to accumulate in plasma, but there is no increase in the levels of 1-methyl-2-pyridone-5-carboxamide and 1-methyl-4-pyridone-5-carboxamide, major metabolites of nicotinic acid.

Trigonelline is absorbed, and after consumption of a coffee containing 81 μmol, reached a peak plasma concentration (C_{max}) in humans of ∼6 μM after ∼3 hours from a baseline of 0.16 μM. The major metabolite of trigonelline, 1-methylpyridinium, had a C_{max} of ∼0.8 μM. In contrast, coffee consumption did not impact significantly on plasma concentrations of 1-methyl-2-pyridone-5-carboxamide and 1-methyl-4-pyridone-5-carboxamide (Figure 5.4), major metabolites of nicotinic acid. Urinary excretion, 0–8 hours post-coffee ingestion, of 57.4 ± 6.9% of trigonelline and 69.1 ± 6.2% of 1-methylpyridinium was observed with males, whereas females excreted slightly less, with 46.2 ± 7.4% and 61.9 ± 12.2%, respectively (Lang et al. 2010). In vitro trigonelline has shown unexpected oestrogen-like effects in MCF-7 cells by an incompletely defined stimulation of the oestrogen receptor (Allred et al. 2009).

A range of diterpenes are present, and some – such as atractyligenin (Figure 5.5) – occur as glycosides, and are thought to contribute to a bitter taste of the beverage (Clifford 1985a). Cafestol, kahweol and 16-O-methylcafestol (Figure 5.5) occur as fatty acyl esters and enter the brew when coffee is prepared by extended boiling and the beverage is not filtered through paper or a bed of coffee grounds (Speer and Kölling-Speer 2001). 16-O-Methylcafestol and 16-O-methylkahweol are of restricted occurrence, found in robusta beans and some parts of the arabicas, such as leaves, but not the beans.

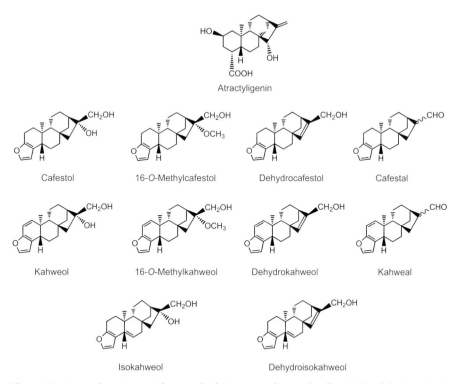

Figure 5.5 Atractyligenins contribute to the bitter taste of roasted coffee. Cafestol, kahweol and 16-O-methylcafestol occur as fatty acyl esters in unfiltered coffee, consumption of which can result in elevated plasma LDL cholesterol.

Accordingly, the presence of the comparatively heat-stable 16-O-methylcafestol can be used to detect robusta beans as an adulterant of roasted arabica beans. Roasting causes some destruction and transformation of the diterpenes, with the products including trace quantities of dehydrocafestol, cafestal, dehydrokahweol, kahweol, isokahweol and dehydroisokahweol (Figure 5.5; Speer and Kölling-Speer 2006). These diterpenes are responsible for the observed reversible elevation of plasma low-density lipoprotein (LDL) cholesterol seen in some populations, notably in Scandinavia and Italy (Urgert et al. 1996, 1997; Urgert and Katan 1997). Education and encouragement of alternative brewing procedures has seen a lowering of plasma LDL cholesterol in some Scandinavian populations.

Cafestol is also hepatoprotective and is subject to epoxidation on the furan ring, with epoxy and epoxy–glutathione conjugates having been characterised and a pathway for their biosynthesis proposed (Figure 5.6; van Cruchten et al. 2010a). Cafestol is absorbed in mice, and a glucuronide metabolite, which is probably subject to enterohepatic circulation, is excreted in bile (van Cruchten et al. 2010b).

Instant coffee powder also contains ~1% mannose and ~2% galactose. These sugars, which rarely occur free in foods, are formed by the hydrolysis of structural arabino-galactan and storage mannan polysaccharides during the high-temperature commercial extraction process (Clifford 1985a).

Roasted coffee is prized for its unique and very attractive aroma. The odour complex contains in excess of 800 known substances, many of which are heterocycles that are not found to any degree elsewhere in the diet. Fourteen of these ~800 volatiles, 2-furfuryl-

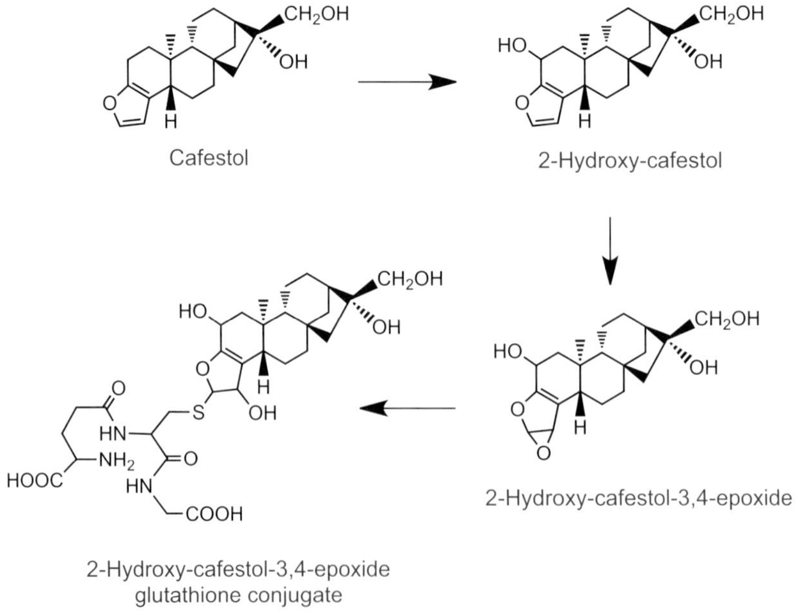

Figure 5.6 Proposed metabolism of cafestol via hydroxylation prior to epoxidation of the furan moiety of cafestol and the formation of a glutathione conjugate of 2-hydroxycafestrol-3,4-epoxide. (After van Cruchten et al. 2010b.)

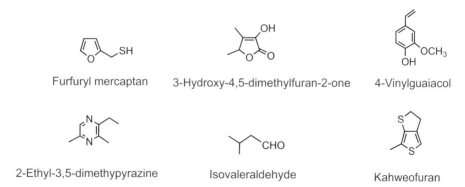

Figure 5.7 Volatile compounds from coffee having a major impact on aroma. Kahweofuran is reputed to have a coffee-like odour even in isolation.

thiol, 4-vinylguaiacol, three alkylpyrazines, four furanones and five aliphatic aldehydes, have been identified as particularly important determinants of odour (Grosch 2001). Kahweofuran is another important volatile as it is reputed to have a coffee-like aroma. Selected structures are shown in Figure 5.7. These volatiles, present in roasted coffee at concentrations in the μg/kg to mg/kg range, are produced during the high-temperature roasting process by a complex series of reactions referred to as the Maillard reaction, in which sugars and amino acids are key reactants (Nursten 2005). Coffee bitterness is only partially due to caffeine, terpenoids and chlorogenic acids, and there is evidence that Maillard products (melanoidins), including cyclic diketopiperazines formed from proline and other amino acids, are also important (Figure 5.8; Ginz and Engelhardt 2000). It has been suggested that the coffee melanoidins should be thought of as 'Maillardised dietary fibre' (Silvan et al. 2010). Studies in vitro suggest that the high-molecular-mass Maillard products formed during coffee roasting are at least partially degraded by the gut microflora yielding acetate and propionate (Reichardt et al. 2009).

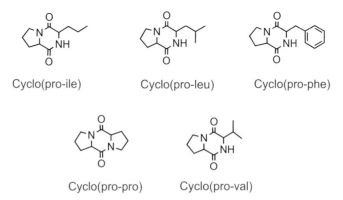

Figure 5.8 Cyclic diketopiperazines formed from proline and other amino acids contribute to the bitter taste of coffee.

5.3.2 Chlorogenic acid intake and coffee consumption

For many consumers, coffee beverage must be the major dietary source of chlorogenic acids (Clifford 1999). Regular coffee drinkers will almost certainly have a greater intake of chlorogenic acids than flavonoids (Gosnay *et al.* 2002; Woods *et al.* 2003).

Coffee, in different preparations, is widely consumed throughout the world, and a single serving can provide between 20 and 675 mg of chlorogenic acids, depending on the type of roast and the volume consumed. Regular consumers can easily have an intake in excess of 1 g/day (Clifford 1999; Stalmach *et al.* 2006). This is confirmed by the data presented in Table 5.3 on CQA levels in espresso coffee purchased from 21 coffee shops, some well-known international chains and others local outlets that are favourite haunts of students at the University of Glasgow. There is a marked variation in the total CQA dosage per serving, with quantities ranging from 423 mg (1195 μmol) to a mere 24 mg (68 μmol) in a cup of Starbucks espresso. Two instant coffees used in bioavailability studies to be discussed in Section 5.4.1 contained 136 mg (384 μmol)

Table 5.3 Levels of CQA isomers in espresso coffees purchased from 21 coffee shops in Glasgow, United Kingdom

Source	Volume of coffee (mL)	3-CQA (mg)	4-CQA (mg)	5-CQA (mg)	Total CQA per serving mg	μmol
Pattiserie Francoise	52	95	112	216	423	1195
S'mug	32	62	78	160	300	847
Costa Coffee	25	48	61	118	227	641
Little Italy	23	37	59	121	217	613
Paperino's	50	65	52	99	216	610
Peckhams	70	65	52	99	216	610
Chapter 1	26	45	58	112	215	607
University Cafe	49	40	54	93	187	528
Baguette Express	45	30	40	74	144	407
Kember & Jones	43	37	46	92	141	398
Heart Buchanan	24	22	37	67	126	356
Working Lunch	100	24	32	57	113	319
Jellyhill	63	26	29	56	111	313
Coffee @ 491	49	23	31	55	109	308
Beanscene	48	19	25	49	93	263
Tinderbox	25	19	24	46	89	251
Café Cinnamon	59	17	23	41	81	229
Crepe á Croissant	34	17	23	41	81	229
Morton's	35	13	16	27	56	158
Antipasti	36	8	15	21	44	124
Starbucks	27	5	7	12	24	68
Range	23–100	5–95	7–112	12–216	24–423	68–1195
Median	43	36	37	67	126	362

T. Crozier, A. Stalmach and A. Crozier, unpublished.
Data are mean values ($n = 3$), with in all instances, the standard error being <5% of the mean.

Table 5.4 Comparison of the chlorogenic acid content of espresso and instant coffees with that of (poly)phenols in selected beverages, fruit and vegetables in which a serving size is standardised to either 200 mL or 100 g

Product	Serving	Compounds	Dose (μmol)	Reference
Espresso coffee	27 mL	Chlorogenic acids	68	Table 5.3
	52 mL	Chlorogenic acids	1195	Table 5.3
Instant coffee	200 mL	Chlorogenic acids	412	Stalmach et al. (2009a)
Globe artichokes	100 g	Chlorogenic acids	762	Padino et al. (2010)
Apple juice	200 mL	Chlorogenic acids	31	Mullen et al. (2007)
		Dihydrochalcones	9	
Green tea	200 ml	Flavan-3-ol monomers	259	Stalmach et al. (2009c)
Unfermented rooibos tea	200 mL	Dihydrochalcones	42	Stalmach et al. (2009b)
Cocoa	200 mL	Flavan-3-ol monomers	36	Mullen et al. (2009)
Fried onions	100 g	Flavonols	102	Mullen et al. (2006)
Strawberries	100 g	Anthocyanins	121	Mullen et al. (2008)
Raspberries	100 g	Anthocyanins	68	González-Barrio et al. (2010)
		Ellagitannins	62	
Pomegranate juice	200 mL	Anthocyanins	69	Borges et al. (2010)
		Ellagitannins	82	
Orange juice	200 mL	Flavanones	11	Mullen et al. (2007)
Purple grape juice	200 mL	Anthocyanins	54	Mullen et al. (2007)
		Hydroxycinnamates	32	
Tomato juice	200 mL	Flavonols	8	Jaganath et al. (2006)

and 146 mg (412 μmol) per serving, which is close to the median CQA content for the espresso coffees.

Even without considering that an intake of four to five servings of coffee per day is not unusual for some people, the higher end of CQA intake from espresso coffee, 1195 μmol per cup, is massive compared with an estimated chlorogenic acid content of apple juice of 31 μmol/200 mL. Although globe artichokes are rich in chlorogenic acids, 762 μmol/100 g (Table 5.4), they are rarely – if ever – consumed in sufficient quantity to compete with coffee beverage. Only maté, providing some 270–320 μmol CGA per cup, is a realistic competitor in some populations (Clifford and Ramìrez-Martìnez 1990).

Regardless of the identity of the polyphenols, few other commodities (Table 5.4) can compete with coffee. Black tea can supply some 100 mg thearubigins per cup (see Chapter 3), and it is now apparent that the proanthocyanidins intake can be substantial, especially in diets rich in berries, grapes and cocoa.

5.4 Bioavailability of coffee chlorogenic acids in humans

The literature describing the fate of chlorogenic acids in humans after the consumption of coffee is scarce and, in some instances, contradictory. A study by Nardini et al. (2002) observed an increase of conjugated caffeic acid in plasma after the ingestion of 200 mL of coffee, while Rechner et al. (2001) detected ferulic acid, isoferulic acid, dihydroferulic

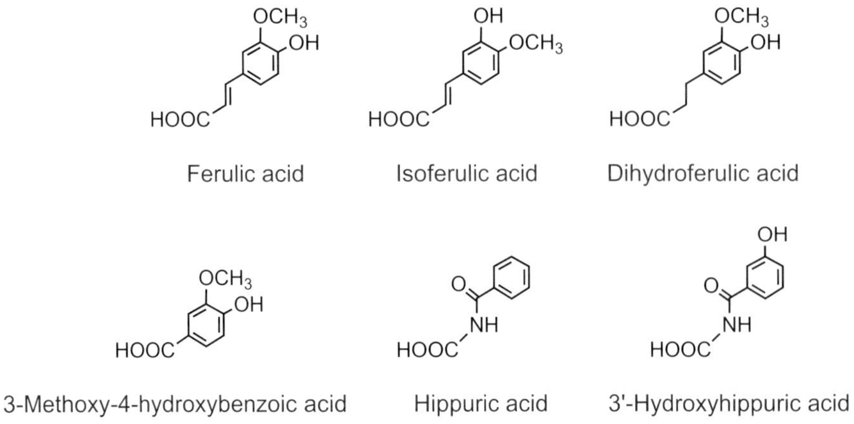

Figure 5.9 Chlorogenic acid catabolites detected in glucuronidase/sulphatase-treated human urine collected after the consumption of coffee (Rechner et al. 2001).

acid, 3-methoxy-4-hydroxybenzoic acid, hippuric acid and 3′-hydroxyhippuric acid (Figure 5.9) in urine excreted by humans after three ingestions of two cups of coffee at 4-hour intervals.

Monteiro et al. (2007) reported the presence of unmetabolised CQA in human plasma with a C_{max} of 7.7 μM and a T_{max} at 2.3 hours after acute ingestion of coffee containing 3395 μmol of CQA. Despite the high-mean C_{max} of the CQA, chlorogenic acids were not detected in urine collected 0–24 hours after coffee intake. However, in a subsequent study by the same group, in which volunteers consumed a coffee containing a much lower 451 μmol of chlorogenic acids, 4- and 5-CQA were detected in sulphatase/glucuronidase-treated urine from some, but not all, subjects (Farah et al. 2008). The chlorogenic acids were also detected in plasma although there were unusually marked variations in the plasma concentration–time profiles of the individual subjects, with – in two instances – an exceedingly high C_{max} of >20 μM being recorded.

The most detailed research on the fate of chlorogenic acids after the ingestion of coffee is that of Stalmach et al. (2009a, 2010) in feeding studies with healthy humans and ileostomists who have had their colon removed surgically for medical reasons. In these investigations, in which analysis of samples was based on HPLC-MS^2 methodology without the use of glucuronidase/sulphatase treatments prior to analysis, it was observed that during passage through the body, extensive metabolism of chlorogenic acids occurs, with some compounds being absorbed in the small intestine and others in the colon.

5.4.1 Studies involving volunteers with and without a functioning colon

Healthy humans with a functioning colon consumed a 200-mL serving of instant coffee, containing 412 μmol (146 mg) of chlorogenic acids, with CQA comprising

Table 5.5 Quantities of chlorogenic acids in a 200-mL serving of instant coffee fed to volunteers[a]

Chlorogenic acids	μmol
3-O-Caffeoylquinic acid	72 ± 1.3
4-O-Caffeoylquinic acid	78 ± 1.5
5-O-Caffeoylquinic acid	119 ± 2.1
Total caffeoylquinic acids	269 ± 4.9
3-O-Feruloylquinic acid	20 ± 1.6
4-O-Feruloylquinic acid	22 ± 1.8
5-O-Feruloylquinic acid	25 ± 1.9
Total feruloylquinic acids	67 ± 5.3
3-O-Caffeoylquinic acid lactone	34 ± 2.9
4-O-Caffeoylquinic acid lactone	23 ± 2.0
Total caffeoylquinic acid lactones	57 ± 4.9
3,4-O-Dicaffeoylquinic acid	5.8 ± 0.6
3,5-O-Dicaffeoylquinic acid	2.8 ± 0.3
4,5-O-Dicaffeoylquinic acid	4.0 ± 0.4
Total dicaffeoylquinic acids	12.6 ± 1.2
4-O-p-Coumaroylquinic acid	3.8 ± 0.1
5-O-p-Coumaroylquinic acid	3.1 ± 0.1
Total p-coumaroylquinic acids	6.8 ± 0.1
Total chlorogenic acids	**412 ± 9.3**

[a] Data expressed as mean values ± SE ($n = 3$).

65% of the total (Table 5.5), after which plasma and urine samples were collected over a 24-hour period (Stalmach et al. 2009a). A total of 12 hydroxycinnamate derivatives, mostly metabolites, were identified and quantified in plasma. Their pharmacokinetic parameters are summarised in Table 5.6 and pharmacokinetic profiles are illustrated in Figure 5.10. Maximum post-ingestion C_{max} values ranged from 2.2 nM for 5-CQA to 385 nM for dihydroferulic acid, with the duration for T_{max} extending from 0.6 (ferulic acid-4'-O-sulphate, 3-CQL-O-sulphate) to 5.2 hours (dihydroferulic acid). The compounds detected in highest concentrations in plasma were free and sulphated conjugates of dihydroferulic acid and dihydrocaffeic acid with C_{max} values ranging from 41 to 385 nM. The T_{max} for these compounds was in a narrow range from 4.7 to 5.2 hours, indicating absorption in the large intestine. Much shorter T_{max} values of 0.6–1.0 hour, indicative of small intestine absorption, were obtained with 5-CQA, two CQL-O-sulphates and three FQA, all of which had relatively low C_{max} values (Table 5.6, Figure 5.10).

As noted by Stalmach et al. (2009a), most of the chlorogenic-derived compounds were rapidly removed from the circulatory system, with elimination half-life ($T_{1/2}$) values of 0.3–1.9 hours. The only compounds with an extended $T_{1/2}$ were dihydroferulic acid-4'-O-sulphate (4.7 hours), dihydroferulic acid-3'-O-sulphate (3.1 hours) and ferulic acid-4'-O-sulphate (which had an unusual biphasic plasma profile with dual T_{max} values at 0.6 and 4.3 hours). It is of note that the free acid, dihydroferulic acid, as

Table 5.6 Pharmacokinetic parameters of chlorogenic acid derivatives and metabolites circulating in plasma of healthy volunteers, 0–24 hours following the ingestion of 412 μmol of chlorogenic acids and derivatives contained in a 200-mL serving of instant coffee[a]

Chlorogenic acids and metabolites	C_{max} (nM)	T_{max} (h)	$T_{1/2}$ (h)
5-O-Caffeoylquinic acid	2.2 ± 1.0	1.0 ± 0.2	0.3 ± 0.3
3-O-Caffeoylquinic lactone-O-sulphate	27 ± 3	0.6 ± 0.1	0.5 ± 0.1
4-O-Caffeoylquinic lactone-O-sulphate	21 ± 4	0.7 ± 0.1	0.4 ± 0.1
3-O-Feruloylquinic acid	16 ± 2	0.7 ± 0.1	0.9 ± 0.1
4-O-Feruloylquinic acid	14 ± 2	0.8 ± 0.1	0.9 ± 0.1
5-O-Feruloylquinic acid	6.0 ± 1.5	0.9 ± 0.1	0.8 ± 0.1
Caffeic acid-3'-O-sulphate	92 ± 11	1.0 ± 0.2	1.9 ± 0.4
Ferulic acid-4'-O-sulphate [b]	76 ± 9	0.6 ± 0.1	4.9 ± 1.0
	46 ± 13	4.3 ± 0.3	
Dihydroferulic acid	385 ± 86	4.7 ± 0.3	1.4 ± 0.4
Dihydroferulic acid-4'-O-sulphate	145 ± 53	4.8 ± 0.5	4.7 ± 0.8
Dihydrocaffeic acid	41 ± 10	5.2 ± 0.5	1.0 ± 0.4
Dihydrocaffeic acid-3'-O-sulphate	325 ± 99	4.8 ± 0.6	3.1 ± 0.3

After Stalmach et al. 2009a.
[a] Data expressed as mean values ± SE ($n = 11$).
[b] Double C_{max} and T_{max} values are due to the biphasic absorption profile of ferulic acid-4-O-sulphate (see Figure 5.5).

opposed to the more typical glucuronide and sulphate metabolites, was the principal component to accumulate in plasma, which also contained dihydrocaffeic acid in a lower concentration (Figure 5.10). Other unmetabolised compounds detected in plasma were three FQA (Figure 5.10) and trace concentrations of 5-CQA (Table 5.6).

Ileostomists drank an instant coffee with a very similar 385-μmol (136-mg) chlorogenic acid profile to that ingested by the healthy subjects (Stalmach et al. 2010). Plasma was not investigated, but data on the amounts of chlorogenic acids and their metabolites in ileal fluid collected over a 0- to 24-hour period after the ingestion of coffee are presented in Table 5.7. There was a 59% recovery of unmetabolised chlorogenic acids, with 3.6% appearing as a mixture of sulphate and glucuronide CQA metabolites. A 77% recovery of parent FQA was observed together with 8.6% as sulphate and glucuronide metabolites. In contrast, only 6.4% of the two ingested CQL appeared in ileal fluid compared to CQL metabolites that corresponded to 56% of the CQL intake. There was a 46% recovery of both p-CoQA and diCQA, and neither of these chlorogenic acids was converted to detectable quantities of metabolites.

Of the 385 μmol of chlorogenic acids ingested by the ileostomists, 274 μmol (71%) was recovered in the 0- to 24-hour ileal fluid as the parent compounds and glucuronide and sulphate metabolites (Table 5.7). This indicates that ∼30% of intake is absorbed in the small intestine and that in subjects with a functioning colon, ∼70% of the ingested chlorogenic acids pass from the small to large intestine, where they are subjected to the action of the colonic microflora. These observations are in line with the findings of Olthof et al. (2001), who fed 2.8 mmol of 5-CQA to humans with an ileostomy and recovered ∼70% of the chlorogenic acid intake in ileal fluid. The data from both

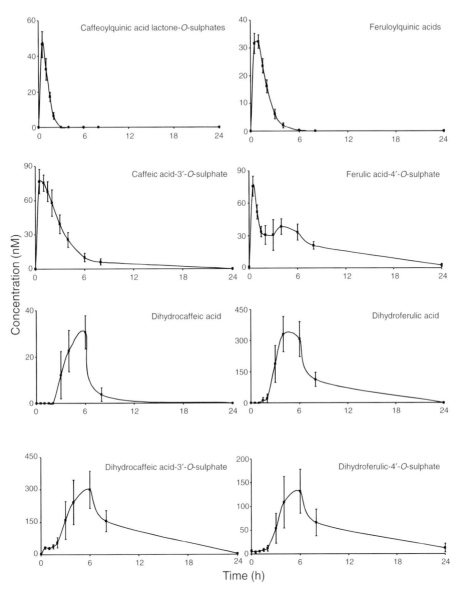

Figure 5.10 Plasma pharmacokinetic profiles of circulating chlorogenic acids and metabolites, following the ingestion of 200 mL of coffee by healthy human subjects (Stalmach et al. 2009a).

studies, albeit at substantially different doses, imply that around one-third of ingested chlorogenic acids in foods are absorbed and enter the bloodstream from the small intestine. *In vitro* studies support this conclusion as 5-CQA is stable when incubated with gastric juice, duodenal fluid and ileostomy effluent (Olthof et al. 2001; Rechner et al. 2001).

Table 5.7 Quantities of chlorogenic acids and metabolites recovered in ileal fluid collected 0–24 hours after the consumption of 200 mL of instant coffee containing 385 μmol of chlorogenic acids by humans with an ileostomy

Chlorogenic acids and metabolites	Ileal fluid (μmol)	Recovery of amount ingested (%)
3-O-Caffeoylquinic acid	55 ± 6	
4-O-Caffeoylquinic acid	45 ± 6	
5-O-Caffeoylquinic acid	64 ± 10	
Total caffeoylquinic acids	164 ± 22	59 ± 8
3-O-Caffeoylquinic acid-O-sulphate	3.5 ± 0.5	
4-O-Caffeoylquinic acid-O-sulphate	5.1 ± 0.6	
5-O-Caffeoylquinic acid-O-sulphate	0.7 ± 0.2	
3-O-Caffeoylquinic acid-O-glucuronide	0.7 ± 0.2	
4-O-Caffeoylquinic acid-O-glucuronide	0.4 ± 0.2	
Total caffeoylquinic acid metabolites	10 ± 1	3.6 ± 0.4
3-O-Feruloylquinic acid	12 ± 1	
4-O-Feruloylquinic acid	12 ± 1	
5-O-Feruloylquinic acid	13 ± 1	
Total feruloylquinic acids	37 ± 4	77 ± 8
3-O-Feruloylquinic acid-O-sulphate	0.7 ± 0.2	
4-O- Feruloylquinic acid-O-sulphate	0.8 ± 0.2	
5-O- Feruloylquinic acid-O-sulphate	0.3 ± 0.1	
3-O- Feruloylquinic acid-O-glucuronide	0.4 ± 0.1	
4-O- Feruloylquinic acid-O-glucuronide	2.0 ± 0.3	
Total feruloylquinic acid metabolites	4.2 ± 0.6	8.8 ± 1.3
3-O-Caffeoylquinic lactone	1.0 ± 0.3	
4-O-Caffeoylquinic lactone	1.5 ± 0.1	
Total caffeoylquinic lactones	2.5 ± 0.3	6.4 ± 0.8
3-O-Caffeoylquinic lactone-O-sulphate	13 ± 2.4	
4-O-Caffeoylquinic lactone-O-sulphate	8.2 ± 1.7	
3-O-Caffeoylquinic lactone-O-glucuronide	0.6 ± 0.2	
4-O-Caffeoylquinic lactone-O-glucuronide	0.4 ± 0.1	
Total caffeoylquinic lactone metabolites	22 ± 4.4	56 ± 11
4-O-p-Coumaroylquinic acid	1.1 ± 0.3	
5-O-p-Coumaroylquinic acid	1.9 ± 0.2	
Total p-coumaroylquinic acids	3.0 ± 0.2	46 ± 3
3,4-O-Dicaffeoylquinic acid	3.0 ± 0.3	
3,5-O-Dicaffeoylquinic acid	1.3 ± 0.2	
4,5-O-Dicaffeoylquinic acid	2.2 ± 0.2	
Total dicaffeoylquinic acids	6.5 ± 0.7	46 ± 5
Caffeic acid	9.0 ± 3.1	
Caffeic acid-4′-O- and 3′-O-sulphates	11 ± 2	
Ferulic acid	0.8 ± 0.3	
Ferulic acid-4′-O-sulphate	4.2 ± 1.0	
Total caffeic and ferulic acids	25 ± 5	—
Total chlorogenic acids and metabolites	**274 ± 28**	**71 ± 7**

After Stalmach et al. 2010.
Data presented as mean values ± standard error ($n = 5$); n.d., not detected.

In the small intestine, 5-CQA, other CQA and FQA are probably subjected to the action of esterases (Buchanan *et al.* 1996; Andreasen *et al.* 2001), resulting in the respective release of caffeic acid and ferulic acid, which after absorption undergo partial conversion to sulphates. A portion of the sulphate metabolites enter the circulatory system (Table 5.6, Figure 5.10), while the remainder, along with the residual caffeic and ferulic acids, is excreted in ileal fluid (Table 5.7), indicating that in subjects with an intact gastrointestinal tract, they would pass from the small to the large intestine.

The quantities of chlorogenic acids and their metabolites excreted in urine by healthy subjects and ileostomists over a 24-hour period post-ingestion of coffee are summarised in Table 5.8. The healthy volunteers excreted a total of 120.2 μmol, which corresponds to 29.2% of intake, while urine from ileostomists contained 30.8 μmol, which equates with only 8.0% of the ingested chlorogenic acids. This highlights the importance of the colon in the bioavailability of dietary chlorogenic acids.

The data presented in Table 5.8 show that absence of a colon had minimal impact on urinary excretion of CQL-O-sulphates and FQA, as well as on caffeic, ferulic and

Table 5.8 Urinary excretion of chlorogenic acid metabolites in 0- to 24-hour urine of healthy subjects ($n = 11$) and ileostomists ($n = 5$) following the ingestion of 200 mL of coffee

Chlorogenic acid and metabolites	Subjects without a colon (385 μmol ingested)	Subjects with a colon (412 μmol ingested)
3-O-Caffeoylquinic lactone-O-sulphate	0.6 ± 0.1	1.1 ± 0.1
4-O-Caffeoylquinic lactone-O-sulphate	0.4 ± 0.1	1.0 ± 0.1
3-O-Feruloylquinic acid	0.9 ± 0.2	1.2 ± 0.1
4-O-Feruloylquinic acid	0.9 ± 0.2	1.1 ± 0.1
5-O-Feruloylquinic acid	1.1 ± 0.2	1.0 ± 0.2
Ferulic acid-4'-O-sulphate	9.9 ± 1.9	11.1 ± 1.6
Feruloylglycine	2.1 ± 0.3[a]	20.7 ± 3.9[b]
Dihydroferulic acid	n.d.[a]	9.7 ± 2.0[b]
Dihydroferulic acid-4'-O-sulphate	0.8 ± 0.2	12.4 ± 3.4
Dihydroferulic acid-4'-O-glucuronide	n.d.	8.4 ± 1.9
Isoferulic acid-3'-O-sulphate	0.2 ± 0.0	0.4 ± 0.1
Isoferulic acid-3'-O-glucuronide	3.9 ± 0.8	4.8 ± 0.5
Dihydroisoferulic acid-3'-O-glucuronide	n.d.[a]	2.5 ± 0.4[b]
Caffeic acid-3'-O-sulphate	6.2 ± 1.2	6.4 ± 0.8
Caffeic acid-4'-O-sulphate	0.6 ± 0.1	0.6 ± 0.1
Dihydrocaffeic acid-3'-O-sulphate	3.2 ± 0.9[a]	37.1 ± 8.2[b]
Dihydrocaffeic acid-3'-O-glucuronide	n.d.[a]	0.7 ± 0.2[b]
Total	**30.8 ± 4.3** *(8.0%)*[a]	**120.2 ± 17.0** *(29.2%)*[b]

After Stalmach *et al.* 2009a, 2010.
Data represent mean values in μmol ± standard error; n.d., not detected. Different superscripts within rows indicate a statistical difference between the two sets of volunteers (Two-sample *t*-test, *P*-value < 0.05). Figures in bold, italicised, within parentheses indicate excretion as a percentage of chlorogenic acid intake.

isoferulic acid-O-sulphates. Furthermore, the data indicate that the small intestine is most probably the site for:

- the cleavage of quinic acid from CQA and FQA, releasing caffeic acid and ferulic acid;
- the metabolism of caffeic acid to its 3′- and 4′-O-sulphates, and ferulic acid to ferulic acid-4′-O-sulphate;
- the methylation of caffeic acid to form isoferulic acid and its subsequent 3′-O-sulphation and glucuronidation.

With the ileostomists there were major reductions in urinary excretion of dihydrocaffeic acid-3′-O-sulphate, dihydrocaffeic acid-3′-O-glucuronide, dihydroferulic acid and its glucuronide and sulphated derivatives, dihydroisoferulic acid-3′-O-glucuronide and feruloylglycine. This demonstrates that the colon is the site for:

- the conversion of ferulic acid to feruloylglycine and dihydroferulic acid;
- the metabolism of caffeic acid to dihydrocaffeic acid, which is further metabolised to dihydroisoferulic acid.

Despite its dual plasma T_{max} at 0.6 and 4.3 hours in healthy subjects (Figure 5.10), urinary excretion of ferulic acid-4′-O-sulphate was unaffected by the absence of a colon (Table 5.8), indicating that its secondary plasma T_{max} is not a consequence of colonic absorption.

Stalmach et al. (2009a, 2010) proposed the data obtained in their coffee-feeding studies with healthy volunteers and ileostomists are in keeping with the metabolic routes illustrated in Figure 5.11, which conveniently summarise their findings.

The data in Table 5.7 show that after coffee consumption a total of 46.2 μmol of ferulic acid-based compounds (ferulic acid, ferulic acid-4′-O-sulphate, FQAs, FQA-O-glucuronides and FQA-O-sulphates) were present in the 0- to 24-hour ileal fluid, and in healthy subjects, they would pass to the large intestine. The quantity of ferulic acid metabolites formed in the large intestine of healthy subjects and excreted in urine (feruloylglycine, and dihydroferulic acid and its 4′-O-glucuronide and 4′-O-sulphate) totalled 51.2 μmol (Table 5.8). As this figure is not greatly in excess of the 46.2 μmol of ferulic acid-based compounds entering the large intestine (Table 5.7), it is not possible to assess to what extent, if any, caffeic acid derivatives, such as CQA and CQL, are methylated and converted to their ferulic acid and feruloyl equivalents within the colon.

The findings of Stalmach et al. (2009a, 2010) with ileostomists and healthy subjects who consumed instant coffee are in line with data obtained in a recent study, where – after an acute intake of coffee – plasma and urine were analysed in extracts that had been subjected to glucuronidase/sulphatase treatment (Renouf et al. 2010). The virtual absence of chlorogenic acids in the plasma of healthy volunteers reported in both this study and studies by Stalmach et al. (2009a, 2010) is in marked contrast to the very high chlorogenic acid plasma C_{max} values, >20 μM in two instances, reported by Monteiro et al. (2007) and Farah et al. (2008). Further investigations are required in order to explain these unexpectedly high plasma chlorogenic acid levels especially as low nM concentrations of CQAs were detected in plasma by Matsui et al. (2007) after the consumption of a green coffee containing 842 μmol of the hydroxycinnamate esters.

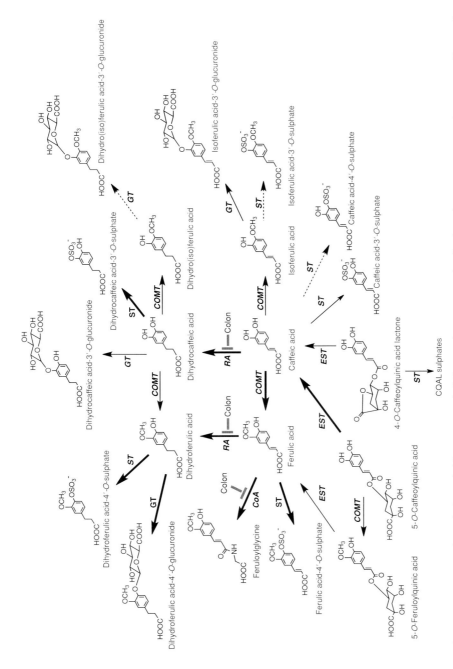

Figure 5.11 Proposed metabolism of chlorogenic acids following the ingestion of coffee by volunteers. 5-CQA and 5-FQA are illustrated structures, but their respective 3'- and 4'-isomers would be metabolised in a similar manner and likewise with 4-CQL and 3-CQL. *COMT*, catechol-*O*-methyltransferase; *EST*, esterase; *RA*, reductase; *GT*, UDP-glucuronyltransferase; *ST*, sulphuryltransferase; *Co-A*, co-enzyme A. Bold arrows indicate major routes, dotted arrows minor pathways. Steps blocked in subjects with an ileostomy and hence occurring in the colon are indicated (Stalmach *et al.* 2010).

It should also be noted that it is difficult to compare the data on 0- to 24-hour urinary metabolites excreted by healthy subjects after drinking a single 200-mL serving of coffee containing 412 µmol of chlorogenic acids (Table 5.8), with the findings of Olthof *et al.* (2003) who collected urine over a 24-hour period after volunteers had consumed a 5.5 mmol 5-CQA supplement on a daily basis for a period of 7 days. After enzyme hydrolysis, the phenolic acid content of urine was analysed by GC–MS. Although this facilitated the separation of a large number of phenolic acids, which if detected in urine would probably be derived from colonic catabolism of 5-CQA, the only compounds of relevance to the 17 urinary metabolites detected by HPLC–MS^2 by Stalmach *et al.* (2009a) (Table 5.8) are caffeic acid and ferulic acid, released by sulphatase hydrolysis of caffeic acid-3'- and 4'-O-sulphate and ferulic acid-4-O-sulphate.

The data presented in Table 5.8 on urinary excretion of hydroxycinnamate metabolites after ingestion of coffee by healthy volunteers clearly show that dihydrocaffeic acid-3'-O-sulphate and feruloylglycine would serve as very sensitive biomarkers for the consumption of relatively small amounts of coffee, the main source of chlorogenic acids in the human diet. A more detailed fingerprint could be obtained by the additional analysis of dihydroferulic acid-4'-O-sulphate, ferulic acid-4'-O-sulphate, dihydroferulic acid and dihydroferulic acid-4'-O-glucuronide.

5.5 Conclusions

Coffee is unique. In terms of global production, it must be the major source of chlorogenic acids, and for many consumers, also the major source of dietary polyphenols. The green bean provides some unique phytochemicals, and yet others are generated during roasting to give coffee its distinctive aroma.

After consumption of coffee by humans, ~30% of the chlorogenic acids are subject to the action of esterases in the small intestine and most of the released caffeic acid and ferulic acid are absorbed into the circulatory system, principally as sulphate metabolites, along with trace quantities of the parent CQA. The remaining ~70% of chlorogenic acids, together with some of the metabolites formed in the small intestine, pass into the colon where they are further metabolised, with the products being absorbed into the portal vein prior to the excretion of a diversity of metabolites, including feruloylglycine, and sulphate and glucuronide conjugates of caffeic acid, dihydrocaffeic acid, ferulic acid, dihydroferulic acid, isoferulic acid and dihydroisoferulic acid. Overall, coffee chlorogenic acids are highly bioavailable, with 30% of intake being excreted as metabolites with 24 hour of ingestion. *In vitro* studies have shown that at low µM concentrations some of these metabolites have antiglycative and neuroprotective activity (Verzelloni *et al.* 2011).

References

Allred, K.F., Yackley, K.M., Vanamala, J. *et al.* (2009) Trigonelline is a novel phytoestrogen in coffee beans. *J. Nutr.*, **139**, 1833–1838.

Alonso-Salces, R.M., Guillou, C. and Berrueta, L.A. (2009a) Liquid chromatography coupled with ultraviolet absorbance detection, electrospray ionization, collision-induced dissociation and tandem mass spectrometry on a triple quadrupole for the on-line characterization of polyphenols and methylxanthines in green coffee beans. *Rapid Commun. Mass Spectrom.*, **23**, 363–383.

Alonso-Salces, R.M., Serra, F., Reniero, F. et al. (2009b) Botanical and geographical characterization of green coffee (*Coffea arabica* and *Coffea canephora*): chemometric valuation of phenolic and methylxanthine contents. *J Agric. Food Chem.*, **57**, 4224–4235.

Andreasen, M.F., Kroon, P.A., Williamson, G. et al. (2001) Esterase activity able to hydrolyze dietary antioxidant hydroxycinnamates is distributed along the intestine of mammals. *J. Agric. Food Chem.*, **49**, 5679–5684.

Anthony, F., Clifford, M.N. and Noirot, M. (1993) Biochemical diversity on the genus *Coffea L.*: chlorogenic acids, caffeine and mozambioside contents. *Gen. Res. Crop Evol.*, **40**, 61–70.

Ashihara, H. and Crozier, A. (1999) Biosynthesis and degradation of caffeine and related purine alkaloids. In J.A. Callow (ed), *Advances in Botanical Research*, Vol. 30. Academic Press, London, pp. 117–205.

Bennat, C. Engelhardt, U.H. Kiehne, A. et al. (1994) HPLC analysis of chlorogenic acid lactones in roasted coffee. *Z. Lebens.-Unters. Forschung*, **199**, 17–21.

Borges, G., Mullen, W., Crozier, A. (2010) Comparison of the polyphenolic composition and antioxidant activity of European commercial fruit juices. *Food Funct.*, **1**, 73–83.

Buchanan, C., Wallace, G. and Fry, S.C. (1996) *In vivo* release of ^{14}C-labelled phenolic groups from intact dietary spinach cell walls during passage through the rat intestine. *J. Sci. Food. Agric.*, **71**, 459–469.

Clarke, R.J. (1985a) Green coffee processing. In M.N. Clifford and K.C. Willson (eds), *Coffee: Botany, Biochemistry and Production of Beans and Beverage*. Croom Helm, London, pp. 230–250.

Clarke, R.J. (1985b) The technology of converting green coffee into the beverage. In M.N. Clifford and K.C. Willson (eds), *Coffee: Botany, Biochemistry and Production of Beans and Beverage*. Croom Helm, London, pp. 375–393.

Clifford, M.N. (1985a) Chemical and physical aspects of green coffee and coffee products. In M.N. Clifford and K.C. Willson (eds), *Coffee: Botany, Biochemistry and Production of Beans and Beverage*. Croom Helm, London, pp. 305–374.

Clifford, M.N. (1985b) Chlorogenic acids. In R.J. Clarke and R. Macrae (eds), *Coffee 1. Chemistry*. Elsevier Applied Science, London, pp. 153–202.

Clifford, M.N. (1999) Chlorogenic acids and other cinnamates – nature, occurrence and dietary burden. *J. Sci. Food Agric.*, **79**, 362–372.

Clifford, M.N. (2006) Chlorogenic acids – their characterisation, transformation during roasting, and potential dietary significance. In *Proc. 21st Int. Conf. Coffee Sci.*, Montpellier, ASIC, Paris.

Clifford, M.N. and Jarvis, T. (1988) The chlorogenic acids content of green robusta coffee beans as a possible index of geographic origin. *Food Chem.*, **9**, 291–298.

Clifford, M.N. and Kazi, T. (1987) The influence of coffee bean maturity on the content of chlorogenic acids, caffeine and trigonelline. *Food Chem.*, **26**, 59–69.

Clifford, M.N. and Knight, S. (2004) The cinnamoyl–amino acid conjugates of green robusta coffee beans. *Food Chem.*, **87**, 457–463.

Clifford, M.N. and Ohiokpehai, O. (1983) Coffee astringency. *Analyt. Proc.*, **29**, 83–86.

Clifford, M.N. and Ramìrez-Martìnez, J.R. (1990) Chlorogenic acids and purine alkaloid content of Maté (*Ilex paraguariensis*) leaf and beverage. Food Chem., **35**, 13–21.

Clifford M.N., Williams, T. and Bridson, D. (1989a) Chlorogenic acids and caffeine as possible taxonomic criteria in *Coffea* and *Psilanthus*. *Phytochemistry*, **28**, 829–838.

Clifford, M.N., Kellard, B. and Ah-Sing, E. (1989b) Caffeoyl-tyrosine from green robusta coffee beans. *Phytochem.*, **28**, 1989–1990.

Clifford, M.N., Johnston, K.L., Knight, S. et al. (2003) A hierarchical scheme for LC-MS[n] identification of chlorogenic acids. *J. Agric. Food Chem.*, **51**, 2900–2911.

Clifford, M.N., Knight, S. and Kuhnert, N. (2005) Discriminating between the six isomers of dicaffeoylquinic acid by LC–MS^n. *J. Agric. Food Chem.*, **53**, 3821–3832.

Clifford, M.N., Knight, S., Suruca, B. et al. (2006a) Characterization by LC-MS^n of four new classes of chlorogenic acids in green coffee beans: dimethoxy-cinnamoylquinic acids, diferuloylquinic acids, caffeoyl-dimethoxycinnamoylquinic acids and feruloyl-dimethyoxycinnamolyquinic acids. *J. Agric. Food Chem.*, **54**, 1957–1969.

Clifford, M.N., Marks, S., Knight, S. et al. (2006b) Characterization by LC–MS^n of four novel classes of p-coumaric acid-containing diacyl chlorogenic acids in green coffee beans. *J. Agric. Food Chem.*, **54**, 4095–4101.

Clifford, M.N., Wu, W., Kirkpatrick, J. et al. (2007) Profiling the chlorogenic acids and other caffeic acid derivatives of herbal Chrysanthemum by LC–MS^n. *J. Agric. Food Chem.*, **55**, 929–936.

Crozier, A. and Ashihara, H. (2006) The cup that cheers. Caffeine biosynthesis: biochemistry and molecular biology. *The Biochemist*, **28**, 23–26.

Farah, A., de Paulis, T., Trugo, L.C. (2005) Effect of roasting on the formation of chlorogenic acid lactones in coffee. *J. Agric. Food Chem.*, **53**, 1505–1513.

Farah, A., de Paulis, T., Moreira, D.P. et al. (2006) Chlorogenic acids and lactones in regular and water-decaffeinated arabica coffees. *J. Agric. Food Chem.*, **54**, 374–381.

Farah, A., Monteiro, M., Donangelo, C.M. et al. (2008) Chlorogenic acids from green coffee extract are highly bioavailable in humans. *J. Nutr.*, **138**, 2309–2315.

Fischer, H.O.L. and Dangschat, G. (1932) Konstitution der chlorogensäure (3. Mitteil. über chinasäure und derivate). *Berichte*, **65**, 1037–1040.

Frank, O., Zehentbauer, G. and Hofmann, T. (2006) Bioresponse-guided decomposition of roast coffee beverage and identification of key bitter taste compounds. *Eur. Food Res. Technol.*, **222**, 492–508.

Ginz, M. (2001) Bittere Diketopiperazine und Chlorogensäure Derivate in Röstkaffee. PhD thesis, 208 pp., Universität Braunschweig.

Ginz, M. and Engelhardt, U.H. (2000) Identification of proline-based diketopiperazines in roasted coffee. *J. Agric. Food Chem.*, **48**, 3528–3532.

González-Barrio, R., Borges, G., Mullen, W. et al. (2010). Bioavailability of raspberry anthocyanins and ellagitannins following consumption of raspberries by healthy humans and subjects with an ileostomy. *J. Agric. Food Chem.*, **58**, 3933–3939.

Gosnay, S.L., Bishop, J.A., New, S.A. et al. (2002) Estimation of the mean intakes of fourteen classes of dietary phenolics in a population of young British women aged 20–30 years. *Proc. Nutr. Soc.*, **61**, 125A.

Grosch, W. (2001) Chemistry III. Volatile compounds. In R.J. Clarke and O.G. Vitzthum (eds), *Coffee: Recent Developments*. Blackwell Publishing Ltd., Oxford, pp. 68–89.

International Coffee Organisation (2010) Total production of exporting countries. http://www.ico.org.

IUPAC (1976) IUPAC nomenclature of cyclitols. *Biochem. J.*, **153**, 23–31.

Jaganath, I.B., Mullen, W., Edwards, C.A. et al. (2006). The relative contribution of the small and large intestine to the absorption and metabolism of rutin in man. *Free Rad. Res.*, **40**, 1035–1046.

Jaiswal, R. and Kuhnert, N. (2010) Hierarchical scheme for liquid chromatography/multi-stage spectrometric identification of 3,4,5-triacyl chlorogenic acids in green robusta coffee beans. *Rapid Commun. Mass Spectrom.*, **24**, 2283–2294.

Jaiswal, R., Patras, M.A., Eravuchira, P.J. et al. (2010) Profile and characterization of the chlorogenic acids in green robusta coffee beans by LC-MS^n: identification of seven new classes of compounds. *J. Agric. Food Chem.*, **58**, 8722–8737.

Lang, R., Wahl, A., Skurk, T. et al. (2010) Development of a hydrophilic liquid interaction chromatography-high-performance liquid chromatography-tandem mass spectrometry based stable isotope dilution analysis and pharmacokinetic studies on bioactive pyridines in human plasma and urine after coffee consumption. *Anal. Chem.*, **82**, 1486–1497.

Matsui, Y., Nakamura, S., Kondou, N. *et al.* (2007) Liquid chromatography–electrospray ionization–tandem mass spectrometry for simultaneous analysis of chlorogenic acids and their metabolites in human plasma. *J. Chromatogr. B*, **858**, 97–105.

Monteiro, M., Farah, A., Perrone, D. *et al.* (2007) Chlorogenic acid compounds from coffee are differentially absorbed and metabolized in humans. *J. Nutr.*, **137**, 2196–2201.

Moon, J.K., Yoo, H.S. and Shibamoto, T. (2009) Role of roasting conditions in the level of chlorogenic acid content in coffee beans: correlation with coffee acidity. *J. Agric. Food Chem.*, **57**, 5365–5369.

Mullen, W., Edwards, C.A., Crozier, A. (2006). Absorption, excretion and metabolic profiling of methyl-, glucuronyl-, glucosyl and sulpho-conjugates of quercetin in human plasma and urine after ingestion of onions. *Br. J. Nutr.*, **96**, 107–116.

Mullen, W., Marks, S.C., Crozier A. (2007). An evaluation of flavonoids and phenolic compounds in commercial fruit juices and juice drinks. *J. Agric. Food Chem.*, **55**, 3148–3157.

Mullen, W., Edwards, C.A., Serafini, M. *et al.* (2008). Bioavailability of pelargonidin-3-glucoside and its metabolites in humans following the ingestion of strawberries with and without cream. *J. Agric. Food Chem.*, **56**, 713–719.

Mullen, W., Borges, G., Donovan J.L. *et al.* (2009) Milk decreases urinary excretion but not plasma pharmacokinetics of cocoa flavan-3-ol metabolites in humans. *Am. J. Clin. Nutr.*, **89**, 1784–1791.

Nardini, M.E., Cirillo, F., Natella, C. *et al.* (2002) Absorption of phenolic acids in humans after coffee consumption. *J. Agric. Food Chem.*, **50**, 5735–5741.

Nursten, H.E. (2005) *The Maillard Reaction: Chemistry, Biology and Implications.* Springer, Heidelberg.

Olthof, M.R., Hollman, P.C. and Katan, M.B. (2001) Chlorogenic acid and caffeic acid are absorbed in humans. *J Nutr.*, **131**, 66–71.

Olthof, M.R., Hollman, P.C.H., Buijsman, M.N.C.P. *et al.* (2003) Chlorogenic acid and caffeic acid are absorbed in humans. *J Nutr.*, **133**, 1806–1814.

Padino, G., Courts, F.L., Lombado, S. *et al.* (2010) Caffeoylquinic acids and flavonoids in the immature inflorescence of globe artichoke, wild cardoon and cultivated cardoon. *J. Agric. Food Chem.*, **58**, 1026–1031.

Payen, S. (1846a) Memoire sur le café (3° part). *Comptes Rendus*, **23**, 244–251.

Payen, S. (1846b) Untersuchung des Kaffees. *Annalen*, **60**, 286–294.

Rechner, A.R., Spencer, J.P.E., Kuhnle, G. *et al.* (2001) Novel biomarkers of the metabolism of caffeic acid derivatives *in vivo*. *Free Rad. Biol. Med.*, **30**, 1213–1222.

Reichardt, N., Gniechwitz, D., Steinhart, H. *et al.* (2009) Characterization of high molecular weight coffee fractions and their fermentation by human intestinal microbiota. *Mol. Nutr. Food Res.*, **53**, 287–299.

Renouf, M., Guy, P.A., Marmet, C. *et al.* (2010) Measurement of caffeic and ferulic acid equivalents in plasma after coffee consumption: small intestine and colon are key sites for coffee metabolism. *Mol. Nutr. Food Res.*, **54**, 760–766.

Scholz, B.M., Maier, H.G. (1990) Isomers of quinic acid and quinide in roasted coffee. *Z. Lebens.-Unters. Forschung*, **190**, 132–134.

Schrader, K., Kiehne, A., Engelhardt, U.H. *et al.* (1996). Determination of chlorogenic acids with lactones in roasted coffee. *J. Sci. Food Agric.,* **71**, 392–398.

Silvan, J.M., Morales, F.J. and Saura-Calixto, F. (2010) Conceptual study on Maillardized dietary fiber in coffee. *J. Agric. Food Chem.*, in press.

Smith, R.F. (1985) A history of coffee. In M.N. Clifford and K.C. Willson (eds), *Coffee: Botany, Biochemistry and Production of Beans and Beverage.* Croom Helm, London, pp. 1–12.

Speer, K. and Kölling-Speer, I. (2006) The lipid fraction of the coffee bean. *Brazil. J. Plant Physiol.*, **18**, 201–216.

Speer, K. and Kölling-Speer, I. (2001) Chemistry IC. Lipids. In R.J. Clarke and O.G Vitzthum (eds), *Coffee: Recent Developments*. Blackwell Publishing Ltd., Oxford, pp. 33–49.

Stadler, R.H., Welti, D.H., Stämpfli, A.A. *et al.* (1996) Thermal decomposition of caffeic acid in model systems. Identification of novel tetraoxygenated phenylindan isomers and their stability in aqueous solution. *J. Agric. Food Chem.*, **44**, 898–905.

Stadler, R.H., Varga, N., Milo, C. *et al.* (2002) Alkylpyrdiniums. 2. Isoation and quantification in roasted and ground coffees. *J. Agric. Food Chem.*, **44**, 898–905.

Stalmach, A., Mullen, W., Nagai, C. *et al.* (2006) On-line HPLC analysis of the antioxidant activity of phenolic compounds in brewed paper-filtered coffee. *Brasil. J. Plant Physiol.*, **18**, 253–262.

Stalmach, A., Mullen, W., Barron, D. *et al.* (2009a) Metabolite profiling of hydroxycinnamate derivatives in plasma and urine after the ingestion of coffee by humans: identification of biomarkers of coffee consumption. *Drug Metab. Dispos.*, **37**, 1749–1758.

Stalmach, A., Mullen, W., Pecorari, M. *et al.* (2009b) Bioavailability of C-linked dihydrochalcone and flavanone glucosides in humans following ingestion of unfermented and fermented rooibos teas. *J. Agric. Food Chem.*, **57**, 7104–7171.

Stalmach, A., Troufflard, S., Serafini M. *et al.* (2009c). Absorption, metabolism and excretion of Choladi green tea flavan-3-ols by humans. *Mol. Nutr. Food Res.*, **53**, S44–S53.

Stalmach, A., Steiling, H., Williamson, G. *et al.* (2010) Bioavailability of chlorogenic acids following acute ingestion of coffee by humans with an ileostomy. *Arch. Biochem. Biophys.*, **501**, 98–105.

Stark, T. and Hofmann, T. (2005) Isolation, structure determination, synthesis, and sensory activity of *N*-phenylpropenoyl-L-amino acids from cocoa (*Theobroma cacao*). *J. Agric. Food Chem.*, **53**, 5419–5428.

Stark, T., Justus, H. and Hofmann, T. (2006) Quantitative analysis of *N*-phenylpropenoyl-L-amino acids in roasted coffee and cocoa powder by means of a stable isotope dilution assay. *J. Agric. Food Chem.*, **54**, 2859–2867.

Urgert, R. and Katan, M.B. (1997) The cholesterol-raising factor from coffee beans. *Ann. Rev. Nutr.*, **17**, 305–324.

Urgert, R., Meyboom, S., Kuilman, M. *et al.* (1996) Comparison of effect of cafetiere and filtered coffee on serum concentrations of liver aminotransferases and lipids: six month randomised controlled trial. *Br. Med. J.*, **313**, 1362–1366.

Urgert, R., Weusten-van der Wouw, M.P., Hovenier, R. *et al.* (1997) Diterpenes from coffee beans decrease serum levels of lipoprotein(a) in humans: results from four randomised controlled trials. *Eur. J. Clin. Nutr.*, **51**, 431–436.

van Cruchten, S.T., de Haan, L.H., Mulder, P.P. *et al.* (2010a) The role of epoxidation and electrophile-responsive element-regulated gene transcription in the potentially beneficial and harmful effects of the coffee components cafestol and kahweol. *J. Nutr. Biochem.*, **21**, 757–763.

van Cruchten, S.T., de Waart, D.R., Kunne, C. *et al.* (2010b) Absorption, distribution, and biliary excretion of cafestol, a potent cholesterol-elevating compound in unfiltered coffees, in mice. *Drug Metab Dispos.*, **38**, 635–640.

Verzelloni, E., Pellacani, C., Tagliazucchi, D. *et al.* (2011) Antiglycative and neuroprotective activity of colon-derived polyphenol catabolites. *Mol. Nutr. Food Res.*, **54**, in press.

Willson, K.C. (1999) *Coffee, Cocoa and Tea*. CABI Publishing, Wallingford, UK.

Woods, E., Clifford M.N., Gibbs, M. *et al.* (2003), Estimation of mean intakes of 14 classes of dietary phenols in a population of male shift workers. *Proc. Nutr. Soc.*, **62**, 60A.

Chapter 6
Coffee and Health

Gary Williamson
School of Food Science and Nutrition, University of Leeds, Leeds LS2 9JT, UK

6.1 Introduction

Coffee is made from roasted coffee beans, which are the fruit of the coffee plant belonging to the Rubiaceae family. Two main species, *Coffea arabica* L. and *Coffea canephora*, are used for arabica and robusta coffees, respectively. Coffee is one of the world's most commonly consumed beverages, after water and tea, and is a complex mixture of compounds, including carbohydrates, lipids, nitrogenous compounds, vitamins, minerals, alkaloids, melanoidins and phenolic compounds. The biologically active classes of compounds in coffee are usually considered to be caffeine, chlorogenic acids, melanoidins and diterpenes (cafestol and kahweol).

Caffeine, an alkaloid naturally present in coffee beans, exhibits numerous and well-studied activities (Heckman *et al.* 2010). The specific effects of caffeine are not covered, since many reviews are already available on this subject (Heckman *et al.* 2010). However, in studies on coffee, it is often not possible to dissect out the contribution of individual components, especially caffeine, unless decaffeinated coffee has been used in the study.

One of the classes of compounds that has received much recent interest are the (poly)phenols, specifically phenolic acids, such as the class of chlorogenic acids, which are esters between *trans*-cinnamic acids (e.g. caffeic or ferulic acid) and quinic acid. The chlorogenic acid content of a 'typical' cup of coffee ranges from 70 to 350 mg (Clifford 1998). Examples of chlorogenic acids include caffeoylquinic acids, feruloylquinic acids, *p*-coumaroylquinic acids and dicaffeoylquinic acids (Clifford 2000). Free hydroxycinnamates, such as caffeic acid, ferulic acid and *p*-coumaric acids, are present only in trace amounts in normal coffees. The chlorogenic acid content of coffee varies with species and maturity, and with processing conditions, but less with geographical origin (Clifford 1998). Decaffeination has little effect on chlorogenic acid levels, but roasting has a significant effect. Clifford estimates a loss of chlorogenic acids of 8–10% for every 1% loss of dry matter, and the chlorogenic acid-to-caffeine ratio is an estimate of the degree of roasting (Clifford 1998).

Melanoidins are complex polymers of mostly undefined composition formed during the roasting process of coffee beans via Maillard reactions, and possess

---

Teas, Cocoa and Coffee: Plant Secondary Metabolites and Health, First Edition.
Edited by Professor Alan Crozier, Professor Hiroshi Ashihara and Professor F. Tomás Barberán.
© 2012 Blackwell Publishing Ltd. Published 2012 by Blackwell Publishing Ltd.

in vitro antioxidant activity (Borrelli *et al.* 2002). They are typically 20–30% of dry soluble coffee.

The diterpenes, cafestol and kahweol, are abundant in coffee (Dias *et al.* 2010), but are removed if the coffee is filtered. Boiled coffee especially has a high content of these compounds (Higdon and Frei 2006).

This chapter covers the effects of coffee on human health, with particular reference to the phenolic acid antioxidants. Human studies *in vivo* (summarised in Table 6.1) and epidemiological studies are covered. There have been hundreds of epidemiological studies on coffee consumption and risk of type 2 diabetes, coronary heart disease and various cancers, and the reader is referred to several reviews and meta-analyses on the subject (Noordzij *et al.* 2005; George *et al.* 2008; Bravi *et al.* 2009; Je *et al.* 2009; Tang *et al.* 2009, 2010; Wu *et al.* 2009; Arab 2010). Several earlier studies suggested that coffee consumption could cause a marked increase in the risk of coronary heart disease as well as several types of cancer. In contrast, more recently, prospective cohort studies, less prone to selection and information bias, have not confirmed these findings. In a summary of epidemiological data up to 2008, van Dam states, 'In sum, the currently available evidence on coffee and risk of cardiovascular diseases and cancer is largely reassuring' (van Dam 2008b).

6.2 Antioxidant status

Many researchers have reported results on coffee using various antioxidant assays, which are simple *chemical* tests for characterising an antioxidant activity. They do not indicate the biological activity or bioavailability of a compound, extract or food, but are often an indication of the (poly)phenolic content since these are chemically the dominant antioxidants in most plant-based foods and beverages. Chemical antioxidant assays are based on hydrogen atom transfer reactions, mostly where antioxidant and substrate compete for thermally generated peroxyl radicals through the decomposition of azo compounds, such as induced low-density lipoprotein auto-oxidation, oxygen radical absorbance capacity (ORAC), total radical-trapping antioxidant parameter (TRAP) and crocin bleaching assays. Antioxidant assays are also based on electron transfer, such as the total phenol assay using the Folin–Ciocalteu reagent, Trolox equivalence antioxidant capacity (TEAC), ferric ion-reducing antioxidant power (FRAP), 'the total antioxidant potential' assay using a Cu(II) complex as an oxidant and the 2,2-diphenyl-1-picrylhydrazyl ($DPPH^+$) radical (Huang *et al.* 2005). These antioxidant assays have also been used to measure the 'antioxidant' activity of biological fluids, such as blood. Generally, these changes are not well defined, since the concentrations of (poly)phenols in plasma are too low to directly influence the plasma antioxidant capacity, which is dominated by powerful endogenous 'antioxidants', especially albumin, uric acid, vitamin C, vitamin E, cysteine and glutathione (Sies 2007). Nevertheless, several studies on coffee have measured 'antioxidant' activity *in vivo*, and although this is not well defined, clearly this is a consequence of a biological change. Therefore, these data are discussed in what follows and summarised in Table 6.1. In addition, there are several endogenous antioxidant defences, which include enzymes such as superoxide dismutase, glutathione peroxidase, glutathione-*S*-transferase and paraoxonase,

Table 6.1 Summary of human intervention studies on coffee consumption and health biomarkers

Compound or food given	Dose per day, comparisons	Approximate total phenolic acid (mg/kg/day)[a]	Days	Study design	$n =$ (per group)	Biomarkers significantly affected	Biomarkers not significantly affected	Reference
Decaffeinated coffee vs. mannitol placebo	12 g		1	Randomised, crossover	15 overweight men (BMI = 25–35)	No effect	No effect on OGTT insulin and glucose AUC	van Dijk et al. (2009)
Chlorogenic acid vs. mannitol placebo	1 g	14	1	Randomised, crossover	15 overweight men (BMI = 25–35)	Lower blood glucose and insulin 15 min after oral glucose load	No effect on OGTT insulin and glucose AUC	van Dijk et al. (2009)
Coffee with increased chlorogenic acid	10 g coffee containing 900 mg chlorogenic acid + 25 g sucrose	13	1	Double-blind, randomised, crossover	6 male, 6 female	Glucose AUC reduced by 7%		Thom (2007)
Instant coffee	10 g coffee with about 300–400 mg chlorogenic acid + 25 g sucrose	13	1	Double-blind, randomised, crossover	6 male, 6 female	Lowered blood glucose after 15–30 min, higher at 45–60 min (not stated if significant)	No effect on glucose AUC	Thom (2007)
Decaffeinated coffee	10 g decaffeinated coffee with about 300–400 mg chlorogenic acid + 25 g sucrose	13	1	Double-blind, randomised, crossover	6 male, 6 female	Lowered blood glucose after 15–30 min, higher at 45–60 min (not stated if significant)	No effect on glucose AUC	Thom (2007)

(continued)

Table 6.1 (Continued)

Compound or food given	Dose per day, comparisons	Approximate total phenolic acid (mg/kg/day)[a]	Days	Study design	n = (per group)	Biomarkers significantly affected	Biomarkers not significantly affected	Reference
Coffee with increased chlorogenic acid	11 g coffee containing 1000 mg chlorogenic acid vs. 300–400 mg	13	49	Parallel, randomised, double-blind	15 slightly overweight men (BMI = 27–32)	Reduction in weight of 5.4 kg cf. 1.7 kg; reduction in body fat of 3.6% (from 27.2 to 23.6%) cf. 0.7%		Thom (2007)
Decaffeinated instant coffee vs. glucose control	400 mL coffee containing 1 mmol chlorogenic acid	5	1	Randomised, crossover	9 healthy volunteers	Decreased GIP and increased GLP-1 over 2 h	Plasma glucose AUC	Johnston et al. (2003)
Green coffee extract	140 mg chlorogenic acid from green coffee extract	2	84	Double-blind, placebo-controlled, randomised	14 with mild essential hypertension	Reduction in SBP and DBP	BMI, pulse rate, serum biochemical markers, plasma Fe, Mg, Cu, Zn, vitamin B_1	Watanabe et al. (2006)
Filtered coffee	60 g/L ground coffee equivalent to 200, 5, 35 mg caffeic, coumaric and ferulic acids, respectively	3.5	1	Subjects as own controls	10 24- to 35-year-old healthy regular coffee drinkers	Increase in LDL lag time from 55 to 65 min up to 18% at 30 and 60 min, incorporation of phenolic acids into LDL		Natella et al. (2007)
Green coffee extract	140 mg chlorogenic acid from green coffee extract	2	120	Parallel, placebo-controlled	10 healthy	Reactive hyperaemia ratio increased, decrease in plasma total homocysteine	Pulse wave velocity, acceleration plethysmograms	Ochiai et al. (2004)

Product	Intervention	Dose (approx.)	Duration (days)	Design	Subjects	Results	Reference
Instant arabica coffee	3 × 8 g portions (no analysis) and 1 week washout		7	Subjects as own controls	11 healthy	Increase in LDL lag time of 8%, decrease in total cholesterol, LDL cholesterol and malondialdehyde	Yukawa et al. (2004)
Instant coffee	120 mg chlorogenic acid per serving	1.7	1	Comparison to other groups	10 healthy	Reduction in iron absorption (less strong than tea)	Hurrell et al. (1999)
Instant coffee plus milk	120 mg chlorogenic acid per serving	1.7	1			Reduction in iron absorption (no effect of milk)	Hurrell et al. (1999)
Hydroxyhydroquinone-free coffee	No chlorogenic acid (80 mg caffeine) vs. no coffee		28	Double-blind, randomised controlled study, dose–response	37 healthy	No change in body weight, heart rate, triglyceride, LDL cholesterol, GOT, GPT, γ-GTP, SBP, DBP	Yamaguchi et al. (2008)
Hydroxyhydroquinone-free coffee	82 mg chlorogenic acid (80 mg caffeine) vs. no coffee	1.2	28	Double-blind, randomised controlled study, dose–response	37 healthy	Decrease in GPT, SBP −2.7 mm Hg, DBP −2.7 mm Hg	Yamaguchi et al. (2008)
Hydroxyhydroquinone-free coffee	172 mg chlorogenic acid (80 mg caffeine) vs. no coffee	2.5	28	Double-blind, randomised controlled study, dose–response	35 healthy	Slight reduction in body weight, small increase in total and HDL cholesterol, SBP −2.8 mm Hg, DBP −2.3 mm Hg	Heart rate, triglyceride, LDL cholesterol, GOT, GPT, γ-GTP	Yamaguchi et al. (2008)

(continued)

Table 6.1 (Continued)

Compound or food given	Dose per day, comparisons	Approximate total phenolic acid (mg/kg/day)[a]	Days	Study design	n = (per group)	Biomarkers significantly affected	Biomarkers not significantly affected	Reference
Hydroxyhydroquinone-free coffee	299 mg chlorogenic acid (80 mg caffeine) vs. no coffee	4.3	28	Double-blind, randomised controlled study, dose–response	37 healthy	Slight reduction in body weight, SBP −3.3 mm Hg, DBP −2.8 mm Hg	Heart rate, triglyceride, LDL cholesterol, GOT, GPT, γ-GTP	Yamaguchi et al. (2008)
Coffee	Three (365 mg total phenolics) or six cups (730 mg)	5.2	21	Two-dose, parallel	43 healthy non-smoking men		Plasma LDL or HDL cholesterol, plasma homocysteine, LDL-conjugated dienes, plasma hydroxyfatty acids, plasma F_2-isoprostanes, plasma glutathione peroxidase, serum paraoxonase, plasma folate or vitamin B_{12}	Mursu et al. (2005)
Coffee	Three (365 mg total phenolics) or six cups (730 mg)	5.2	1	Two-dose, parallel	35 healthy non-smoking men		LDL-conjugated dienes, plasma hydroxyfatty acids, plasma F_2-isoprostanes	Mursu et al. (2005)
Chlorogenic acid	Pure compound, 2 g (5.5 mmol)[b]	28.5	7	Four-treatment, randomised	20 healthy men and women	Total homocysteine in fasting plasma 4–5 h after intake increased 12%		Olthof et al. (2001)

Coffee	1 L unfiltered coffee	5	Single group	10	Threefold induction of glutathione-S-transferase P in blood; small change in total activity using chloro-dinitrobenzene	No change in plasma glutathione-S-transferase A	Steinkellner et al. (2005)
Coffee	1 L unfiltered coffee	5	Single group	7	45% reduction BPDE-induced DNA migration in COMET assays		
Coffee	1 L unfiltered compared to paper filtered coffee	5	Single group	7	Threefold induction of glutathione-S-transferase P in blood; small change in total glutathione-S-transferase activity using chloro-dinitrobenzene		
Coffee	600 mL coffee (30 g ground coffee)	5	Single group	8	DNA damage from H_2O_2 decreased 17%, from 3-amino-1-methyl-5H-pyrido[4,3-bisindole-acetate] decreased 35%; superoxide dismutase activity increased 38%	Glutathione peroxidase activity	Bichler et al. (2007)
Coffee	200 mL (60 g roasted coffee brewed in 1 L water)	1	One-dose, parallel	10	Increased crocin activity at 1–2 h; increased TRAP by 6% at 1–2 h		Natella et al. (2002)

(continued)

Table 6.1 (Continued)

Compound or food given	Dose per day, comparisons	Approximate total phenolic acid (mg/kg/day)[a]	Days	Study design	n = (per group)	Biomarkers significantly affected	Biomarkers not significantly affected	Reference
Weak coffee[c]	One 'tablespoon' coffee grounds in 300 mL water		1	Randomised, crossover	5	Reduced variability of the glycaemic index	No change in magnitude of glycaemic index when consumed with white bread, cheese puffs or 'fruit leather'	Aldughpassi and Wolever (2009)
Filter coffee	30 days of no coffee, 1 month of four cups filter coffee per day, and then 1 month of eight cups filter coffee per day		60	Single-group	47 habitual coffee drinkers	Eight cups cf. zero cups: serum IL-18 decreased 8%, 8-isoprostane decreased 16%, adiponectin increased 6%, total cholesterol increased 12%, HDL cholesterol increased 7%, apolipoprotein B increased 4%, LDL to HDL cholesterol decreased by 8%, apolipoprotein B/A-1 decreased 9%	Markers of glucose metabolism in OGTT	Kempf et al. (2010)
Filter coffee	60 g roasted and ground coffee in 1 L[d]		1	Three-arm ex vivo post-consumption (compared to water or pure caffeine)	10 healthy regular coffee drinkers	Inhibition of collagen and arachidonic acid-induced platelet aggregation in blood obtained 30 and 60 min after coffee drinking. Increased in caffeic acid derivatives in platelets	Caffeine content in platelets	Natella et al. (2008)

176

Filter decaffeinated coffee	Decaffeinated coffee (compared to water) with carbohydrate meal and then OGTT after 3 h	1	Single-blind, randomised, crossover	10 healthy males 18–50 years	Blood glucose AUC increased	Blood: insulin AUC, C-peptide AUC	Moisey et al. (2010)
Filter decaffeinated coffee	Decaffeinated coffee (compared to caffeinated coffee) with carbohydrate meal and then OGTT after 3 h	1	Single-blind, randomised, crossover	10 healthy males 18–50 years		No impairment of insulin sensitivity compared to a 40% reduction with caffeinated coffee; no change in serum free fatty acids (caffeinated coffee gave increase)	Moisey et al. (2008)
Coffee	284 mg total chlorogenic acids	4	Crossover	10 adults	Plasma FRAP increased 2.6%; TRAP increased 7.6% (90 min): these were not due to changes in uric acid, albumin, bilirubin, ascorbate or tocopherol		Moura-Nunes et al. (2009)
Decaffeinated coffee	Italian espresso coffee, 25 mL from 20 s extraction from 7 g compared to 50 mL	1	Single-blind, crossover, no control	15	Increase in FMD	Blood pressure	Buscemi et al. (2009)

(*continued*)

Table 6.1 (Continued)

Compound or food given	Dose per day, comparisons	Approximate total phenolic acid (mg/kg/day)[a]	Days	Study design	n = (per group)	Biomarkers significantly affected	Biomarkers not significantly affected	Reference
Coffee	Decaffeinated compared to caffeinated		1	Double-blind, crossover	20	Caffeinated increased blood pressure, but not decaffeinated; FMD decreased by caffeinated coffee, but not decaffeinated	No change in blood glucose by decaffeinated	Buscemi et al. (2010)
Decaffeinated coffee			1	Double-blind, randomised	11 healthy	Glucose AUC was 50% of that of an equi-glucose control	No change in insulin or C-peptide	Battram et al. (2006)
Unfiltered coffee	1 L		14	Parallel, two-group, randomised	64 healthy in two groups	Fasting plasma homocysteine increased by 10%	Serum cholesterol, triacylglycerol, vitamin B_{12}, B_6 or folate	Grubben et al. (2000)
Italian-style mocca and espresso	Five cups		7	Parallel, controlled (n = 5 in control group)	18 with normal homocysteine	Plasma glutathione increased by 16%	Plasma homocysteine	Esposito et al. (2003)
Decaffeinated coffee	Compared to warm water		1	Randomised, crossover, controlled	11 young men	For the OGTT, increased glucose and insulin; less than caffeinated coffee		Greenberg et al. (2010)

SBP, systolic blood pressure; DBP, diastolic blood pressure; AUC, area under the curve; GIP, glucose-dependent insulinotropic polypeptide; GLP-1, glucagon-like peptide 1; BMI, body mass index; OGTT, oral glucose tolerance test; LDL, low-density lipoprotein; HDL, high-density lipoprotein; FMD, flow-mediated dilation; TRAP, total radical-trapping antioxidant parameter; FRAP, ferric ion-reducing antioxidant power; GOT, glutamic oxaloacetic transaminase; GPT, glutamic pyruvic transaminase; γ-GTP, γ-glutamyl transpeptidase.
[a]Calculated based on an average weight of 70 kg.
[b]Similar effect from black tea; Rutin alone had no effect.
[c]Weak tea gave the same effect.
[d]No effect of caffeine alone.

which act together with small-molecule reductants such as glutathione. Some studies have examined the effect of coffee on these defences, which may be most commonly mediated through gene expression. It is usually considered that boosting endogenous antioxidant defences reflects a beneficial effect.

6.2.1 Effect of coffee consumption on antioxidant status: epidemiological and cohort studies

In a cross-sectional study of 12,687 Japanese individuals (7398 men and 5289 women, aged 40–69 years), increased coffee consumption was strongly and independently associated with decreased γ-glutamyltransferase activity among males; the inverse association between coffee and serum γ-glutamyltransferase was stronger among heavier alcohol consumers and was absent among non-alcohol drinkers. Among females, however, coffee was only weakly related to lower γ-glutamyltransferase level. Similar inverse associations with coffee and interactions between coffee and alcohol intake were observed for serum aspartate aminotransferase and alanine aminotransferase. These results were not modified when considering intake of green tea (Tanaka *et al.* 1998). Also, coffee consumption may affect circulating vitamin B levels at high levels of vitamin consumption, possibly by increasing excretion in the urine (Ulvik *et al.* 2008), and may weakly affect absorption of thyroxine T4 (Benvenga *et al.* 2008).

6.2.2 Effect of coffee consumption on antioxidant status: intervention studies

Several studies have examined the effect of coffee consumption on the antioxidant capacity of plasma. Ten adults were tested in a crossover design before and 90 minutes after coffee consumption compared to water. After coffee consumption, FRAP and TRAP significantly increased by 2.6% and 7.6%, but decreased by 2.5% and 1.0%, respectively, after water consumption. These changes did not correlate with changes in uric acid, albumin, ascorbate, tocopherol or bilirubin (Moura-Nunes *et al.* 2009). The increase in plasma TRAP was also seen after consumption of both coffee and tea (+6% and +4%, respectively, at peak time), but in this study, both produced an increase in plasma uric acid (+5% and +7%, respectively). An alternative antioxidant assay, the crocin test, showed an increase in plasma only in subjects consuming coffee (+7% at peak time) and not tea (Natella *et al.* 2002). After ingestion of unfiltered coffee each day over a period of 2 weeks by healthy subjects, there was a significant increase in [glutathione] both in colorectal mucosa (8%, $p = 0.01$) and in plasma (15%, $p = 0.003$). There was, however, no induction of glutathione-S-transferase enzyme activity or isoenzymes in the colorectal mucosa (Grubben *et al.* 2000). Similarly, after ingestion of five cups of coffee daily for 7 days by healthy subjects, there was a significant increase (16%) in plasma [glutathione] (Esposito *et al.* 2003). There was a weak, but significant, increase in the activity of glutathione-S-transferase activity and a pronounced increase in the P isoform after consumption of unfiltered coffee (Steinkellner *et al.* 2005). When healthy subjects consumed 600 mL of coffee per day over 5 days, there was a significant 38% increase in superoxide dismutase activity in the cytosol of human peripheral lymphocytes, but not in glutathione peroxidase activity (Bichler *et al.* 2007). The latter is in agreement with Mursu *et al.* (2005), who measured the effect of coffee consumption on glutathione peroxidase and paraoxonase activity. After consumption of no coffee, 450 or 900 mL of filtered coffee over 3 weeks, the activities of neither plasma glutathione peroxidase nor paraoxonase were significantly changed. In this study, however, there was no randomisation and the subjects were free to choose how much coffee they drank. A metabolomics study of coffee consumption by 284 men showed that coffee intake is positively associated with the plasma content of two classes of sphingomyelins, one containing a hydroxy group and one with an additional carboxy group. In contrast, long- and medium-chain acylcarnitine species decreased with increasing coffee consumption. Total cholesterol also increased with increased coffee intake (Altmaier *et al.* 2009). Black tea, cocoa and coffee are all well known to modulate iron absorption, although the overall effect on iron status depends on the population group studied (Hurrell *et al.* 1999).

6.3 Diabetes

Diabetes mellitus is a group of diseases characterised by an elevated blood glucose level (hyperglycaemia) resulting from defects in insulin secretion and/or in insulin action, and is a combination of aetiologically different metabolic defects. Common symptoms are lethargy from marked hyperglycaemia, blurred vision and susceptibility to infection. Severe hyperglycaemia may lead to insulin deficiency, and chronic hyperglycaemia causes long-term damage, dysfunction and failure of various cells, tissues and organs. Tests to assess pre-diabetes and type 2 diabetes are fasting blood glucose (levels of 5.5–7.0 mmol/L are above normal, i.e. pre-diabetic with impaired fasting glucose (IFG)) and the oral glucose tolerance test (OGTT), which determines the efficiency of glucose metabolism. The OGTT distinguishes metabolically healthy individuals with impaired glucose tolerance (IGT), a combination of an impaired secretion of insulin and reduced insulin sensitivity (insulin resistance), from those with diabetes.

6.3.1 Effect of coffee consumption on diabetes risk: epidemiological and cohort studies

There is a robust inverse relation between regular coffee consumption and risk of type 2 diabetes based on meta-analyses of epidemiological studies (van Dam and Hu 2005; van Dam, 2006a, 2006b). Decaffeinated coffee is also associated with decreased risk of diabetes (Battram *et al.* 2006; van Dam *et al.* 2006c); substances other than caffeine are responsible for this activity (van Dam 2006b), possibly the chlorogenic acids (Ranheim and Halvorsen 2005; Tunnicliffe and Shearer 2008). Chlorogenic acids affect the absorption and utilisation of glucose in human studies (Johnston *et al.* 2003; Thom 2007), consistent with animal and *in vitro* studies on glucose regulation (mainly inhibition of glucose-6-phosphatase and regulation of glucose-dependent insulinotropic polypeptide/glucagon-like peptide 1 (GIP/GLP-1)). Oxidative stress may be one of the mechanisms linking insulin resistance with β-cell and endothelial dysfunction, eventually leading to type 2 diabetes and cardiovascular diseases (Ceriello and Motz 2004). The results are not always clear cut, and there are some caveats. For example, in the Puerto Rico Heart Health Program sub-cohort, comprising 4685 middle-aged men (aged 35–79 years) participants, there was support for a protective effect of coffee intake on diabetes risk, but abstainers may be at reduced risk (Fuhrman et al. 2009). Sometimes studies show some unexpected relationships. In a study on British men ($n = 4055$) and women ($n = 1768$) from the Whitehall II cohort with 11.7 years follow-up, there were 387 incident cases of diabetes. Intake of more than three cups of coffee and tea per day were not prospectively associated with the incidence of type 2 diabetes, although there was evidence of a combined effect. There was an inverse association between coffee intake and 2-hour post-load glucose concentration at the baseline assessment. The authors try to explain their results by referring to the limited range of exposure and varied beverage consumption according to socio-economic class (Hamer et al. 2008).

Despite some studies that show no effect, generally epidemiological (mainly cohort) studies have shown consistently that coffee consumption reduces the risk of type 2 diabetes (Ranheim and Halvorsen 2005), often equivalent to a risk reduction of 30–60%, in the same range as observed with pharmacological approaches (Legrand and Scheen 2007). Meta-analyses on data from 18 studies with information on 457,922 participants for the association between coffee, decaffeinated coffee, and tea consumption with risk of diabetes showed an association between coffee consumption and diabetes. Six ($n = 225,516$) and seven studies ($n = 286,701$) also reported estimates of the association between decaffeinated coffee and tea with diabetes, respectively. There was an inverse log–linear relationship between coffee consumption and subsequent risk of diabetes: every additional cup of coffee consumed in a day was associated with a 7% reduction in the excess risk of diabetes relative risk. Similar significant and inverse associations were observed with decaffeinated coffee and tea and risk of diabetes (Huxley et al. 2009). At least 14 out of the above 18 cohort studies showed a substantially lower risk of type 2 diabetes mellitus with frequent coffee intake, with often a dose–response relationship (Pimentel et al. 2009). There is some evidence that this is due to the (poly)phenol intake, since in a prospective cohort study of 40,011 participants with 918 follow-up cases of type 2 diabetes over a 10-year period showed that total daily consumption of at least three cups of coffee and/or tea reduced the risk of type 2 diabetes by approximately 42%. This could not be explained by magnesium, potassium, caffeine or blood pressure effects (van Dieren et al. 2009). A total of 21,826 Finnish men and women who were 35–74 years of age and without any history of diabetes at baseline were prospectively followed up for onset of type 2 diabetes ($n = 862$ cases). Habitual coffee consumption was associated with lower incidence of type 2 diabetes, particularly in those with higher baseline serum GGT levels (Bidel et al. 2008). In a study on American Indian men and women 45–74 years of age, participants with normal glucose tolerance ($n = 1141$) at baseline were followed up for 7.6 years. In this population, a high level of coffee consumption was associated with a reduced risk of deterioration of glucose metabolism over the follow-up compared to those who did not drink coffee, and participants who drank 12 or more cups of coffee daily had a 67% less risk of developing diabetes during the follow-up (Zhang et al. 2011). In a prospective cohort study on 69,532 French women, an 11-year follow-up of 1415 new cases of diabetes supported an inverse association between coffee consumption and diabetes, but suggested that the time of drinking coffee plays a distinct role in glucose metabolism. At lunchtime, this inverse association was observed for both regular and decaffeinated coffee and for filtered and black coffee, with no effect of sweetening. Total caffeine intake was also associated with a statistically significantly lower risk of diabetes. Neither tea nor chicory consumption was associated with diabetes risk (Sartorelli *et al.* 2010). In the Singapore Chinese Health Study on 36,908 female and male participants 45–74 years of age, participants who consumed four or more cups of coffee per day showed a 30% reduction in the risk of diabetes compared with those who reported little consumption, indicating that regular coffee consumption is associated with lower risk of type 2 diabetes in Asian men and women in Singapore (Odegaard *et al.* 2008).

6.3.2 Effect of coffee consumption on diabetes risk: intervention studies

The epidemiological association is supported by intervention studies, which have shown effects of coffee consumption on blood glucose response (Johnston *et al.* 2003; Battram *et al.* 2006; Thom 2007). There are several intervention studies on coffee that show an effect of coffee on diabetic markers. In a randomised, crossover, placebo-controlled trial of the effects of decaffeinated coffee, caffeinated coffee, and caffeine on glucose, insulin and GIP levels during an OGTT in 11 young men, glucose and insulin were higher for decaffeinated coffee than for placebo within the first hour of the OGTT. During the whole OGTT, decaffeinated coffee yielded higher insulin than placebo, and lower glucose and a higher insulin sensitivity index than caffeine. Changes in GIP could not explain any beverage effects on glucose and insulin (Greenberg *et al.* 2010). In a randomised, crossover design, ten healthy males ingested caffeinated coffee, decaffeinated coffee or water co-ingested with a high glycaemic index cereal followed 3 hours later by an OGTT. After the initial meal, insulin area under the curve (AUC) and insulin sensitivity index did not differ for both coffees, although the glucose AUC for both was greater than water. After the second carbohydrate load, insulin AUC for caffeinated coffee was 49% and 57% greater than for decaffeinated coffee and water, respectively. Despite the greater insulin response, glucose AUC for caffeinated coffee was greater than both decaffeinated coffee and water. Insulin sensitivity index after the second meal was lower after caffeinated coffee compared with both decaffeinated coffee and water. Co-ingestion of caffeinated coffee with one meal resulted in insulin insensitivity during the postprandial phase of a second meal in the absence of further caffeinated coffee ingestion (Moisey *et al.* 2010).

In one study, drinking coffee or tea with test meals did not affect the mean glycemic index value obtained, but reduced variability in the measurements (Aldughpassi and Wolever 2009). This study illustrates that there is a small effect of acute doses of coffee phenolics on glucose and insulin, but that because the effect is small, it reaches significance only in some studies. No changes were seen for markers of glucose metabolism after coffee consumption in an OGTT (Kempf *et al.* 2010). On separate occasions in random order, lean, young, healthy subjects ($n = 8$) consumed a potato-based meal

1 hour after consumption of test substances. Compared to water, black coffee caused a 28% increase in postprandial glycaemia. In contrast, black coffee with sucrose decreased glycaemia compared with either black coffee or water, but had no effect on insulin responses. Decaffeinated coffee, black tea and sucrose had no significant effects on postprandial responses. Sucrose and decaffeinated coffee reduced the absolute glucose concentration at the start of the meal, demonstrating that only sweetened coffee significantly reduced postprandial glycaemia (Louie *et al.* 2008).

Other studies have produced different results. Ten healthy men underwent four trials in a randomised order, ingesting caffeinated coffee or the same volume of decaffeinated coffee followed 1 hour later by either a high or low glycaemic index cereal mixed meal tolerance test. Caffeinated coffee with the high glycaemic index meal resulted in 147%, 29% and 40% greater AUCs for glucose, insulin and C-peptide, respectively, compared with the values for decaffeinated coffee. Similar results were obtained from a low-glycaemic-index treatment. Insulin sensitivity was significantly reduced with the high-glycaemic-index treatment after caffeinated coffee was ingested compared with decaffeinated coffee. The ingestion of caffeinated coffee with either a high or low GI meal significantly impairs acute blood glucose management and insulin sensitivity compared with ingestion of decaffeinated coffee (Moisey *et al.* 2008). Despite several studies on coffee, very few studies have looked at the effect of purified coffee phenolic acids in intervention studies. One gram chlorogenic acid and 500 mg trigonelline ingestion significantly reduced glucose and insulin concentrations 15 minutes following an OGTT compared with placebo (1 g mannitol) in 15 overweight men in a randomised, crossover trial, although total AUC values were not affected. Decaffeinated coffee (12 g) had no effect on insulin or glucose (van Dijk *et al.* 2009).

6.4 Cardiovascular disease

6.4.1 Effect of coffee consumption on cardiovascular risk: epidemiological and cohort studies

Coffee ingestion has been traditionally related to an increase in cardiovascular disease risk. However, the results are conflicting, showing discrepancies between cohort and case–control results. In more recent investigations, the controversial harmful effects were likely exacerbated by association between high intakes of coffee and unhealthy lifestyle habits (smoking, alcohol drinking, physical inactivity, high-fat diets, etc.). The risk associated with cardiovascular disease was later shown to be related to the presence of coffee diterpenes (cafestol and kahweol) in unfiltered boiled coffee (typically, 6–12 mg per cup). Coffee seems to have distinct acute and long-term effects on cardiovascular health, some beneficial and some harmful. Caffeine also shows effects, including an increase in blood pressure in a few studies (Geleijnse 2008). This could explain discrepancies in the results observed between cohort and case–control studies (Sofi *et al.* 2007). A recent meta-analysis including 21 cohort studies (15,599 cases from 407,806 participants were included in pooling the overall effects) showed that coffee ingestion could even be related to a lower risk of cardiovascular disease in women who were moderate drinkers (Wu *et al.* 2009). Until recently, few epidemiological studies collected information about the brewing process used to prepare coffee: the overall consumption of boiled (unfiltered) coffee is now lower than that when some of the older epidemiological studies were carried out, and it is possible that coffee phenolics may partially or fully overcome the harmful effects of diterpenes, which are present only in low amounts (0.2–0.6 mg per cup) in filtered or instant coffee (Kleemola *et al.* 2000). In fact, filtered or instant coffee may be beneficial for cardiovascular health (Higdon and Frei 2006; Cornelis and El-Sohemy 2007; van Dam 2008a).

The risk of stroke has been studied in some detail. For 83,076 women in the Nurses' Health Study without history of stroke, coronary heart disease, diabetes or cancer at baseline, 2280 strokes – of which 426 were haemorrhagic, 1224 were ischaemic and 630 were undetermined – were documented. Long-term coffee consumption was not associated with a modestly reduced risk of stroke. Other drinks containing caffeine, such as tea and caffeinated soft drinks, were not associated with stroke. Again, caffeine may be playing a negative role since decaffeinated coffee was associated with a lower risk of stroke after adjustment for caffeinated coffee consumption (Lopez-Garcia *et al.* 2009). There may also be gender differences since, using prospective data from the Amsterdam Growth and Health Longitudinal Study, associations between coffee consumption over a period of 15 years and the components of the metabolic syndrome (at age 42 years) showed that consumption of two or more cups of coffee per day was significantly associated with lower high-density lipoprotein (HDL) in women. However, for men, coffee consumption was not associated with any of the components of the metabolic syndrome (Balk *et al.* 2009). Prospective data from the Alpha-Tocopherol, Beta-Carotene Cancer Prevention Study, a cohort study of 26,556 male Finnish smokers aged 50–69 years without a history of stroke at baseline showed that consumption of both coffee and tea was inversely associated with the risk of cerebral infarction, but not with intracerebral or subarachnoid haemorrhage (Larsson *et al.* 2008). Coffee consumption was not associated with metabolic syndrome in a healthy sample followed up for 9 years (Driessen *et al.* 2009).

6.4.2 Effect of coffee consumption on cardiovascular risk: intervention studies

Interventions studies on coffee and cardiovascular risk biomarkers show mixed results. It appears that the beneficial phenolics may be counteracted by cafestol/kahweol, hydroxyhydroquinones (formed on roasting) and possibly by caffeine. The beneficial effects are often seen in decaffeinated coffee or with chlorogenic acid. Patients with mild hypertension ($n = 28$) in a placebo-controlled, randomised clinical trial were randomised to receive treatment with chlorogenic acid (140 mg per day) from green coffee extract or placebo. In the chlorogenic acid group, but not the placebo group, blood pressure (systolic and diastolic) decreased significantly during the ingestion period. There was no difference in body mass index and pulse rate between groups (Watanabe *et al.* 2006).

In a double-blind, randomised controlled trial on 203 mildly hypertensive men and women consuming control, zero-dose, low-dose, middle-dose and high-dose coffees with reduced levels of hydroxyhydroquinones for 4 weeks, there was a dose-dependent reduction in blood pressure only in the low hydroxyhydroquinone group, which the authors ascribe to chlorogenic acid (Yamaguchi et al. 2008). Very high doses of chlorogenic acid may, of course, produce deleterious effects. In a crossover study, 20 healthy men and women ingested molar equivalent amounts of chlorogenic acid, black tea solids, quercetin-3-O-rutinoside or a placebo daily for 7 days in random order. Total homocysteine in plasma collected 4–5 hours after supplement intake was 12% higher after chlorogenic acid and 11% higher after black tea than after placebo. Total homocysteine in fasting plasma collected 20 hours after supplement intake was 4% higher after chlorogenic acid and 5% higher after black tea than after placebo. Quercetin-3-O-rutinoside did not significantly affect homocysteine concentrations (Olthof et al. 2001).

Some studies show a positive beneficial effect. Instant coffee increased LDL oxidation lag time by 8% and decreased total cholesterol, LDL cholesterol, malondialdehyde and thiobarbituric acid-reactive substances in healthy subjects compared to water consumption (Yukawa et al. 2004). Phenolic acids were incorporated into LDL, and the lag time of LDL susceptibility to oxidative stress increased significantly. However, the proportion of oxidised LDL remained unaffected after coffee consumption, despite the potential effect of coffee on resistance of LDL to oxidation (Natella et al. 2007). Habitual coffee drinkers ($n = 47$) consumed four cups of filtered coffee per day for 1 month and then eight cups of filtered coffee per day for a further month. For the highest dose compared to none, significant changes were observed for serum concentrations of interleukin-18, 8-isoprostane, adiponectin, total cholesterol, HDL cholesterol and apolipoprotein AI, whereas the ratios of LDL to HDL cholesterol and apolipoprotein B to apolipoprotein AI decreased significantly by 8% and 9%, respectively (Kempf et al. 2010).

Some studies show a negative effect of caffeinated coffee, but not of decaffeinated coffee. Twenty healthy non-obese subjects ingested one cup of caffeinated or decaffeinated Italian espresso coffee in random order at 5- to 7-day intervals in a double-blind, crossover study. Following ingestion of caffeinated Italian espresso coffee, flow-mediated dilation (FMD) decreased progressively and significantly, but it did not change after decaffeinated Italian espresso coffee ingestion. Similarly, caffeinated Italian espresso coffee significantly increased both systolic and diastolic blood pressure; this effect was not observed after decaffeinated Italian espresso coffee ingestion. Blood glucose concentrations remained unchanged after ingestion of both, but blood insulin and C-peptide decreased significantly only after caffeinated coffee (Buscemi et al. 2009, 2010). In a crossover study on ten healthy subjects, subjects drank 200 mL coffee, containing 180 mg caffeine, or a capsule of caffeine (180 mg) with 200 mL water. Platelets were separated from plasma at baseline and 30 and 60 minutes after coffee drinking. Platelet aggregation was induced with three different agonists: collagen, arachidonic acid and ADP. Coffee drinking inhibited collagen and arachidonic acid-induced platelet aggregation. Caffeine intake alone did not affect platelet aggregation induced by the three agonists. The antiplatelet effect of coffee is independent from caffeine and could be a result of the interaction of coffee phenolic acids with the intracellular signalling network leading to platelet aggregation since some phenolic acids were found in platelets (Natella et al. 2008).

Mursu et al. (2005) measured lipid peroxidation after ingestion of 450 and 900 mL of filtered coffee per day for 3 weeks or 150 and 300 mL of filtered coffee as a single ingestion. When compared to baseline, no acute or chronic effect of coffee consumption on plasma triglycerides, plasma cholesterol, serum lipid peroxidation, LDL diene conjugated or plasma F_2-isoprostane could be seen. The study was not randomised, and the subjects were free to choose whether they wanted to consume zero, three or six cups of coffee and baseline characteristics differed between study groups.

6.5 Effect of coffee on inflammation

Several studies have looked directly or indirectly at the effect of coffee on inflammation, which is related to many diseases including cardiovascular disease, diabetes or carcinogenesis. In 982 diabetic and 1058 non-diabetic women without cardiovascular disease from the Nurses' Health Study, high consumption of caffeine-containing coffee was associated with higher adiponectin and lower inflammatory marker concentrations as well as previously reported inverse associations of coffee consumption with inflammatory markers, C-reactive protein and tumour necrosis factor-α receptor II (Williams *et al.* 2008). In another study, subjects were 10,325 men and women, 49–76 years of age, living in Fukuoka City, who participated in a baseline survey of a cohort study on lifestyle-related diseases. Coffee and green tea consumption and other lifestyle characteristics were assessed using a structured questionnaire. Anthropometric measurements and venous blood samples were also included. C-reactive protein concentrations were progressively lower with increasing levels of coffee consumption, after adjustment for smoking and other co-variates, in men (primarily limited to those with a high alcohol consumption), but not in women. Green tea consumption showed no measurable relationship with C-reactive protein concentrations in either men or women (Maki *et al.* 2010).

6.6 Effect of coffee consumption on cancer risk

Impact of coffee on colorectal cancer is interesting to consider as a significant portion of (poly)phenols (e.g. chlorogenic acids) present in coffee is not absorbed in the intestine and reaches the colon (Lafay *et al.* 2006). There are many animal studies on the effect of coffee on carcinogenesis. For example, CGA supplementation significantly reduced the incidence of chemical-induced pre-neoplastic lesions and carcinogenicity in different animal models (Shimizu *et al.* 1999; Matsunaga *et al.* 2002). Moreover, chlorogenic acid and caffeic acid significantly induce both liver and intestinal detoxifying enzyme activities in rodents (Kitts and Wijewickreme 1994; Somoza *et al.* 2003). However, intervention studies on coffee consumption and cancer biomarkers are rare, although there are many epidemiological studies.

6.6.1 Effect of coffee consumption on cancer risk: epidemiological and cohort studies

There are over 500 epidemiological papers relating coffee consumption to the risk of cancer at various sites, and many meta-analyses have been reported on these studies.

Table 6.2 Effect of coffee consumption on cancer risk derived from epidemiological studies

Site of cancer	Effect of coffee consumption
Hapatocellular	Strong and consistent protective association
Endometrial	Strong and consistent protective association
Colorectal	Borderline protective association
Breast	No association
Pancreatic	No association
Kidney	No association
Ovarian	No association
Prostate	No association
Gastric	No association
Bladder	Increased with heavy coffee consumption in men
Childhood leukaemia (mother's consumption)	Ambiguous, possibly some risk at high levels of consumption

Adapted from Arab 2010.

The most up-to-date review of all studies is informative since it includes summaries of meta-analyses (Arab 2010) and is provided in Table 6.2. The risk of most cancers is not modified by coffee consumption, but for specific cancers or in specific populations or individuals, there is some effect on the risk.

The relationship between coffee consumption and cancer risk at other sites has also been covered. In a meta-analysis of lung cancer incidence, five prospective studies and eight case–control studies involving 110,258 individuals with 5347 lung cancer cases indicated a significant positive association between highest coffee intake and lung cancer. An increase in coffee consumption of two cups per day was associated with a 14% increased risk of developing lung cancer. In stratified analyses, the highest coffee consumption was significantly associated with increased risk of lung cancer in prospective studies, studies conducted in America and Japan, but borderline significantly associated with decreased risk of lung cancer in non-smokers. In addition, decaffeinated coffee drinking was associated with decreased lung cancer risk, although the number of studies on this topic was relatively small. The authors state that because the residual confounding effects of smoking or other factors may still exist, these results should be interpreted with caution (Tang *et al.* 2010). An additional study on colorectal cancer is also noteworthy. In the Singapore Chinese Health Study, during the first 12 years of follow-up, 961 colorectal cancer cases occurred in the cohort of over 60,000 middle-aged or older Chinese men and women living in Singapore. No overall association between coffee intake and colorectal cancer was observed. However, analysis by subsite for smokers showed that coffee may protect against smoking-related advanced colon cancer (Peterson *et al.* 2010).

In addition, the role of other added substances and the brewing method may influence the relationship between coffee intake and cancer risk. The role of tea and coffee and substances added (sugar/honey, creamers and milk) on endometrial cancer risk in a population-based case–control study in six counties in New Jersey, including 417 cases and 395 controls, indicates that sugars and milk/cream added to coffee and tea should be considered in future studies evaluating coffee/tea and endometrial cancer risk (Bandera *et al.* 2010). In a study on subjects from the Vasterbotten Intervention

Project (64,603 participants, including 3034 cases), with up to 15 years of follow-up, the potential relevance of the brewing method in investigations of coffee consumption was suggested to affect cancer risk (Nilsson *et al.* 2010).

6.6.2 Effect of coffee consumption on cancer risk: intervention studies

There are only a limited number of intervention studies on coffee and cancer risk, which are not directly concerned with caffeine. Healthy, non-smoking subjects consumed 600 mL of coffee for 5 days. DNA damage caused by H_2O_2 and by 3-amino-1-methyl-5H-pyrido[4,3-b]indole acetate was significantly reduced by 17% and 35%, respectively, after coffee consumption (Bichler *et al.* 2007). When subjects ingested 1 L of unfiltered coffee per day for 5 days, coffee consumption reduced the 7b,8a-dihydroxy-9a,10a-epoxy-7,8,9,10-tetrahydrobenzo[a]pyrene-induced DNA damage in peripheral lymphocytes in $\geq$90% of the participants by 45% (Steinkellner *et al.* 2005).

6.7 Summary

Coffee consumption is often linked with unhealthy lifestyles, which confounded older epidemiological studies. Changes in consumer habits have meant that consumption no longer has this association, which has clarified more recent epidemiological studies. There is now good epidemiological evidence that coffee consumption could reduce the risk of type 2 diabetes as shown in the majority of epidemiological studies and supported by some intervention studies. This is currently the most convincing effect of coffee consumption. There is generally no effect of coffee consumption on cancer risk, with the exception of a protective effect against hepatocellular and endometrial cancers, and a negative effect on bladder cancer with high consumption in men. The well-established protective effect of (poly)phenols on cardiovascular risk may be confounded by other components in coffee, such as caffeine and diterpenes (in some coffees). Further research needs to examine the mechanism by which the phenolic acids in coffee might exert these protective effects.

References

Aldughpassi, A. and Wolever, T.M. (2009) Effect of coffee and tea on the glycaemic index of foods: no effect on mean but reduced variability. *Br. J. Nutr.*, **101**, 1282–1285.

Altmaier, E., Kastenmuller, G., Romisch-Margl, W. *et al.* (2009) Variation in the human lipidome associated with coffee consumption as revealed by quantitative targeted metabolomics. *Mol. Nutr. Food Res.*, **53**, 1357–1365.

Arab, L. (2010) Epidemiologic evidence on coffee and cancer. *Nutr. Cancer*, **62**, 271–283.

Balk, L., Hoekstra, T. and Twisk, J. (2009) Relationship between long-term coffee consumption and components of the metabolic syndrome: the Amsterdam Growth and Health Longitudinal Study. *Eur. J. Epidemiol.*, **24**, 203–209.

Bandera, E.V., Williams-King, M.G., Sima, C. et al. (2010) Coffee and tea consumption and endometrial cancer risk in a population-based study in New Jersey. *Cancer Causes Control*, **21**, 1467–1473.

Battram, D.S., Arthur, R., Weekes, A. et al. (2006) The glucose intolerance induced by caffeinated coffee ingestion is less pronounced than that due to alkaloid caffeine in men. *J. Nutr.*, **136**, 1276–1280.

Benvenga, S., Bartolone, L., Pappalardo, M.A. et al. (2008) Altered intestinal absorption of L-thyroxine caused by coffee. *Thyroid*, **18**, 293–301.

Bichler, J., Cavin, C., Simic, T. et al. (2007) Coffee consumption protects human lymphocytes against oxidative and 3-amino-1-methyl-5H-pyrido[4,3-b]indole acetate (Trp-P-2) induced DNA-damage: results of an experimental study with human volunteers. *Food Chem. Toxicol.*, **45**, 1428–1436.

Bidel, S., Silventoinen, K., Hu, G. et al. (2008) Coffee consumption, serum γ-glutamyltransferase and risk of type II diabetes. *Eur. J. Clin. Nutr.*, **62**, 178–185.

Borrelli, R.C., Visconti, A., Mennella, C. et al. (2002) Chemical characterization and antioxidant properties of coffee melanoidins. *J. Agric. Food Chem.*, **50**, 6527–6533.

Bravi, F., Scotti, L., Bosetti, C. et al. (2009) Coffee drinking and endometrial cancer risk: a metaanalysis of observational studies. *Am. J. Obstet. Gynecol.*, **200**, 130–135.

Buscemi, S., Verga, S., Batsis, J.A. et al. (2009) Dose-dependent effects of decaffeinated coffee on endothelial function in healthy subjects. *Eur. J. Clin. Nutr.*, **63**, 1200–1205.

Buscemi, S., Verga, S., Batsis, J.A. et al. (2010) Acute effects of coffee on endothelial function in healthy subjects. *Eur. J. Clin. Nutr.*, **64**, 483–489.

Ceriello, A. and Motz, E. (2004) Is oxidative stress the pathogenic mechanism underlying insulin resistance, diabetes, and cardiovascular disease? The common soil hypothesis revisited. *Arterioscler. Thromb. Vasc. Biol.*, **24**, 816–823.

Clifford, M.N. (1998) The nature of chlorogenic acids – are they advantageous compounds in coffee? In *Proceedings of the 17th International Conference on Coffee Science*, Nairobi, ASIC, Paris, pp. 79–91.

Clifford, M.N. (2000) Chlorogenic acids and other cinnamates – nature, occurrence, dietary burden, absorption and metabolism. *J. Sci. Food Agric.*, **80**, 1033–1043.

Cornelis, M.C. and El-Sohemy, A. (2007) Coffee, caffeine, and coronary heart disease. *Curr. Opin. Lipidol.*, **18**, 13–19.

Dias, R.C., Campanha, F.G., Vieira, L.G. et al. (2010) Evaluation of kahweol and cafestol in coffee tissues and roasted coffee by a new high-performance liquid chromatography methodology. *J. Agric. Food Chem.*, **58**, 88–93.

Driessen, M.T., Koppes, L.L., Veldhuis, L. et al. (2009) Coffee consumption is not related to the metabolic syndrome at the age of 36 years: the Amsterdam Growth and Health Longitudinal Study. *Eur. J. Clin. Nutr.*, **63**, 536–542.

Esposito, F., Morisco, F., Verde, V. et al. (2003) Moderate coffee consumption increases plasma glutathione but not homocysteine in healthy subjects. *Aliment. Pharmacol. Ther.*, **17**, 595–601.

Fuhrman, B.J., Smit, E., Crespo, C.J. et al. (2009) Coffee intake and risk of incident diabetes in Puerto Rican men: results from the Puerto Rico Heart Health Program. *Public Health Nutr.*, **12**, 842–848.

Geleijnse, J.M. (2008) Habitual coffee consumption and blood pressure: an epidemiological perspective. *Vasc. Health Risk Manag.*, **4**, 963–970.

George, S.E., Ramalakshmi, K. and Mohan Rao, L.J. (2008) A perception on health benefits of coffee. *Crit. Rev. Food Sci. Nutr.*, **48**, 464–486.

Greenberg, J.A., Owen, D.R. and Geliebter, A. (2010) Decaffeinated coffee and glucose metabolism in young men. *Diabetes Care*, **33**, 278–280.

Grubben, M.J., Boers, G.H., Blom, H.J. et al. (2000) Unfiltered coffee increases plasma homocysteine concentrations in healthy volunteers: a randomized trial. *Am. J. Clin. Nutr.*, **71**, 480–484.

Hamer, M., Witte, D.R., Mosdol, A. et al. (2008) Prospective study of coffee and tea consumption in relation to risk of type 2 diabetes mellitus among men and women: the Whitehall II study. *Br. J. Nutr.*, **100**, 1046–1053.

Heckman, M.A., Weil, J. and Gonzalez de Mejia, E. (2010) Caffeine (1,3,7-trimethylxanthine) in foods: a comprehensive review on consumption, functionality, safety, and regulatory matters. *J. Food Sci.*, **75**, R77–R87.

Higdon, J.V. and Frei, B. (2006) Coffee and health: a review of recent human research. *Crit. Rev. Food Sci. Nutr.*, **46**, 101–123.

Huang, D., Ou, B. and Prior, R.L. (2005) The chemistry behind antioxidant capacity assays. *J. Agric. Food Chem.*, **53**, 1841–1856.

Hurrell, R.F., Reddy, M. and Cook, J.D. (1999) Inhibition of non-haem iron absorption in man by polyphenolic-containing beverages. *Br. J. Nutr.*, **81**, 289–295.

Huxley, R., Lee, C.M., Barzi, F. et al. (2009) Coffee, decaffeinated coffee, and tea consumption in relation to incident type 2 diabetes mellitus: a systematic review with meta-analysis. *Arch. Intern. Med.*, **169**, 2053–2063.

Je, Y., Liu, W. and Giovannucci, E. (2009) Coffee consumption and risk of colorectal cancer: a systematic review and meta-analysis of prospective cohort studies. *Int. J. Cancer*, **124**, 1662–1668.

Johnston, K.L., Clifford, M.N. and Morgan, L.M. (2003) Coffee acutely modifies gastrointestinal hormone secretion and glucose tolerance in humans: glycemic effects of chlorogenic acid and caffeine. *Am. J. Clin. Nutr.*, **78**, 728–733.

Kempf, K., Herder, C., Erlund, I. et al. (2010) Effects of coffee consumption on subclinical inflammation and other risk factors for type 2 diabetes: a clinical trial. *Am. J. Clin. Nutr.*, **91**, 950–957.

Kitts, D.D. and Wijewickreme, A.N. (1994) Effect of dietary caffeic and chlorogenic acids on in vivo xenobiotic enzyme systems. *Plant Foods Hum. Nutr.*, **45**, 287–298.

Kleemola, P., Jousilahti, P., Pietinen, P. et al. (2000) Coffee consumption and the risk of coronary heart disease and death. *Arch. Intern. Med.*, **160**, 3393–3400.

Lafay, S., Gil-Izquierdo, A., Manach, C. et al. (2006) Chlorogenic acid is absorbed in its intact form in the stomach of rats. *J. Nutr.*, **136**, 1192–1197.

Larsson, S.C., Mannisto, S., Virtanen, M.J. et al. (2008) Coffee and tea consumption and risk of stroke subtypes in male smokers. *Stroke*, **39**, 1681–1687.

Legrand, D. and Scheen, A.J. (2007) Does coffee protect against type 2 diabetes?. *Rev. Med. Liege*, **62**, 554–559.

Lopez-Garcia, E., Rodriguez-Artalejo, F., Rexrode, K.M. et al. (2009) Coffee consumption and risk of stroke in women. *Circulation*, **119**, 1116–1123.

Louie, J.C., Atkinson, F., Petocz, P. et al. (2008) Delayed effects of coffee, tea and sucrose on postprandial glycemia in lean, young, healthy adults. *Asia Pac. J. Clin. Nutr.*, **17**, 657–662.

Maki, T., Pham, N.M., Yoshida, D. et al. (2010) The relationship of coffee and green tea consumption with high-sensitivity C-reactive protein in Japanese men and women. *Clin. Chem. Lab Med.*, **48**, 849–854.

Matsunaga, K., Katayama, M., Sakata, K. et al. (2002) Inhibitory effects of chlorogenic acid on azoxymethane-induced colon carcinogenesis in male F344 rats. *Asian Pac. J. Cancer Prev.*, **3**, 163–166.

Moisey, L.L., Kacker, S., Bickerton, A.C. et al. (2008) Caffeinated coffee consumption impairs blood glucose homeostasis in response to high and low glycemic index meals in healthy men. *Am. J. Clin. Nutr.*, **87**, 1254–1261.

Moisey, L.L., Robinson, L.E. and Graham, T.E. (2010) Consumption of caffeinated coffee and a high carbohydrate meal affects postprandial metabolism of a subsequent oral glucose tolerance test in young, healthy males. *Br. J. Nutr.*, **103**, 833–841.

Moura-Nunes, N., Perrone, D., Farah, A. *et al.* (2009) The increase in human plasma antioxidant capacity after acute coffee intake is not associated with endogenous non-enzymatic antioxidant components. *Int. J. Food Sci. Nutr.*, in press.

Mursu, J., Voutilainen, S., Nurmi, T. *et al.* (2005) The effects of coffee consumption on lipid peroxidation and plasma total homocysteine concentrations: a clinical trial. *Free Radic. Biol. Med.*, **38**, 527–534.

Natella, F., Nardini, M., Giannetti, I. *et al.* (2002) Coffee drinking influences plasma antioxidant capacity in humans. *J. Agric. Food Chem.*, **50**, 6211–6216.

Natella, F., Nardini, M., Belelli, F. *et al.* (2007) Coffee drinking induces incorporation of phenolic acids into LDL and increases the resistance of LDL to *ex vivo* oxidation in humans. *Am. J. Clin. Nutr.*, **86**, 604–609.

Natella, F., Nardini, M., Belelli, F. *et al.* (2008) Effect of coffee drinking on platelets: inhibition of aggregation and phenols incorporation. *Br. J. Nutr.*, **100**, 1276–1282.

Nilsson, L.M., Johansson, I., Lenner, P. *et al.* (2010) Consumption of filtered and boiled coffee and the risk of incident cancer: a prospective cohort study. *Cancer Causes Control*, **21**, 1533–1544.

Noordzij, M., Uiterwaal, C.S., Arends, L.R. *et al.* (2005) Blood pressure response to chronic intake of coffee and caffeine: a meta-analysis of randomized controlled trials. *J. Hypertens.*, **23**, 921–928.

Ochiai, R., Jokura, H., Suzuki, A. *et al.* (2004) Green coffee bean extract improves human vasoreactivity. *Hypertens. Res.*, **27**, 731–737.

Odegaard, A.O., Pereira, M.A., Koh, W.P. *et al.* (2008) Coffee, tea, and incident type 2 diabetes: the Singapore Chinese Health Study. *Am. J. Clin. Nutr.*, **88**, 979–985.

Olthof, M.R., Hollman, P.C., Zock, P.L. *et al.* (2001) Consumption of high doses of chlorogenic acid, present in coffee, or of black tea increases plasma total homocysteine concentrations in humans. *Am. J. Clin. Nutr.*, **73**, 532–538.

Peterson, S., Yuan, JM., Koh, W.P. *et al.* (2010) Coffee intake and risk of colorectal cancer among Chinese in Singapore: the Singapore Chinese Health Study. *Nutr. Cancer*, **62**, 21–29.

Pimentel, G.D., Zemdegs, J.C., Theodoro, J.A. *et al.* (2009) Does long-term coffee intake reduce type 2 diabetes mellitus risk? *Diabetol. Metab Syndr.*, **1**, 6.

Ranheim, T. and Halvorsen, B. (2005) Coffee consumption and human health–beneficial or detrimental? Mechanisms for effects of coffee consumption on different risk factors for cardiovascular disease and type 2 diabetes mellitus. *Mol. Nutr. Food Res.*, **49**, 274–284.

Sartorelli, D.S., Fagherazzi, G., Balkau, B. *et al.* (2010) Differential effects of coffee on the risk of type 2 diabetes according to meal consumption in a French cohort of women: the E3N/EPIC cohort study. *Am. J. Clin. Nutr.*, **91**, 1002–1012.

Shimizu, M., Yoshimi, N., Yamada, Y. *et al.* (1999) Suppressive effects of chlorogenic acid on N-methyl-N-nitrosourea-induced glandular stomach carcinogenesis in male F344 rats. *J. Toxicol. Sci.*, **24**, 433–439.

Sies, H. (2007) Total antioxidant capacity: appraisal of a concept. *J. Nutr.*, **137**, 1493–1495.

Sofi, F., Conti, A.A., Gori, A.M. *et al.* (2007) Coffee consumption and risk of coronary heart disease: a meta-analysis. *Nutr. Metab. Cardiovasc. Dis.*, **17**, 209–223.

Somoza, V., Lindenmeier, M., Wenzel, E. *et al.* (2003) Activity-guided identification of a chemopreventive compound in coffee beverage using *in vitro* and *in vivo* techniques. *J. Agric. Food Chem.*, **51**, 6861–6869.

Steinkellner, H., Hoelzl, C., Uhl, M. *et al.* (2005) Coffee consumption induces GSTP in plasma and protects lymphocytes against $(+/-)$-anti-benzo[a]pyrene-7,

8-dihydrodiol-9,10-epoxide induced DNA-damage: results of controlled human intervention trials. *Mutat. Res.*, **591**, 264–275.

Tanaka, K., Tokunaga, S., Kono, S. *et al.* (1998) Coffee consumption and decreased serum gamma-glutamyltransferase and aminotransferase activities among male alcohol drinkers. *Int. J. Epidemiol.*, **27**, 438–443.

Tang, N., Zhou, B., Wang, B. *et al.* (2009) Coffee consumption and risk of breast cancer: a metaanalysis. *Am. J. Obstet. Gynecol.*, **200**, 290–299.

Tang, N., Wu, Y., Ma, J. *et al.* (2010) Coffee consumption and risk of lung cancer: a meta-analysis. *Lung Cancer*, **67**, 17–22.

Thom, E. (2007) The effect of chlorogenic acid enriched coffee on glucose absorption in healthy volunteers and its effect on body mass when used long-term in overweight and obese people. *J. Int. Med. Res.*, **35**, 900–908.

Tunnicliffe, J.M. and Shearer, J. (2008) Coffee, glucose homeostasis, and insulin resistance: physiological mechanisms and mediators. *Appl. Physiol. Nutr. Metab.*, **33**, 1290–1300.

Ulvik, A., Vollset, S.E., Hoff, G. *et al.* (2008) Coffee consumption and circulating B-vitamins in healthy middle-aged men and women. *Clin. Chem.*, **54**, 1489–1496.

van Dam, R.M. (2006a) Coffee and type 2 diabetes: from beans to beta-cells. *Nutr. Metab Cardiovasc. Dis.*, **16**, 69–77.

van Dam, R.M. (2006b) Green tea, coffee, and diabetes. *Ann. Intern. Med.*, **145**, 634–635.

van Dam RM (2008a) Coffee consumption and coronary heart disease: paradoxical effects on biological risk factors versus disease incidence. *Clin. Chem.*, **54**, 1418–1420.

van Dam, R.M. (2008b) Coffee consumption and risk of type 2 diabetes, cardiovascular diseases, and cancer. *Appl. Physiol. Nutr. Metab.*, **33**, 1269–1283.

van Dam, R.M. and Hu, F.B. (2005) Coffee consumption and risk of type 2 diabetes: a systematic review. *JAMA*, **294**, 97–104.

van Dam, R.M., Willett, W.C., Manson, J.E. *et al.* (2006c) Coffee, caffeine, and risk of type 2 diabetes: a prospective cohort study in younger and middle-aged U.S. women. *Diabetes Care*, **29**, 398–403.

van Dieren, S., Uiterwaal, C.S., van der Schouw, Y.T. *et al.* (2009) Coffee and tea consumption and risk of type 2 diabetes. *Diabetologia*, **52**, 2561–2569.

van Dijk, A.E., Olthof, M.R., Meeuse, J.C. *et al.* (2009) Acute effects of decaffeinated coffee and the major coffee components chlorogenic acid and trigonelline on glucose tolerance. *Diabetes Care*, **32**, 1023–1025.

Watanabe, T., Arai, Y., Mitsui, Y. *et al.* (2006) The blood pressure-lowering effect and safety of chlorogenic acid from green coffee bean extract in essential hypertension. *Clin. Exp. Hypertens.*, **28**, 439–449.

Williams, C.J., Fargnoli, J.L., Hwang, J.J. *et al.* (2008) Coffee consumption is associated with higher plasma adiponectin concentrations in women with or without type 2 diabetes: a prospective cohort study. *Diabetes Care*, **31**, 504–507.

Wu, J.N., Ho, S.C., Zhou, C. *et al.* (2009) Coffee consumption and risk of coronary heart diseases: a meta-analysis of 21 prospective cohort studies. *Int. J. Cardiol.*, **137**, 216–225.

Yamaguchi, T., Chikama, A., Mori, K. *et al.* (2008) Hydroxyhydroquinone-free coffee: a double-blind, randomized controlled dose-response study of blood pressure. *Nutr. Metab. Cardiovasc. Dis.*, **18**, 408–414.

Yukawa, G.S., Mune, M., Otani, H. *et al.* (2004) Effects of coffee consumption on oxidative susceptibility of low-density lipoproteins and serum lipid levels in humans. *Biochemistry (Mosc.)*, **69**, 70–74.

Zhang, Y., Lee, E.T., Cowan, L.D. *et al.* (2011) Coffee consumption and the incidence of type 2 diabetes in men and women with normal glucose tolerance: the Strong Heart Study. *Nutr. Metab. Cardiovasc. Dis.*, **21**, 418–423.

Chapter 7
Phytochemicals in Cocoa and Flavan-3-ol Bioavailability

Francisco Tomás-Barbéran[1], Gina Borges[2] and Alan Crozier[2]

[1]CEBAS CSIC, PO Box 164, Espinardo 30100 Murcia, Spain
[2]School of Medicine, College of Medical, Veterinary and Life Sciences, University of Glasgow, Glasgow G12 8QQ, UK

7.1 Introduction

Cocoa is a tree that originated in the tropical regions of South America. There are two forms sufficiently distinct as to be considered subspecies. Criollo developed north of the Panama Isthmus and Forastero in the Amazon Basin. A so-called hybrid, Trinitario, developed in Trinidad (Willson 1999). The cocoa plant is now cultivated worldwide, major producers being the Ivory Coast, Ghana, Nigeria, Indonesia, Brazil and Cameroon. The main cultivated form is *Theobroma cacao* var. *Forastero*, which accounts for more than 90% of the world's usage. Criollo and Trinitario are also grown, and some regard these as providing better-flavour qualities to cocoa-based products (Leung and Foster 1996).

Ripe cocoa pods contain about 30–40 seeds that are embedded in a sweet mucilaginous pulp comprised mainly of sugars. The pods are harvested and broken open, and the pulp and seeds are formed into large mounds and covered with leaves. The pulp is fermented for 6–8 days. During this period sucrose is converted to glucose and fructose by invertase, and the glucose is subsequently utilised in fermentation, yielding ethanol that is metabolised to acetic acid. As the tissues of the beans loose cellular integrity and die, storage proteins are hydrolysed to peptides and amino acids, while polyphenol oxidase converts phenolic components to quinones, which polymerise yielding brown, highly insoluble compounds that give chocolate its characteristic colour (Haslam 1998). After fermentation, the seeds are dried in the sun, reducing the moisture content from 55% to 7.5%. The resulting cocoa beans are then packed for the wholesale trade.

Cocoa beans are used extensively in the manufacture of both chocolate and cocoa that is used as a beverage. To produce the cocoa powder used in the beverage, the beans are roasted at 150°C and the shell (hull) and meat of the bean (nib) are mechanically separated. The nibs, which contain about 55% cocoa butter, are then finely ground while hot to produce a liquid 'mass' or 'liquor'. This sets on cooling and is then pressed to express the 'butter' that is used in the manufacture of chocolate. The residual cake is pulverised to produce the cocoa powder. An alkalisation process is often also employed to modify the dispersibility and flavour of cocoa powders. This involves the exposure of the nibs prior to processing to a warm solution of caustic soda, a process known as Dutching, which darkens the cocoa, resulting in its characteristic brown colour (Bixler and Morgan 1999).

7.2 Phytochemicals in cocoa

Cocoa powder and chocolate are rich dietary sources of phytochemicals. The main ones, in terms of quantitative composition, are proanthocyanidins and purine alkaloids. Other minor compounds include flavonols, anthocyanins, phenolic acid amides and stilbenoids.

7.2.1 Purine alkaloids, theobromine and caffeine

Cocoa beans contain significant amounts of theobromine (2.2–2.8% of dry weight), together with smaller amounts of caffeine (0.6–0.8%) (Figure 7.1; Ashihara et al. 2008). These alkaloids, which are discussed in detail in Chapter 2, are preserved in cocoa powder and easily extracted during the preparation of cocoa beverages. In addition, the phenolic amine salsolinol (S-form) has been reported (Figure 7.1; Buckingham 1994).

7.2.2 Flavan-3-ols

Cocoa beans are a source of flavan-3-ols in the form of the monomers (−)-epicatechin and (+)-catechin (Figure 7.2; Aron and Kennedy 2008). As well as simple monomers, flavan-3-ols also exist as proanthocyanidins at high levels in cocoa and some dark chocolates (Gu et al. 2006). Type B proanthocyanidins are formed from (+)-catechin and (−)-epicatechin, with the oxidative coupling occurring between the C_4 of the heterocycle and the C_6 or C_8 positions of the adjacent unit to create oligomers and

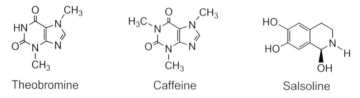

Figure 7.1 Structures of cocoa purine alkaloids and the phenolic amine, salsolinol.

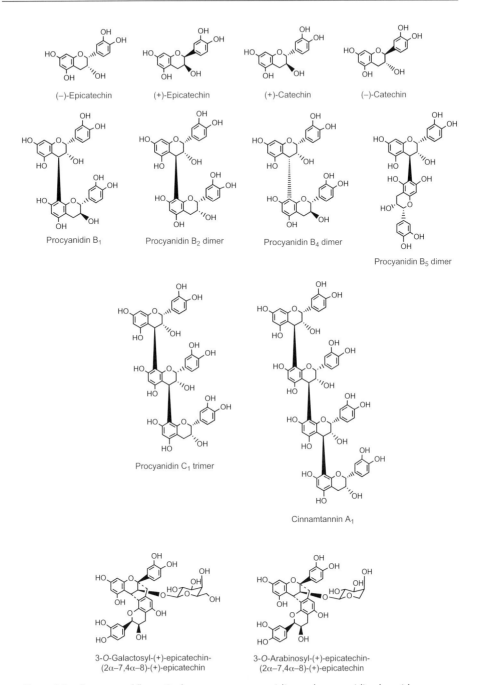

Figure 7.2 Structures of flavan-3-ol monomers, procyanidins and procyanidin glycosides.

polymers. Individual procyanidins that have been identified in cocoa include the B_1, B_2, B_4 and B_5 dimers, the C_1 trimer and tetramers such as cinnamtannin A_1 (Figure 7.2; Haslam 1998), although longer chain polymers comprising seven or more monomer units seemingly predominate (Haslam 1998; Hammerstone et al. 1999; Gu et al. 2006; Miller et al. 2008). Some dimeric epicatechin glycosides as the 3-O-arabinoside and the 3-O-galactoside conjugates of (+)-epicatechin-(2α→7,4α→8)-(+)-epicatechin (Figure 7.2) have also been reported in *T. cacao* (Buckingham 1994).

Flavan-3-ols are lost during processing, with declines in levels being reported to occur during fermentation (Kim and Keeney 1984) and Dutching (Miller et al. 2008). Losses are likely to occur at other steps in the processing chain, including roasting. The total flavan-3-ol content of commercial cocoa can, therefore, vary by tenfold or more (Miller et al. 2006). Processing can also result in some epimerisation of (−)-epicatechin to form (−)-catechin (Figure 7.2), and as a result, the predominant form of catechin in chocolate products is the (−)-isomer rather than the naturally occurring (+)-form (Gotti et al. 2006; Cooper et al. 2007). Because of all these events, the flavan-3-ol content of most commercial cocoas and chocolate products, which is monitored rarely, if at all, by most manufacturers, is highly variable. To complicate matters further there is little or no information on the fate of procyanidins during processing.

7.2.3 Phenolic acid derivatives

Cocoa also contains *N*-caffeoyl-3-O-hydroxytyrosine (clovamide) and *N*-*p*-coumaroyl-tyrosine (deoxyclovamide) (Figure 7.3; Sanbongi et al. 1998). These compounds, along with the proanthocyanidins, contribute to the astringent taste of unfermented cocoa beans and roasted cocoa nibs, but not to the same degree as other amides, in particular cinnamoyl-L-aspartic acid and caffeoyl-L-glutamic acid (Figure 7.3; Stark and Hofmann (2005). During fermentation, the conversion of many of the phenolic components to insoluble brown polymeric compounds takes place, and as a consequence, the level of soluble polyphenols falls by ∼90%.

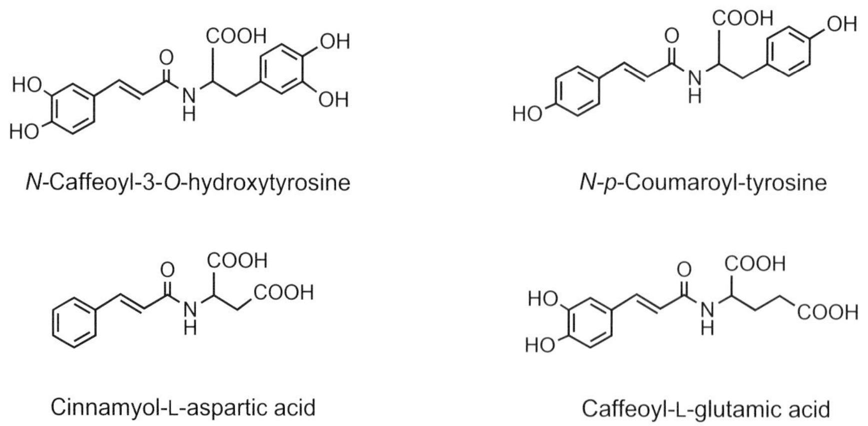

N-Caffeoyl-3-O-hydroxytyrosine

N-*p*-Coumaroyl-tyrosine

Cinnamyol-L-aspartic acid

Caffeoyl-L-glutamic acid

Figure 7.3 Phenolic acid derivatives found in cocoa.

7.2.4 Minor phytochemicals

7.2.4.1 Anthocyanins

The pigments cyanidin-3-O-galactoside and cyanidin-3-O-arabinoside (Figure 7.4) have been reported in cocoa beans. Their occurrence in cocoa powder and chocolate is, however, unlikely due to degradation during the manufacturing process. Cocoa powders obtained under mild, controlled conditions to avoid phytochemical degradation can, however, contain small amounts of anthocyanins.

7.2.4.2 Flavonols

Several quercetin-O-glycosides, including quercetin-3-O-galactoside, quercetin-3-O-arabinoside (Tomás-Barberán *et al.* 2007) and quercetin-3-O-glucoside (Hammerstone *et al.* 1999), have been identified in cocoa and cocoa powder (Figure 7.4). They are generally present in much smaller amounts than flavan-3-ol monomers and proanthocyanidins.

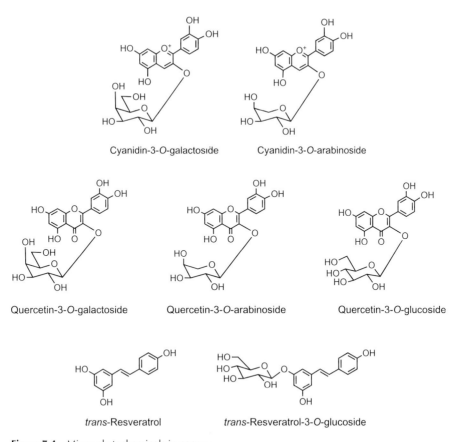

Figure 7.4 Minor phytochemicals in cocoa.

7.2.4.3 Stilbenes

The stilbene *trans*-resveratrol and its 3-*O*-glucoside, piceid (Figure 7.4), have been detected in cocoa powders, dark chocolate and cocoa liquor (Counet *et al.* 2006; Hurst *et al.* 2008). Stilbenoids have also recently been determined in 22 cocoa liquors from 11 different countries (Jerkovic *et al.* 2010), with resveratrol and piceid acid being found in concentrations of 0.4 and 2.6 µg/g in the richest samples. Another *trans*-piceid-*O*-hexoside was also detected, this being most probably the galactoside.

7.3 Bioavailability of cocoa flavan-3-ols

The biological activity of cocoa products, particularly those with potential beneficial effects on cardiovascular disease, is associated with their phytochemical content, especially to that of flavan-3-ols (Rasmussen *et al.* 2005). The biological activity of this flavonoid subclass depends on their bioavailability and metabolism, and the effects on cardiovascular disease have been associated with (−)-epicatechin circulating in plasma (Schroeter *et al.* 2006). Information on flavan-3-ol bioavailability and metabolism is therefore essential for our understanding of the role of cocoa phytochemicals in human health and for the development of food products with improved bioavailability and health benefits.

Most of the available studies focus on the bioavailability and pharmacokinetics of cocoa flavan-3-ols, both the epicatechin and catechin monomers, and procyanidins. There are no publications on the bioavailability of phenolic acid amides and the minor phytochemicals of cocoa powder and chocolate, although there are reports of studies on the bioavailability of stilbenes, anthocyanins and flavonols from other food sources (see reviews by Crozier *et al.* 2009, 2010; Cottart *et al.* 2010; Del Rio *et al.* 2010).

7.3.1 Background

The post-consumption fate of flavan-3-ol monomers from cocoa has been the subject of study for some years, although, arguably, there has been more research on the fate of green tea flavan-3-ols, which, as well as (−)-epicatechin and (+)-catechin, include gallocatechins and their 3-*O*-galloylated derivatives (see Chapter 3). Typically, human volunteers are placed on a low-polyphenol diet for 24 or 48 hours, and after an overnight fast, baseline blood and/or urine are collected before acute supplementation and the subsequent collection of blood and urine at various time points over a period typically of 24 hours or less. Studies with rats indicate that (−)-epicatechin is absorbed more readily than (+)-catechin (Baba *et al.* 2001), which in turn is more bioavailable than (−)-catechin (Donovan *et al.* 2006), the main catechin stereoisomer in most cocoa products. Human bioavailability studies with chocolate products have monitored the fate of (−)-epicatechin, which is absorbed principally in the small intestine, appearing in the circulatory system within 30 minutes of ingestion. During passage from the small intestine to the portal vein, flavan-3-ols are subject to the action of sulphotransferases, uridine-5′-diphosphate glucuronosyltransferases and catechol-O-methyltransferases, and as a consequence, it is principally sulphate, glucuronide and methylated metabolites that appear in the circulatory system rather than the parent flavan-3-ols found in cocoa. Once in the bloodstream, the glucuronidated, sulphated and/or methylated metabolites may undergo further phase II metabolism, with conversions occurring in the liver, where enterohepatic transport in the bile may result in some recycling back to the small intestine.

In vivo human studies have shown that cocoa flavan-3-ol monomers and procyanidins are stable while in the stomach (Rios *et al.* 2002). *In vitro* incubations simulating conditions in the stomach and small intestine suggest that (−)-epicatechin is stable under such conditions (Auger *et al.* 2008), and in keeping with these observations, in feeding studies in which ileostomists ingested Polyphenon E and green tea, as opposed to cocoa, sizeable amounts of epicatechin, catechin and their metabolites, equivalent to >20% of intake, were recovered in ileal fluid (Auger *et al.* 2008; Stalmach *et al.* 2010). Feeding studies with apple juice have revealed that 90% of ingested procyanidins were recovered in ileal fluid, albeit with a reduced degree of polymerisation (Kahle *et al.* 2007). This indicates that in healthy subjects with a functioning colon, these compounds will pass from the small to the large intestine where they will be subject to the action of the colonic microbiota, resulting in their breakdown to phenolic catabolites.

7.3.1.1 Subtleties in the analysis of monomeric flavan-3-ols and their metabolites

Although it is comparatively easy to resolve by reverse-phase chromatography the relatively rapidly eluting (+)-catechin from the more hydrophobic (−)-epicatechin, it should be noted that analysis by HPLC-MS of the associated mammalian metabolites is more complex than is generally appreciated. Firstly, in the absence of enantiomerically pure standards, these chromatographically resolvable compounds cannot be distinguished by their mass fragmentation. Secondly, even when the identity of the consumed flavan-3-ol diastereomer has been established, there is evidence for the gut microbial conversion of (+)-catechin to (+)-epicatechin by C_2 epimerisation (Tzounis *et al.* 2008). While (+)-catechin and (+)-epicatechin can be resolved by reverse-phase chromatography, this procedure does not separate (+)-epicatechin metabolites from (−)-epicatechin metabolites that would also be expected after consumption of cocoa products (Donovan *et al.* 2006). In addition, as noted in the preceding text, there is evidence for the epimerisation of flavan-3-ols at C_2 during processing of cocoa, and accordingly, the main flavan-3-ol monomers to be ingested when cocoa products are consumed are (−)-epicatechin and its C_2 epimer, (−)-catechin. For these reasons when a cocoa flavan-3-ol metabolite is detected in human plasma or urine, it can only be assigned as an (epi)catechin metabolite and not to a particular enantiomeric form, and this comparatively non-specific terminology is used throughout this chapter.

7.3.2 Flavan-3-ol monomers

7.3.2.1 Early studies

Since reference compounds of flavan-3-ol glucuronide and sulphate metabolites were unavailable, the initial human studies on the post-ingestion fate of cocoa, treated plasma and urine samples with β-glucuronidase/sulphatase prior to the analysis of the released (epi)catechin monomer by reverse-phase HPLC, typically with fluorescence (Ho et al. 1995; Richelle et al. 1999) or electrochemical detection (Rein et al. 2000; Wang et al. 2000). With this methodology, Richelle et al. (1999) showed that following the consumption of 40 g of dark chocolate containing 282 μmol of (−)-epicatechin, the (epi)catechin levels rose rapidly and reached a peak plasma concentration (C_{max}) of 355 nmol/L after 2.0 hours (T_{max}). With double the chocolate intake, the C_{max} increased to 676 nmol/L, while the T_{max} was extended to 2.6 hours, which was attributed to the *ad libitum* consumption of bread by the volunteers rather than the increased intake of chocolate. Wang et al. (2000) also carried out a dose study in which varying amounts of chocolate were served with 40 g of bread. The data obtained, which are summarised in Table 7.1, show a positive relationship between intake and (epi)catechin plasma concentration. This study also showed that the rise in plasma (epi)catechin contributed to the ability of the plasma to inhibit lipid peroxidation and scavenge free radicals.

In a further study, Baba et al. (2000) fed a chocolate containing 760 μmol of (−)-epicatechin and 214 μmol of catechin, most probably the (−)-isomer, to human subjects and collected plasma and urine over the ensuing 24-hour period. By selective incubation of samples with glucuronidase and sulphatase, it was possible to distinguish between glucuronide, sulphate and sulphoglucuronide metabolites of the released (epi)catechin and methyl-(epi)catechin aglycones. The main metabolites in plasma were sulphates and sulphoglucuronides of (epi)catechin and methyl-(epi)catechin with lower levels of (epi)catechin-glucuronide and (epi)catechin. No flavan-3-ols were detected in either the 0-hour baseline or 24-hour plasma (Figure 7.5). The combined C_{max} of the metabolites was 4.8 ± 0.9 μmol/L and the T_{max} was 2 hours. The same metabolites were present in urine, with most being excreted in the initial 8-hour period after consumption of the chocolate. The total 0- to 24-hour urinary excretion of the (epi)catechin and methyl-(epi)catechin metabolites was 227 ± 39 μmol that

Table 7.1 Concentration of (epi)catechin metabolites in plasma of human volunteers after the ingestion of chocolate containing 159, 312 and 417 μmol of (−)-epicatechin

(−)-Epicatechin intake (μmol)	0 h	2 h	6 h
0	1 ± 1[a]	19 ± 14[a]	1 ± 1[a]
159	2 ± 2[a]	133 ± 27[c]	26 ± 8[b]
312	4 ± 2[a]	258 ± 29[c]	66 ± 8[b]
471	4 ± 3[a]	355 ± 49[c]	103 ± 16[b]

After Wang et al. 2000.
Data expressed as mean values in nmol/L ± standard error ($n = 9$–13). Mean values with a different superscript are significantly different ($P < 0.05$).

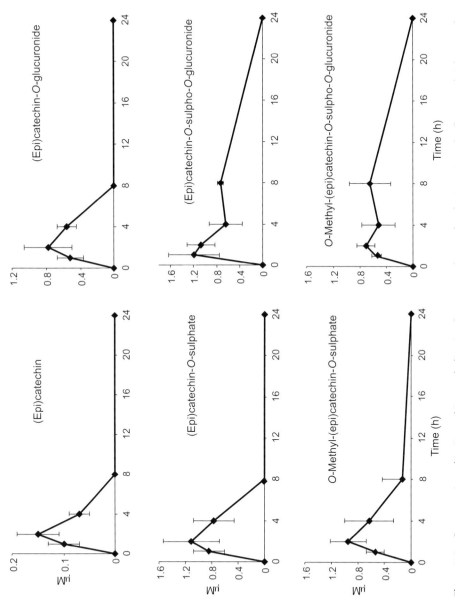

Figure 7.5 Concentration of (epi)catechin metabolites in plasma 0–24 hours after the ingestion of a flavan-3-ol-rich cocoa by human volunteers. Data presented as mean values ± standard error ($n = 5$). (After Baba et al. 2000.)

Table 7.2 Total (epi)catechin metabolites excreted in urine by human volunteers 0–24 hours after the consumption of chocolate containing 760 μmol of (–)-epicatechin

Excretion period (h)	Excretion (μmol)	Excretion (% of intake)
0–8	188 ± 33	24.7 ± 4.3
8–24	39 ± 19	5.1 ± 2.5
Total (0–24)	**227 ± 39**	**29.9 ± 5.2**

After Baba et al. 2000.
Data expressed as mean values ± standard error ($n = 5$).

corresponds to 29.9% of the ingested (–)-epicatechin (Table 7.2). Baba et al. (2000) obtained broadly similar data when the same dose of flavan-3-ols was consumed as a cocoa drink rather than as chocolate.

7.3.2.2 More recent investigations

Some cocoa flavan-3-ol bioavailability studies are now analysing plasma and urine samples by HPLC with multistage MS (i.e. MS^2 and MS^3) without recourse to enzyme hydrolysis. The advantage of this method of detection is that even in the absence of reference compounds, low ng quantities of O-glucuronide, O-sulphate and O-sulpho-O-glucuronides of (epi)catechin and O-methyl-(epi)catechin can be partially identified and quantified on a routine basis. As a consequence, much more information is obtained on post-consumption metabolism than when samples are subjected to β-glucuronidase/sulphatase treatment prior to analysis. However, the absence of reference compounds can impact on the accuracy of quantitative estimates, but as not all metabolites are cleaved by glucuronidase/sulphatase treatment, this also applies to quantitative estimates based on enzyme hydrolysis.

HPLC-MS^2 methodology was used in a recent human bioavailability study in which volunteers drank a beverage made with 10 g of commercial cocoa powder and 250 mL of hot water (Mullen et al. 2009). The drink contained 22.3 μmol of catechin, almost all of it as the less bioavailable (–)-isomer, and 23.0 μmol of (–)-epicatechin along with 70 mg of procyanidins. Two flavan-3-ol metabolites were detected in plasma, an O-methyl-(epi)catechin-O-sulphate and a (epi)catechin-O-sulphate. Both had a C_{max} below 100 nM and a T_{max} of <1.5 hours, indicative of absorption in the small rather than the large intestine. The two sulphated flavan-3-ols were also the main metabolites in urine, which, in addition, contained smaller quantities of an (epi)catechin-O-glucuronide and an additional (epi)catechin-O-sulphate. The amount of flavan-3-ol metabolites excreted in urine over the 0- to 24-hour collection period was 7.32 ± 0.82 μmol (Table 7.3).

The urinary excretion of (epi)catechin metabolites in μmol amounts appears to provide a more realistic quantitative guide to bioavailability than the nmol/L C_{max} concentrations in plasma. It is interesting to note that in humans, excretion of flavan-3-ol metabolites, as a percentage of intake, is substantially higher than that

Table 7.3 Quantities of flavan-3-ol metabolites excreted in urine by human subjects 0–24 hours after the consumption of 250 mL of a water-based cocoa drink, containing 45 μmol of flavan-3-ol monomers

Flavan-3-ol metabolites	0–2 h	2–5 h	5–8 h	8–24 h	Total (0–24 h)
(Epi)catechin-O-sulphate	0.93 ± 0.11	0.54 ± 0.06	0.16 ± 0.02	0.13 ± 0.06	1.76 ± 0.12
(−)-Epicatechin-O-glucuronide	0.41 ± 0.04	0.25 ± 0.04	0.08 ± 0.02	0.02 ± 0.01	0.76 ± 0.07
(Epi)catechin-O-sulphate	1.13 ± 0.20	0.92 ± 0.19	0.17 ± 0.04	0.09 ± 0.05	2.31 ± 0.30
O-Methyl-(epi)catechin-O-sulphate	1.15 ± 0.23	0.90 ± 0.17	0.23 ± 0.04	0.21 ± 0.15	2.49 ± 0.42
Total	**3.62 ± 0.54**	**2.61 ± 0.41**	**0.64 ± 0.11**	**0.45 ± 0.17**	**7.32 ± 0.82**

After Mullen et al. 2009.
Data expressed as mean values in μmol ± standard error ($n = 9$).

of many other flavonoids. In the Mullen et al. (2009) cocoa study, the 0- to 24-hour urinary excretion of 7.32 ± 0.82 μmol of (epi)catechin metabolites (Table 7.3) is equivalent to 16.3 ± 1.8% of intake. Considering that the ~50% of the flavan-3-ol monomer content of the cocoa was (−)-catechin, which has reduced bioavailability (Donovan et al. 2006), this figure is probably nearer 30%, and as such, it is comparable with urinary (epi)catechin excretion levels obtained by Baba et al. (2000) (Table 7.2) as well as a 28.5% excretion of (epi)catechin metabolites observed after consumption of green tea (Stalmach et al. 2009b) and even higher levels after ingestion of Polyphenon E (Auger et al. 2008).

Varying human (epi)catechin metabolite profiles have been obtained by different research groups. In the Mullen et al. (2009) investigation, an (epi)catechin-O-sulphate and an O-methyl-(epi)catechin-O-sulphate were detected in plasma. These sulphates were also the main metabolites in urine, which contained an additional (epi)catechin-O-sulphate and an (epi)catechin-O-glucuronide (Table 7.3). Stalmach et al. (2009b, 2010) detected a similar spectrum of (epi)catechin metabolites in plasma and urine collected after ingestion of green tea. These findings are also in keeping with the data of Baba et al. (2000), discussed earlier, in which post-enzyme hydrolysis indicated, albeit indirectly, the presence of sulphate and sulphoglucuronides of (epi)catechin and methyl-(epi)catechin as the main metabolites in both plasma (Figure 7.5) and urine. However, in other cocoa studies in which plasma and urine samples were analysed by HPLC-MS^2, an (epi)catechin-O-glucuronide was the main metabolite and sulphates were either absent or minor components (Roura et al. 2005, 2007a, 2007b, 2008; Tomás-Barberán et al. 2007). Initially, the reason for these varying metabolite profiles, especially the absence of sulphated metabolites in plasma in some studies and not others, was unclear, although it was thought that the amount of (−)-epicatechin ingested could be a factor (Mullen et al. 2009). However, recent research has shown that the discrepancy may be due to losses of (epi)catechin-O-sulphates as

(−)-epicatechin-3′-O-sulphate is unstable and breaks down readily during processing of plasma and urine prior to analysis (Schroeter, unpublished data).

7.3.2.3 Matrix effects

Schramm et al. (2003) investigated the effects of carbohydrate on the absorption of cocoa flavan-3-ols by humans and reported that the presence of carbohydrate resulted in an increased plasma C_{max} of epicatechin, released by enzyme hydrolysis, in a dose-dependent manner following consumption of cocoa, while the T_{max} appeared to be similar regardless the carbohydrate content.

Neilson et al. (2009) investigated the influence of the chocolate matrix on the bioavailability of cocoa (epi)catechins. In a crossover study, volunteers ingested a dark chocolate and a similar chocolate containing either high sucrose or milk protein. In addition, two cocoa drinks containing milk protein and either sucrose at the same level as the dark chocolate or an artificial sweetener were consumed. All the matrices contained 94 μmol of (−)-epicatechin and 32 μmol of catechin, most of which was presumably the (−)-isomer. Plasma collected over a 6-hour post-ingestion period was then subjected to enzyme hydrolysis before analysis of (epi)catechin by HPLC. Pharmacokinetic analysis revealed C_{max} values ranging from 24.7 to 42.6 nmol/L and area-under-the-curve (AUC) figures of 100.7–142.7 nmol/L/hour. Significantly higher C_{max} and AUC values were obtained with the two cocoa drinks compared to the three chocolate products. Although T_{max} values were not significantly different, those with the cocoa drink were 0.9 and 1.1 hours, compared with 2.3, 1.8 and 2.3 hours with the chocolates (Table 7.4). Thus, with the different cocoa matrices there appear to be differences in (epi)catechin bioavailability, but they are relatively small. It is unfortunate that urinary excretion of (epi)catechin metabolites was not investigated as it would have been substantial and, as discussed in Section 7.3.2.2, would almost certainly have

Table 7.4 (Epi)catechin metabolite pharmacokinetic parameters in plasma of human volunteers 0–6 hours after the consumption of different chocolate formulations containing 94 μmol of (−)-epicatechin and 32 μmol of catechin

Product	Composition	C_{max}(nmol/L)	T_{max}(h)	AUC(nmol/L/h)
Chocolate	Sucrose (6.6 g)	31.6 ± 2.8[bc]	2.3 ± 0.8[a]	121.1 ± 12.6[ab]
	Sucrose (14.6 g)	34.0 ± 3.3[ab]	1.8 ± 0.5[a]	128.1 ± 13.1[ab]
	Sucrose (6.6 g), milk protein (6.0 g)	24.7 ± 1.9[c]	2.3 ± 0.8[a]	107.0 ±11.1[b]
Cocoa beverage	Sucrose (6.6 g), milk protein (6.1 g)	42.5 ± 4.3[a]	0.9 ± 0.1[a]	132.1 ±14.8[a]
	Artificial sweetener, milk protein (6.1 g)	41.6 ± 2.1[a]	1.1 ± 0.3[a]	142.7 ± 9.3[a]

After Neilson et al. 2009.
Data expressed as mean values ± standard error ($n = 6$). Common superscripts in the same column indicate no significant difference ($P > 0.05$) between formulations.

provided evidence of more clear-cut differences in (epi)catechin bioavailability with the five cocoa products than the comparison of plasma pharmacokinetics.

In a further investigation into the influence of the chocolate matrix on the bioavailability of cocoa (epi)catechins with rats, Neilson et al. (2010) showed that plasma concentrations of O-methyl-(epi)catechin-O-glucuronides and (epi)catechin-O-glucuronides, analysed by HPLC-MS, were highest with a high-sucrose chocolate and lowest with milk chocolate, while a reference dark chocolate yielded intermediate concentrations.

There is much interest in whether or not milk reduces the bioavailability of cocoa flavan-3-ols. Serafini and co-workers (2003) reported that although consumption of 100 g of dark chocolate brought about an increase in human plasma antioxidant capacity, this effect was reduced substantially when the chocolate was ingested with 200 mL of milk and no increase in antioxidant capacity was observed after eating milk chocolate. They also showed that the absorption of (−)-epicatechin from chocolate was reduced when consumed with milk or as milk chocolate. It was hypothesised that proteins in the milk bind to the flavan-3-ols and limit their absorption from the gastrointestinal tract. This report generated much controversy with subsequent studies producing conflicting data on the impact of milk on cocoa flavan-3-ol absorption.

The Mullen et al. (2009) investigation referred to in the previous section was a randomised crossover study with human volunteers who drank 250 mL of a commercial cocoa made not just with hot water but also with hot milk. Both drinks contained 45 μmol of flavan-3-ol monomers. Milk did not have a significant effect on either the plasma C_{max} or T_{max} of sulphated and methylated (epi)catechin metabolites (Figure 7.6), but did bring about a significant reduction in (epi)catechin metabolites excreted 0–2 and 2–5 hours after ingestion (Figure 7.7), with the overall amounts being excreted over a 24-hour period declining from 16.3% to 10.4% of intake. This was not due to effects of milk either on gastric emptying or on the time for the head of the meal to reach the colon, ruling out the possibility that milk slowed the rate of transport of the meal through the gastrointestinal tract. The reduced excretion of the flavan-3-ol metabolites is, therefore, probably a consequence of components in the milk that either bind directly to flavan-3-ols or interfere with the mechanism involved in their transport across the wall of the small intestine into the portal vein.

The findings of this study contrast with reports that milk does not affect the absorption of flavan-3-ols. These include a study by Roura et al. (2007a), who monitored flavan-3-ol metabolites in plasma collected 2 hours after drinking of cocoa containing 128 μmol of flavan-3-ol monomers, a threefold higher quantity than the 45 μmol ingested in the Mullen et al. (2009) study. It is, however, interesting to note that although not statistically significant, urinary excretion in the study of Roura et al. (2008) was 20% lower with cocoa milk compared with cocoa water. Keogh et al. (2007), who analysed plasma 0–8 hours after the consumption of a flavan-3-ol-rich cocoa drink, also reported that milk had no effect on the absorption of catechin and epicatechin. In this instance, the ingested dose of flavan-3-ol monomers was 2374 μmol, which is 53-fold higher than the quantity consumed by volunteers in the Mullen et al. study. This high dose is reflected in a C_{max} of ∼12 μmol/L compared with ∼150 nmol/L in the Mullen et al. (2009) study. Schroeter et al. (2003) also reported that milk did not influence plasma epicatechin after consumption of a cocoa beverage, which in this

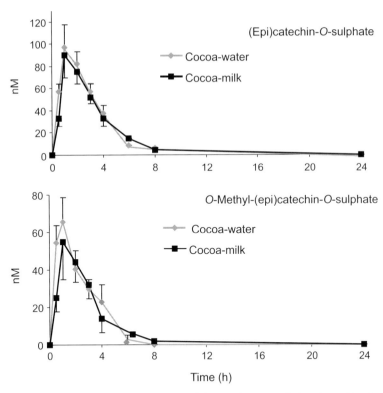

Figure 7.6 Concentration of a (epi)catechin-O-sulphate and an O-methyl-(epi)catechin-O-sulphate in plasma 0–24 hours after the ingestion of a cocoa, prepared with water and containing 45 μmol of (epi)catechins, by human volunteers. Data presented as means vaules ± standard error ($n = 8$). (After Mullen et al. 2008.)

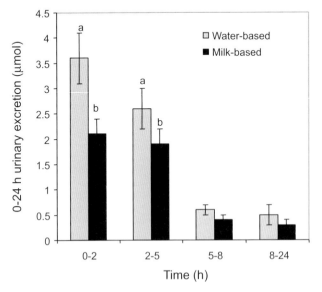

Figure 7.7 Urinary excretion of (epi)catechin metabolites 0–2, 2–5, 5–8 and 8–24 hours after the consumption of cocoas, made with milk and water and containing 45 μmol of (epi)catechins, by human subjects ($n = 8$). Different letters above paired histograms indicate values that are significantly different ($P > 0.05$). (After Mullen et al. 2008.)

instance was consumed at a dose of 1314 μmol of flavan-3-ol monomers for a 70-kg human.

There is an explanation for these seemingly contradictory reports. It would appear that with high flavan-3-ol cocoas, which are principally research products, the factors in milk that reduce absorption have minimal overall impact. In the case of drinks with a lower flavan-3-ol content, such as the one used in the Mullen *et al.* (2009) study, which is typical of many commercial cocoas that are found on supermarket shelves (Miller *et al.* 2008) and available to the general public, milk does have the capacity to interfere with absorption. An additional factor that may impact on the effect of milk on flavan-3-ol bioavailability could be the physical form of the cocoa product, whether it is consumed as a cocoa drink or as a solid chocolate with varying contents of carbohydrate and milk protein as implied by the findings of Neilson *et al.* (2009, 2010).

7.3.2.4 Degradation of (−)-epicatechin in the large intestine

As noted in Section 5.3.1, even though absorption occurs in the small intestine, in excess of 20% of (−)-epicatechin intake appears to pass from the small intestine to the colon (Auger *et al.* 2008; Stalmach *et al.* 2010). To mimic the events occurring in the colon, 50 μmol of (−)-epicatechin was incubated under anaerobic conditions *in vitro* with human faecal slurries and degradation to phenolic acids and related compounds by the microbiota monitored (Roowi et al. 2010). GC-MS analysis revealed the appearance of four catabolites. 4′-Hydroxyphenylacetic acid was present in small amounts, while 3-(3′-hydroxyphenyl)propionic acid, 5-(3′,4′-dihydroxyphenyl)-γ-valerolactone and 5-(3′,4′-dihydroxyphenyl)valeric acid appeared, albeit transitorily, after 24- to 48-hour incubation periods, in amounts which on a mole basis correspond to 32–54% of the (−)-epicatechin substrate (Table 7.5). As is typical with this model system, there were substantial differences in the catabolite profiles of the individual subjects presumably reflecting variations in their colonic microflora (Jaganath *et al.* 2006, 2009).

In keeping with studies by Toshiyuki *et al.* (2001), the data of Roowi *et al.* (2010) indicate that the breakdown of (−)-epicatechin by colonic bacteria leads to the production of 5-(3′,4′-dihydroxyphenyl)-γ-valerolactone with 1-(3′,4′)-dihydroxyphenyl)-3-(2″,4″,6″-trihydroxy)propan-2-ol, which did not accumulate in detectable amounts in the faecal slurries, acting as a transient intermediate, as illustrated in Figure 7.8. The dihydroxyvalerolactone is then converted to 5-(3′,4′-dihydroxyphenyl)valeric acid that is further metabolised to 3-(3′-hydroxyphenyl)propionic acid, which, in turn, yields low levels of 4′-hydroxyphenylacetic acid. In a parallel study, additional phenolic compounds in the form of 3-(3′-hydroxyphenyl)hydracrylic acid, 4-hydroxybenzoic acid, hippuric acid and trace levels of 3′-methoxy-4′-hydroxyphenylacetic acid, arguably derived from phase II metabolism in the liver, were detected in urine after green tea consumption (Roowi *et al.* 2010). The potential routes for the formation of these catabolites are also illustrated in Figure 7.8. The catabolism of (−)-epicatechin will, however, be more complex than the routes illustrated, as some phase II urinary catabolites will have escaped detection by GC-MS. This includes the glucuronides and sulphates of 5-(3′,4′-dihydroxyphenyl)-γ-valerolactone and 5-(3′-methoxy-4′-hydroxyphenyl)-γ-valerolactone that have been detected in urine after cocoa consumption (Llorach *et al.* 2009).

Table 7.5 Quantities of catabolites in the faecal slurries from three donors incubated for 4, 6, 24, 30 and 48 hours with 50 µmol of (−)-epicatechin

Substrates	Catabolites	Donor	4 h	6 h	24 h	30 h	48 h
(−)-Epicatechin	4′-Hydroxyphenylacetic acid	I	n.d.	n.d.	2.3 ± 0.5	n.d.	1.4 ± 0.2
		II	0.3 ± 0.0	0.3 ± 0.0	n.d.	0.2 ± 0.0	0.1 ± 0.0
		III	0.2 ± 0.0	0.5 ± 0.0	0.1 ± 0.0	0.4 ± 0.0	0.6 ± 0.1
	3-(3′-Hydroxyphenyl)propionic acid	I	n.d.	n.d.	27.4 ± 4.3	n.d.	n.d.
		II	n.d.	n.d.	n.d.	0.3 ± 0.0	0.2 ± 0.0
		III	n.d.	0.3 ± 0.0	1.2 ± 0.1	4.4 ± 0.8	2.7 ± 0.4
	5-(3′,4′-Dihydroxyphenyl)valeric acid	I	n.d.	n.d.	n.d.	n.d.	n.d.
		II	n.d.	n.d.	n.d.	2.6 ± 0.2	5.3 ± 0.1
		III	n.d.	0.1 ± 0.0	4.1 ± 0.5	16.3 ± 3.0	5.1 ± 0.2
	5-(3′,4′-Dihydroxyphenyl)-γ-valerolactone	I	n.d.	n.d.	2.2 ± 0.5	n.d.	n.d.
		II	n.d.	1.8 ± 0.2	0.5 ± 0.1	19.1 ± 1.1	1.8 ± 0.0
		III	0.3 ± 0.1	1.0 ± 0.1	n.d.	n.d.	n.d.
Blank	4′-Hydroxyphenylacetic acid	I	n.d.	n.d.	0.2 ± 0.3	0.1 ± 0.0	0.2 ± 0.0
		II	n.d.	n.d.	n.d.	n.d.	0.1 ± 0.0
		III	n.d.	n.d.	n.d.	n.d.	0.2 ± 0.0
	3-(3′-Hydroxyphenyl)propionic acid	I	n.d.	n.d.	n.d.	n.d.	n.d.
		II	n.d.	n.d.	n.d.	n.d.	n.d.
		III	n.d.	n.d.	0.7 ± 0.2	0.4 ± 0.1	0.5 ± 0.1

After Roowi et al. 2010.
Data expressed as mean values in µmol ± standard error ($n = 3$); n.d., not detected.

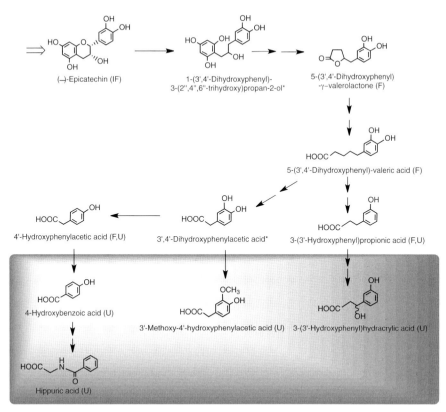

Figure 7.8 Proposed pathways involved in the colonic catabolism and urinary excretion of (−)-epicatechin. When incubated with faecal slurries, (−)-epicatechin is converted to the catabolites labelled (F) by the colonic microflora probably via the pathways illustrated. Some, but not all of these, catabolites are excreted in urine (U) with additional products that are probably produced by phase II metabolism in the liver before being excreted in urine. Potential steps in phase II metabolism are located in the shaded rectangle. Double arrows indicate conversions where the intermediate(s) did not accumulate and are unknown. '∗' indicates a potential intermediate that did not accumulate in detectable quantities in faecal slurries. (After Roowi et al. 2010.)

Similar faecal catabolites to those detected by Roowi *et al.* (2010) were reported by Tzounis *et al.* (2008) after incubations with (−)-epicatechin and (+)-catechin. Seemingly, the formation of catabolites from (+)-catechin required its conversion to (+)-epicatechin. (+)-Catechin incubation affected the growth of selected microflora, resulting in a statistically significant increase in the growth of *Clostridium coccoides*–*Eubacterium rectale* group, *Bifidobacterium* spp. and *Escherichia coli*, as well as a significant inhibitory effect on the growth of the *C. histolyticum* group. In contrast, the effect of (−)-epicatechin was less marked, only significantly increasing the growth of the *C. coccoides*–*E. rectale* group. It was suggested that as both (−)-epicatechin and (+)-catechin were converted to the same catabolites, the more dramatic change in the growth of the distinct microfloral populations produced by (+)-catechin may be linked to its bacterial conversion to (+)-epicatechin. Tzounis *et al.* (2008) also suggested that the consumption of flavan-3-ol-rich foods may support gut health through their ability to exert pre-biotic actions.

It is unclear if urinary phenolic catabolites are excreted predominantly as phase II glucuronide and sulphate conjugates, as is the case with the valerolactones, or as free unconjugated structures. Most, but not all, data point to the latter rather than the former option. In the study by Roowi *et al.* (2010), glucuronidase/sulphatase treatment of urine prior to analysis had minimal effect on the levels of phenolic acids detected and likewise in the investigation by Llorach *et al.* (2009). In contrast, after feeding a cranberry powder to rats, urinary excretion of a diversity of phenolic acids was 65–100% as conjugated forms (Prior *et al.* 2010). Also, the majority of hydroxycinnamates excreted by humans after drinking a chlorogenic acid-rich coffee were identified by HPLC-MS^2 as glucuronide and sulphate conjugates with dihydrocaffeic acid-3′-*O*-sulphate, dihydroferulic acid-4′-*O*-sulphate and dihydroferulic acid-4′-*O*-glucuronide predominating (Stalmach *et al.* 2009a; see Chapter 5), while glucuronide derivatives of phenylacetic acid and phenylpropionic acid have been similarly detected in human urine following ingestion of orange juice flavanones (Tomás-Barbéran, unpublished data). In contrast, feeding [2-^{14}C]quercetin-4′-glucoside to rats resulted in the colonic conversion of quercetin exclusively to unconjugated phenolic acids (Mullen *et al.* 2008). With respect to cocoa, this situation and the operation of minor routes in addition to those illustrated in Figure 7.8 will be clarified only when feeding experiments are carried out with isotopically labelled (–)-epicatechin.

In the context of matrix effects discussed in Section 7.3.2.3, it is of interest that consumption of cocoa with milk had a significant effect on the excretion of microbial degradation products in rats and humans (Urpi-Sarda *et al.* 2010). Nine out of fifteen phenolic catabolites analysed in urine were affected by milk consumption with seven decreasing and two increasing.

7.3.3 Procyanidins

There are numerous feeding studies with animals and humans indicating that the oligomeric and polymeric flavan-3-ols, the proanthocyanidins, are not absorbed to any degree. Most pass unaltered to the large intestine where they are catabolised by the colonic microflora, yielding a diversity of phenolic acids (Manach *et al.*, 2005) including 3-(3′-hydroxyphenyl)propionic acid and 4-*O*-methyl-gallic acid (Déprez *et al.* 2000; Gonthier *et al.* 2003; Ward *et al.* 2004) that are absorbed into the circulatory system and excreted in urine. There is one report based on data obtained in an *in vitro* model of gastrointestinal conditions, that procyanidins degrade yielding more readily absorbable flavan-3-ol monomers (Spencer *et al.* 2000). Subsequent *in vivo* studies have not supported this conclusion (Donovan *et al.*, 2002; Rios *et al.* 2002; Tsang *et al.* 2005). There is a report of minor quantities of procyanidin B$_2$ being detected in enzyme-treated human plasma collected after the consumption of cocoa (Holt *et al.* 2002). The T_{max} and pharmacokinetic profile of the B$_2$ dimer were similar to that of flavan-3-ol monomers, but the C_{max} was ~100-fold lower (Holt *et al.* 2002).

Urpi-Sarda et al. (2009) also detected and quantified procyanidin B_2 in human and rat urine after cocoa intake.

Recent studies using procyanidin B_2 and [^{14}C]procyanidin B_2 have provided information about their *in vitro* gut flora catabolism (Appeldoom et al. 2009; Stoupi et al. 2010a, 2010b) and rodent pharmacokinetics (Stoupi et al. 2010c). Following oral dosing of rats with [^{14}C]procyanidin B_2 dimer, approximately 60% of the radioactivity was excreted in urine after 96 hours, with the vast majority in a form(s) very different from the intact procyanidin dosed (Stoupi et al. 2010c). This observation is consistent with the *in vitro* studies that show extensive catabolism by the gut microflora. The scission of the interflavan bond represents a minor route, and the dominant products are a series of phenolic acids having one or two phenolic hydroxyls and between one and five aliphatic carbons in the side chain (Appeldoom et al. 2009; Stoupi et al. 2010a, 2010b). There are, in addition, some C_6–C_5 catabolites with a side-chain hydroxyl, and associated lactones, and several diaryl-propan-2-ols, most of which are also produced from the flavan-3-ol monomers, with the routes illustrated in Figure 7.9 being proposed by Appeldoom et al. (2009). In this figure, 3′,4′-dihydroxyphenylacetic acid is derived from cleavage of the C-ring of the upper flavan-3-ol unit. However, the findings of Stoupi et al. (2010b) also indicate that a feature of flavan-3-ol catabolism is conversion to C_6–C_5 valerolactones and progressive β-oxidation to C_6–C_3 and C_6–C_1 products, broadly in keeping with the routes illustrated in Figure 7.8.

The *in vitro* fate of procyanidin B_2 in faecal slurries is also much more complex than the routes indicated in Figure 7.9 as the dimer also yields 24 'dimeric' catabolites, i.e. having a mass greater than the constituent monomer (–)-epicatechin (290 amu), and which early in the incubation collectively accounted for some 20% of the substrate (Stoupi et al. 2010b). Clearly, these catabolites retain the interflavan bond. One was identified tentatively as either 6- or 8-hydroxy-procyanidin B_2. Thirteen were characterised as having been microbially reduced in at least one of the epicatechin units. Five contained an apparently unmodified epicatechin unit, but in at least one case, this was shown to consist of the B-ring of the 'upper' epicatechin unit and the A-ring of the 'lower'. It is not known whether these unique catabolites are produced *in vivo*, and if so, whether they are absorbed (Stoupi et al. 2010b).

The potential biological effects of procyanidins are generally attributed to their more readily absorbed colonic breakdown products, the phenolic acids, although there is a lack of detailed study in this area. There is, however, a dissenting view as trace levels of procyanidins, in contrast to (–)-catechin and (+)-epicatechin, inhibit platelet aggregation *in vitro* and suppress the synthesis of the vasoconstriction peptide, endothelin-1 by cultured endothelial cells (Corder 2008). Supporting this view is a study in which individual procyanidins were fed to rats after which dimers through to pentamers were detected in plasma, which was extracted with 8 mol/L urea, rather than the more traditional methanol/acetonitrile, which it was proposed prevented the irreversible binding of procyanidins to plasma proteins (Shoji et al. 2006). The procyanidins were, however, administered by gavage at an extremely high dose, 1 g/kg body weight, and it remains to be determined if procyanidins can be similarly detected in urea-extracted plasma after the ingestion by humans of more nutritionally relevant quantities in cocoa or chocolate products.

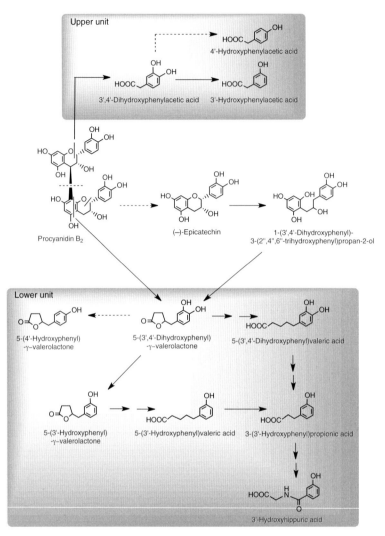

Figure 7.9 Proposed pathways for human microbial degradation of procyanidin B_1 dimer. Main routes are indicated with solid arrows, minor pathways with dotted arrows. Metabolites derived from upper and lower units are grouped within the shaded rectangles. After Appeldoom et al. (2009) and Stoupi et al. (2010a, 2010b and 2010c).

7.4 Conclusions

Cocoa contains many phytochemicals, but from a nutrition perspective, those of most interest are the flavan-3-ol monomers and the oligomeric and polymeric proanthocyanidins. The bioavailability of flavan-3-ol monomers is high, and although plasma concentrations of metabolites are typically only in the nmol/L to low μmol/L range, urinary excretion can be >30% of intake. Absorption is affected by the spatial configuration of the molecules as (−)-epicatechin seems to be much better absorbed than

(+)-catechin. The dimer procyanidin B_2 is only marginally bioavailable, while the rest of oligomers and polymers have not been detected in either plasma or urine. Studies with apple juice and ileostomists suggest that more than 90% of procyanidins pass from the small to large intestine (Kahle *et al.* 2007). This raises the question of the bioavailability of procyanidin oligomers, particularly that of trimers, tetramers and pentamers, which have never been detected in plasma or urine but which exhibit high biological activity *in vitro*. Are these procyanidins really not absorbed? Is there a possibility that despite their size, they are absorbed, but the analytical methods used for sample preparation, detection and quantification are not appropriate for their detection? Are they transformed to non-extractable metabolites after absorption perhaps binding to proteins and lipids? Can they be absorbed via lymphatic system or in combination with lipid particles?

It is clear that the bulk of cocoa procyanidins are catabolised by the colonic microbiota to produce low-molecular-weight phenolics that are then absorbed and detected in plasma and urine. These are hydroxylated valerolactones, phenylpropionic acids, phenylacetic acids and benzoic acids. The degree of conjugation of these metabolites in plasma and urine still remains uncertain. Can these metabolites account for the health effects associated with cocoa consumption? Recent studies have shown that low-molecular-weight phenolic catabolites can act as inhibitors of NADPH oxidase, which could partly explain an indirect effect of cocoa polyphenols in cardiovascular function (Schewe *et al.* 2008). There is also evidence that some of these catabolites have anti-inflammatory effects in experimental models (Larrosa *et al.* 2009) and that *in vitro* they have neuroprotective and antiglycative effects (Verzelloni *et al.* 2011).

References

Appeldoom, M.M., Vincken, J.-P., Aura, A.-M. *et al.* (2009) Procyanidin dimers are metabolized by human microbiota with 2-(3,4-dihydroxyphenyl)acetic acid and 5-(3,4-dihydroxyphenyl)-γ-valerolactone as the major metabolites. *J. Agric. Food Chem.*, **57**, 1084–1092.

Aron, P.M. and Kennedy, J.A. (2008) Flavan-3-ols: nature, occurrence and biological activity. *Mol. Nutr. Food Res.*, **52**, 79–104.

Ashihara, H., Sano, H. and Crozier, A. (2008) Caffeine and related purine alkaloids: biosynthesis, catabolism, function and genetic engineering. *Phytochemistry*, **69**, 841–856.

Auger, C., Hara, Y. and Crozier, A. (2008). Bioavailability of Polyphenon E flavan-3-ols in humans with an ileostomy. *J. Nutr.*, **138**, 1535S–1542S.

Baba, S., Osakabe, N., Yasuda, A. *et al.* (2000) Bioavailability of (−)-epicatechin upon intake of chocolate and cocoa in human volunteers. *Free Rad. Res.*, **33**, 635–641.

Baba, S., Osakabe, N., Natsume, M. *et al.* (2001) *In vivo* comparison of the bioavailability of (+)-catechin, (−)-epicatechin and their mixture in orally administered rats. *J. Nutr.*, **131**, 2885–2891.

Bixler, R.G. and Morgan, J.N. (1999) Cocao bean and chocolate processing. In I. Knight (ed), *Chocolate and Cocoa: Health and Nutrition*. Blackwell Publishing Ltd., Oxford, pp. 43–60.

Buckingham, J. (1994) *Dictionary of Natural Products*. Chapman and Hall, London.

Cooper, K.A., Campos-Gimenez, E., Jimenez Alvarez, D. *et al.* (2007) Rapid reversed phase ultra-performance liquid chromatography analysis of the major coca polyphenols and inter-relationships of their concentrations in chocolate. *J. Agric. Food Chem.*, **55**, 2841–2847.

Corder, R. (2008) Red wine, chocolate and vascular health: developing the evidence base. *Heart*, **94**, 821–823.

Cottart, C.-H., Nivet-Antoine, V., Laguillier-Morizot, C. *et al.* (2010) Resveratrol bioavailability and toxicity in humans. *Mol. Nutr. Food Res.*, **54**, 7–16.

Counet, C., Callemien, D. and Collin, S. (2006) Chocolate and cocoa: new sources of *trans*-resveratrol and *trans*-piceid. *Food Chem.*, **98**, 649–657.

Crozier, A., Jaganath, I.B. and Clifford, M.N. (2009) Dietary phenolics: chemistry, bioavailability and effects on health. *Nat. Prod. Rep.*, 26, 1001–1043.

Crozier, A., Del Rio, D. and Clifford, M.N. (2010) Bioavailability of dietary flavonoids and phenolic compounds. *Mol. Asp. Med.*, **31**, 446–467.

Del Rio, D., Borges, G. and Crozier, A. (2010) Berry flavonoids and phenolics: bioavailability and evidence of protective effects. *Br. J. Nutr.*, **104**, S67–S90.

Déprez, S., Brezillon, C., Rabot, S. *et al.* (2000) Polymeric proanthocyanidins are catabolized by human colonic microflora into low-molecular-weight phenolic acids. *J. Nutr.*, **130**, 2733–2738.

Donovan, J.L., Manach, C., Rios, L. *et al.* (2002) Procyanidins are not bioavailable in rats fed a single meal containing a grape seed extract or the procyanidin dimer B3. *Brit. J. Nutr.*, **87**, 299–306.

Donovan, J.L., Crespy, V., Oliveira, M. *et al.* (2006) (+)-Catechin is more bioavailable than (−)-catechin: relevance to the bioavailability of catechin from cocoa. *Free Rad. Res.*, **40**, 1029–1034.

Gonthier, M.P., Donovan, J.L., Texier, O. *et al.* (2003) Metabolism of dietary procyanidins in rats. *Free Rad. Biol. Med.*, **35**, 837–844.

Gotti, R., Furlanetto, S., Pinzauti, S. *et al.* (2006) Analysis of catechins in *Theobroma cacao* beans by cyclodextrin-modified micellar electrokinetic chromatography. *J. Chromatogr. A*, **1112**, 345–352.

Gu, L., House, S.E., Wu, X. *et al.* (2006) Procyanidin and catechin contents and antioxidant capacity of cocoa and chocolate products. *J. Agric. Food Chem.*, **54**, 4057–4061.

Hammerstone, J.F., Lazarus, S.A., Mitchell, A.E. *et al.* (1999) Identification of procyanidins in cocoa (*Theobroma cacao*) and chocolate using high-performance liquid chromatography/mass spectrometry. *J. Agric. Food Chem.*, **47**, 490–496.

Haslam, E. (1998) *Practical Polyphenols – From Structure to Molecular Recognition and Physiological Action*. Cambridge University Press, Cambridge.

Ho, Y., Lee, Y.-L. and Hsu, K.-Y. (1995) Determination of (+)-catechin in plasma by high-performance liquid chromatography using fluorescence detection. *J. Chromatogr. B*, **665**, 383–389.

Holt, R.R., Lazarus, S.A., Sullards, M.C. *et al.* (2002) Procyanidin dimer B_2 [epicatechin-(4β-8)-epicatechin] in human plasma after the consumption of a flavanol-rich cocoa. *Am. J. Clin. Nutr.*, **76**, 798–804.

Hurst, W.J., Glinski, J.A., Miller, K.B. *et al.* (2008) Survey of the *trans*-resveratrol content of cocoa-containing chocolate products. *J. Agric. Food Chem.*, **56**, 8374–8378.

Jaganath, I.B., Mullen, W., Edwards, C.A. *et al.* (2006). The relative contribution of the small and large intestine to the absorption and metabolism of rutin in man. *Free Rad. Res.*, **40**, 1035–1046.

Jaganath, I.B., Mullen, W., Lean, M.E.J. *et al.* (2009) *In vitro* catabolism of rutin by human fecal bacteria and the antioxidant capacity of its catabolites. *Free Rad. Biol. Med.*, **47**, 1180–1189.

Jerkovic, V., Bröhan, M., Monnart, E. *et al.* (2010) Stilbenic profile of cocoa liquors from different origins determined by RP-HPLC-APCI(+)-MS/MS. Detection of a new resveratrol hexoside. *J. Agric. Food Chem.*, **58**, 7067–7074.

Kahle, K., Huemmer, W., Kempf, M. *et al.* (2007) Polyphenols are extensively metabolized in the human gastrointestinal tract after apple juice consumption. *J. Agric Food Chem.*, **55**, 10695–10614.

Keogh, J.B., McInerney, J. and Clifton, P.M. (2007) The effect of milk protein on the bioavailability of cocoa polyphenols. *J. Food Sci.*, **72**, S230–S233.

Kim, H. and Keeney, P.G. (1984) (−)-Epicatechin content in fermented and unfermented cocoa beans. *J. Food Sci.*, **49**, 1090–1092.

Larrosa, M., Luceri, C., Vivoli, E. *et al.* (2009) Polyphenol metabolites from colonic microbiota exert anti-inflammatory activity on different inflammation models. *Mol. Nutr. Food Res.*, **53**, 1044–1054.

Leung, A.Y. and Foster, S. (1996) *Encyclopedia of Common Natural Ingredients Used in Food, Drugs and Cosmetics*, 2nd edition. John Wiley & Sons, Inc., New York.

Llorach, R., Urpi-Sarda, M., Jauregui, O. *et al.* (2009) An LC-MS-based metabolomics approach for exploring urinary metabolome modifications after cocoa consumption. *J. Prot. Res.*, **8**, 5060–5068.

Manach, C., Williamson, G., Morand, C. *et al.* (2005) Bioavailability and bioefficacy of polyphenols in humans. I. Review of 97 bioavailability studies. *Am. J. Clin. Nutr.*, **81**, 2230S–242S.

Miller, K.B., Stuart, D.A., Smith, N.L. *et al.* (2006) Antioxidant activity and polyphenol and procyanidin contents of selected commercially available cocoa-containing and chocolate products in the United States. *J. Agric. Food Chem.*, **54**, 4062–4068.

Miller, K.B., Hurst, W.J., Payne, M.J. *et al.* (2008) Impact of alkalization on the antioxidant and flavanol content of commercial cocoa powders. *J. Agric. Food Chem.*, **56**, 8527–8533.

Mullen, W., Rouanet, J.-M., Auger, C. *et al.* (2008) The bioavailability of [2–^{14}C]quercetin-4′-glucoside in rats. *J. Agric. Food Chem.*, **56**, 12137–12137.

Mullen, W., Borges, G., Donovan, J.L. *et al.* (2009) Milk decreases urinary excretion but not plasma pharmacokinetics of cocoa flavan-3-ol metabolites in humans. *Am. J. Clin. Nutr.*, **89**, 1784–1791.

Neilson, A.P., George, J.C., Janle, E.M. *et al.* (2009) Influence of chocolate matrix composition of cocoa flavan-3-ol bioaccessibility *in vitro* and bioavailability in humans. *J. Agric. Food. Chem.*, **57**, 9418–9426.

Neilson, A.P., Sapper, T.N., Janle, E.M. *et al.* (2010) Chocolate matrix factors modulate the pharmacokinetic behaviour of cocoa flavan-3-ol phase II metabolites following oral consumption by Sprague-Dawley rats. *J. Agric. Food Chem.*, **58**, 6685–6691.

Prior, R.L., Rogers, T.R., Khanal, R.C. *et al.* (2010) Urinary excretion of phenolic acids in rats fed cranberry. *J. Agric. Food Chem.*, **58**, 3940–3949.

Rasmussen, S.E., Frederiksen, H., Struntze-Kronghol m, K. *et al.* (2005) Dietary proanthocyanidins: occurrence, dietary intake, bioavailability, and protection against cardiovascular disease. *Mol. Nutr. Food Res.*, **49**, 159–174.

Rein, D., Lotito, S., Holt, R.R. *et al.* (2000) Epicatechin in plasma: *in vivo* determination and effect of chocolate consumption on plasma oxidative status. *J. Nutr.*, **130**, 2109S–2114S.

Richelle, M., Tavazzi, I., Enslen, M. *et al.* (1999) Plasma kinetics in man of epicatechin from black chocolate. *Eur. J. Clin. Nutr.*, **53**, 22–26.

Rios, L.Y., Bennett, R.N., Lazarus, S.A. *et al.* (2002) Cocoa procyanidins are stable during gastric transit in humans. *Am. J. Clin. Nutr.*, **76**, 1106–1110.

Roowi, S., Stalmach, A., Mullen, W. *et al.* (2010) Green tea flavan-3-ols: colonic degradation and urinary excretion of catabolites by humans. *J. Agric. Food Chem.*, **58**, 1296–1304.

Roura, E., Andrés-Lacueva, C., Jáuregui, O. *et al.* (2005) Rapid liquid chromatography tandem mass spectrometer assay to quantify plasma (−)-epicatechin metabolites after ingestion of a standard portion of cocoa beverage in humans. *J. Agric. Food Chem.*, **53**, 6190–6194.

Roura, E., Andés-Lacueva, C., Estruch, R. *et al.* (2007a) Milk does not affect the bioavailability of cocoa powder flavonoid in healthy human. *Ann. Nutr. Metab.*, **51**, 493–498.

Roura, E., Almajano, M.P., Mata-Bilbao, M.L. *et al.* (2007b) Human urine: epicatechin metabolites and antioxidant activity after cocoa intake. *Free Rad. Res.*, **41**, 943–949.

Roura, E., Andrés-Lacueva, C., Estruch, R. *et al.* (2008) The effects of milk as a food matrix for polyphenols on the excretion profile of cocoa (−)-epicatechin metabolites in healthy human subjects. *Br. J. Nutr.*, **100**, 846–851.

Sanbongi, C., Osakabe, N., Natsume, M. *et al.* (1998) Antioxidative polyphenols isolated from *Theobroma cacao*. *J. Agric. Food Chem.*, **46**, 454–457.

Schewe, T., Steffen, Y. and Sies, H. (2008) How do dietary flavanols improve vascular function? A position paper. *Arch. Biochem. Biophys.*, **476**, 102–106.

Schramm, D.D., Karim, M., Schrader, H.R. *et al.* (2003) Food effects on the absorption and pharmacokinetics of cocoa flavanols. *Life Sci.*, **73**, 857–869.

Schroeter, H., Holt, R.R., Orozco, T.J. *et al.* (2003) Milk and absorption of dietary flavanols. *Nature*, **426**, 787–788.

Schroeter, H., Heiss, C., Balzer, J. *et al.* (2006) (−)-Epicatechin mediates beneficial effects of flavanol-rich cocoa on vascular function in humans. *Proc. Natl. Acad. Sci.*, **103**, 1024–1029.

Serafini, M., Bugianesi, S., Maiani, G. *et al.* (2003) Plasma antioxidants from chocolate. Dark chocolate may offer its consumers health benefits the milk variety cannot match. *Nature*, **422**, 1013.

Shoji, T., Matsumoto, S., Moriichi, N. *et al.* (2006) Apple procyanidin oligomers absorption in rats after oral administration. Analysis of procyanidins in plasma using the Porter method and high-performance liquid chromatography/tandem mass spectrometry. *J. Agric. Food Chem.*, **54**, 884–892.

Spencer, J.P., Chaudry, F., Pannala, A.S. *et al.* (2000) Decomposition of cocoa procyanidins in the gastric milieu. *Biochem. Biophys. Res. Commun.*, **272**, 236–241.

Stalmach, A., Mullen, W., Barron, D. *et al.* (2009a) Metabolite profiling of hydroxycinnamate derivatives in plasma and urine following the ingestion of coffee by humans: identification of biomarkers of coffee consumption. *Drug Met. Disp.*, **37**, 1759–1768.

Stalmach, A., Troufflard, S., Serafini, M. *et al.* (2009b) Absorption, metabolism and excretion of Choladi green tea flavan-3-ols by humans. *Mol. Nutr. Food Res.*, **53**, S44–S53.

Stalmach, A., Mullen, W., Steiling, H. *et al.* (2010) Absorption, metabolism, efflux and excretion of green tea flavan-3-ols in humans with an ileostomy. *Mol. Nutr. Food Res.*, **54**, 323–334.

Stark, T. and Hofmann, T. (2005) Isolation, structure determination, synthesis, and sensory activity of N-phenylpropenoyl-L-amino acids from cocoa (*Theobroma cacao*). *J. Agric. Food Chem.*, **53**, 5419–5428.

Stoupi, S., Williamson, G., Drynan, J.W. *et al.* (2010a) A comparison of the *in vitro* biotransformation of (−)-epicatechin and procyanidin B2 by human faecal microbiota. *Mol. Nutr. Food Res.*, **54**, 747–759.

Stoupi, S., Williamson, G., Drynan, J.W. *et al.* (2010b) Procyanidin B_2 catabolism by human fecal microflora: partial characterization of 'dimeric' intermediates. *Arch. Biochem. Biophys. Acta*, **501**, 73–78.

Stoupi, S., Williamson, G., Viton, F. *et al.* (2010c) *In vivo* bioavailability, absorption, excretion, and pharmacokinetics of [^{14}C]procyanidin B_2 in male rats. *Drug Met. Disp.*, **38**, 287–291.

Tomás-Barberán, F.A., Cienfuegos-Jovellanos, E., Marín, A. *et al.* (2007) A new process to develop a cocoa powder with higher flavonoid monomer content and enhanced bioavailability in healthy humans. *J. Agric. Food Chem.*, **55**, 3926–3935.

Toshiyuki, K., Natsuki, M., Mana, Y. *et al.* (2001). Metabolic fate of (−)-[4–^{3}H]epigallocatechin gallate in rats after oral administration. *J. Agric. Food. Chem.*, **49**, 4102–4112.

Tsang, C., Auger, C., Bornet, A. *et al.* (2005). The absorption, metabolism and excretion of flavan-3-ols and procyanidins following the ingestion of a grape seed extract by rats. *Brit. J. Nutr.*, **94**, 170–181.

Tzounis, X., Vulevic, J., Kuhnle, G.G. *et al.* (2008) Flavanol monomer-induced changes to the human faecal microflora. *Br. J. Nutr.*, **99**, 782–792.

Urpi-Sarda, M., Monagas, M., Khan, N. *et al.* (2009) Epicatechin, procyanidins and phenolic microbial metabolites after cocoa intake in humans and rats. *Anal. Bioanal. Chem.*, **394**, 1545–1556.

Urpi-Sarda, M., Llorach, R., Khan, N. *et al.* (2010) Effect of milk on the urinary excretion of microbial phenolic acid after cocoa powder consumption in humans. *J. Agric. Food Chem.*, **58**, 4706–4711.

Verzelloni, E., Pellacani, C., Tagliazucchi, D. *et al.* (2011) Antiglycative and neuroprotective activity of colon-derived polyphenol catabloites. *Mol. Nutr. Food Res.*, **55**, S35–S43.

Wang, J.F., Schramm, D.D., Holt, R.R. *et al.* (2000) A dose-response effect from chocolate consumption on plasma epicatechin and oxidative damage. *J. Nutr.*, **130**, 2115S–2119S.

Ward, N.C, Croft, K.D., Puddey, I.B. *et al.* (2004) Supplementation with grape seed polyphenols results in increased urinary excretion of 3-hydroxyphenylpropionic acid, an important metabolite of proanthocyanidins in humans. *J. Agric. Food. Chem.*, **52**, 5545–5549.

Willson, K.C. (1999) *Coffee, Cocoa and Tea*. CABI Publishing, Wallingford, UK.

Chapter 8
Cocoa and Health

Jennifer L. Donovan[1], Kelly A. Holes-Lewis[1], Kenneth D. Chavin[2] and Brent M. Egan[3]

[1] Departments of Psychiatry and Behavioral Sciences, Medical University of South Carolina, Charleston, SC 29401, USA
[2] Department of Surgery, Medical University of South Carolina, Charleston, SC 29401, USA
[3] Department of Medicine, Medical University of South Carolina, Charleston, SC 29401, USA

8.1 Introduction

The medicinal, ritual and nutritional use of cocoa originated in Central and South America as early as 600 BC (Hurst *et al.* 2002). At the time of the Spanish Conquest, cocoa was consumed with most meals and mixed with other ingredients (e.g. honey and maize) to produce a wide variety of foods and drinks. Cocoa became popularly consumed as a beverage in Europe in the mid-1500s (Dillinger *et al.* 2000; see Chapter 1). During most of the nineteenth century, traditional cocoa-containing foods (e.g. chocolate) have been considered indulgent and presumed to have negative health consequences because most cocoa-containing foods are rich in sugar, fat and calories. However, an emerging body of evidence now suggests that many cocoa products may actually have health benefits because of their high concentrations of flavonoids (see Chapter 7) with diverse biological activities.

The majority of scientific research has focused on cocoa's ability to alter pathological processes involved in the development of cardiovascular disease (CVD). Short-term intervention trials indicate positive changes in vascular endothelial function, insulin sensitivity and possibly blood pressure. There are no long-term intervention studies addressing the health benefits of cocoa consumption, but epidemiological and ecological evidence also support beneficial effects on CVD. While scientists and clinicians are not ready to recommend cocoa for the treatment or prevention of any disease, current evidence indicates that cocoa-containing foods can certainly be part of a healthy diet.

Teas, Cocoa and Coffee: Plant Secondary Metabolites and Health, First Edition.
Edited by Professor Alan Crozier, Professor Hiroshi Ashihara and Professor F. Tomás Barberán.
© 2012 Blackwell Publishing Ltd. Published 2012 by Blackwell Publishing Ltd.

8.2 Composition of cocoa products

Cocoa is produced from *Theobroma cacao*, a tree that is native to the tropical region of South America. The majority of the crop is now grown in Africa and South America. The ripe fruits, known as pods, contain 30–40 seeds surrounded by a sweet pulp. The initial processing includes fermentation of the pulp. During this time, the enzyme polyphenol oxidase converts some of naturally occurring phytochemicals to larger insoluble compounds. The seeds or 'beans' are then dried and sold to producers. In the processing of cocoa powder, the beans are roasted and the outer shell, known as the hull, is separated. What remains is referred to as cocoa liquor, which contains the flavonoids and approximately 55% fat (i.e. cocoa butter). Cocoa liquor is pressed to remove much of the fat during the production of cocoa powder, which contains approximately 12% fat. Dark chocolate refers to a solid confection containing variable amounts of cocoa liquor or powder, cocoa butter and sugar. Varying amounts of milk or milk solids are added to some chocolate formulations to create milk chocolate. The manufacturing of commercial chocolate from the nibs (beans after removal of the hull) includes nib grinding, alkalinising (in Dutch processing) and conching (heating in the presence of friction to promote smoothness) (Cooper *et al.* 2007b).

Typical concentrations of polyphenols in commercial cocoa-containing products are shown in Figure 8.1 (Vinson *et al.* 2006), and structures of common constituents in cocoa are shown in Figure 8.2. The polyphenols in chocolate are from the subclass of flavonoids known as flavan-3-ols. Flavan-3-ols (also called flavanols or catechins) occur

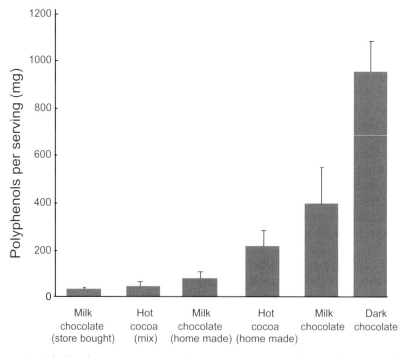

Figure 8.1 Polyphenol concentrations in products containing cocoa. (Based on Vinson *et al.* 2006.)

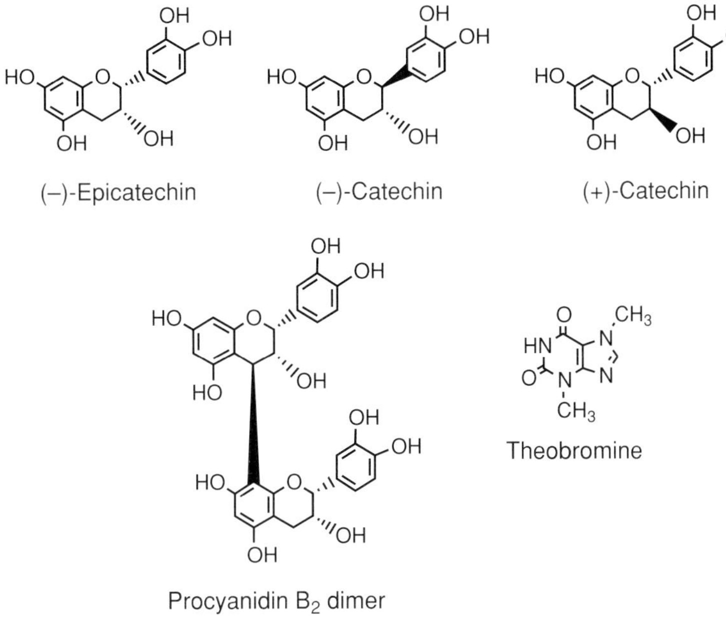

Figure 8.2 Structures of key phytochemical constituents of cocoa.

as monomers, oligomers and polymers in the human diet. Oligomers and polymers are called proanthocyanidins, procyanidins and condensed tannins, and are made up of (+)-catechin or (−)-epicatechin units. The main monomeric flavan-3-ol in cocoa is (−)-epicatechin. The oligomers contain between two and ten monomers connected at the 4 and 8 positions or the 4 and 6 positions of the monomers (see Chapter 5). Unlike most other types of flavonoids, flavan-3-ols are generally not glycosylated (Wollgast and Anklam 2000). Small amounts of flavonol glycosides such as quercetin-3-arabinoside are also present in cocoa. Although most of the purported health effects of cocoa are thought to be due to the flavan-3-ol components, cocoa also contains the methylxanthines, theobromine and caffeine. Despite similarities between theobromine and caffeine, theobromine appears to have little influence on mood and vigilance even at the fairly large doses present in cocoa and chocolate (Judelson et al. 2010). The biological effects of caffeine have been well studied (see Chapter 2), but are unlikely to be clinically very significant because of their low concentrations in chocolate/cocoa in comparison to other sources such as tea and coffee (Donovan and DeVane 2001).

Like most fruit crops, the phytochemical composition of the cocoa beans varies based on cultivar (i.e. genetics), as well as location and seasonal changes in environment. In addition to the choice of particular plant, variable processing techniques can dramatically alter the composition of the final product. The flavonoid content of the fresh cocoa bean is approximately 6–8% of its weight, but all of the processes involved in the manufacture of cocoa and chocolates reduce and/or alter the flavonoid content and chemical composition of the final product (Tomas-Barberán et al. 2007). For example, although chocolate contains the monomers (−)-epicatechin and (−)-catechin, the *T. cacao* plant produces very little (−)-catechin (Gotti et al. 2006). The catechin that is present in processed cocoa products is not the naturally occurring enantiomer (+)-catechin, but (−)-catechin (Figure 8.2) which is thought to be formed during cocoa processing. Because of the variability in chemical composition between processed cocoas, it is important to note that the flavan-3-ol content of the individual products used in studies discussed in this chapter will undoubtedly have varied markedly in composition, and this may have resulted in or explain differences in biological activity.

8.3 Worldwide consumption of cocoa and its contribution to flavonoid intake

The International Cocoa Organization (ICCO) reported that the world production of cocoa beans reached almost 3.7 million tonnes in 2007/2008, an increase of approximately 10% compared with the previous year. Germany, Belgium, Switzerland and the United Kingdom are among the highest consumers (nearly 10 kg per capita). Western Europeans, in general, consume the largest share of chocolate (45%) compared to those in the United States, who consume 20% of the chocolate that is available globally. In the United States the average consumption of chocolate is ∼5 kg per year. Although the ICCO has not reported the specific amount of cocoa consumed by each country, Western Europeans may be consuming even higher quantities of cocoa in comparison to those in the United States because they favour chocolate with higher cocoa concentrations (e.g. dark chocolate) than Americans, who tend to prefer chocolate with lower concentrations of cocoa (e.g. milk chocolate).

Chocolate and cocoa products can contribute substantially to the dietary intake of polyphenols/flavan-3-ols. Vinson *et al.* (2006) estimated an intake of approximately 100 mg per person of polyphenols per day in the United States based on the consumption of cocoa liquor equivalents. Another study, based on self-reported food recall data, reported that cocoa did not contribute substantially to the overall intake of flavan-3-ol monomers in the United States, which was estimated to be 187 mg per day (Chun *et al.* 2007). In the Netherlands, chocolate was reported to be a significant source of dietary flavan-3-ols, contributing to 20% of the total intake (Arts *et al.* 1999). In children, cocoa may be a more substantial source of flavan-3-ols in the diet in countries where chocolate milk and cocoa are consumed regularly (Lamuela-Raventos *et al.* 2005). Clearly, the dietary intake of flavonoids from cocoa varies greatly and some people are known to regularly consume more than five servings of chocolate or cocoa-containing foods each week. These individuals have been referred to as 'chocoholics' (Perkel 2007; Triche *et al.* 2008).

8.4 Epidemiological and ecological studies of cocoa

Ecological studies indicate positive effects of cocoa consumption on blood pressure and CVD risk. The Kuna Islanders are a Native American population that live on the San Blas Islands off the Caribbean coast of Panama. Kuna Islanders consume very large quantities of a beverage they make from cocoa. Most of the islanders consume at least five cups of cocoa 'brews' per day, which has been estimated to provide ~900 mg of flavan-3-ols. This flavan-3-ol intake is thought to be the highest in the world and is around ten times the intake of flavan-3-ols than those living in mainland Panama. In Panama, like most Western countries, CVD is the leading cause of death (83 ± 1 age-adjusted deaths per 100,000) and cancer was second (68 ± 2). In contrast, the rate of CVD and cancer among those living in San Blas is much lower (9 ± 3) and (4 ± 4) respectively. Similarly, deaths attributed to type 2 diabetes mellitus are also more common on the mainland (24 ± 1) than in the San Blas Islands (7 ± 2). Kuna Islanders have a very low prevalence of hypertension despite a high sodium intake and their blood pressure does not appear to rise with age. Those who relocate to Panama from the Islands develop typical age-dependent increases in blood pressure and a higher prevalence of hypertension than those living on the island, and are thus not genetically protected from hypertension (Bayard et al. 2007). Although there are certainly other dietary and lifestyle differences that differ among those living on the Islands and in Panama, these data strongly support potential beneficial cardiometabolic effects of cocoa consumption.

Epidemiological studies also support reduction of CVD risk in those that consume higher amounts of cocoa. The Zutphen Elderly Study, a prospective study of 472 elderly men living in the Netherlands, showed that cocoa intake was inversely associated with 15-year CVD and all-cause mortality. The relative risk for men in the highest intake tertile was 0.50 ($P = 0.04$) for CVD and 0.53 ($P < 0.001$) for all-cause mortality (Buijsse et al. 2006). A cross-sectional analysis of blood pressure performed at baseline and after 5 years showed that the mean systolic blood pressure in the highest intake tertile of cocoa intake was 3.7 mm Hg lower than the lowest intake tertile ($P = 0.03$) (Buijsse et al. 2006). The same group more recently assessed 19,357 men and women aged 35–65 years, and evaluated their chocolate consumption for a period of at least 10 years. They found that those who ate the most chocolate (7.5 g per day) had a 39% lower risk of myocardial infarction (MI) and stroke than individuals who ate much less chocolate (1.7 g per day). The lower risk could be partly explained by blood pressure reduction observed in the study. The risk reduction was stronger for stroke than for MI, a finding that is in agreement with the fact that blood pressure is a higher risk factor for stroke than for MI (Buijsse et al. 2010).

Two studies performed in the United States found that chocolate consumption was either weakly or not related to CVD. In a prospective study of 34,489 women that had a 16-year follow-up period, chocolate consumption was associated with a significant reduction in mortality from CVD ($P = 0.02$) and stroke ($P = 0.01$). In multivariate analysis that included age and energy intake, both trends were no longer statistically significant (Mink et al. 2007). In the Nurses' Health Study of 80,082 women living in the United States, there was no significant association between chocolate bar intake and the incidence of CVD. The categories of consumption (almost never, once per month, once per week, and three to four times per week) had relative risks of 1.0, 1.15, 1.08 and 1.11 (Hu et al. 1999, 2000; Kris-Etherton et al. 2000). It should be noted that the United States has a lower standard for the amount of cocoa used in milk chocolate bars (at least 12% cocoa vs. 24% in the European Union) and that most epidemiological studies do not differentiate between milk and dark chocolate. The lack of differentiation between milk and dark chocolate and the reduced flavan-3-ol content in chocolate consumed in the United States may explain the discrepancy between findings of studies in Europe where chocolate generally contains more cocoa and dark chocolate is more commonly consumed.

The Stockholm Heart Epidemiology Study showed that chocolate consumption was associated with lower cardiac mortality in a dose-dependent manner in patients free of diabetes surviving their first MI. They assessed self-reported chocolate consumption among 1169 non-diabetic patients. When compared with those never eating chocolate, the multivariable-adjusted hazard ratios were 0.73, 0.56 and 0.34 for those consuming chocolate less than once per month, up to once per week and twice or more per week respectively. Of interest, other sweets were not associated with cardiac or total mortality (Janszky et al. 2009). The most recent prospective study demonstrated an inverse relationship between moderate consumption of chocolate and the incidence of hospitalisation or death due to heart failure. They studied chocolate intake of 31,823 Swedish women aged 48–83 years (Mostofsky et al. 2010). Over 9 years, 419 women were hospitalised for incident heart failure ($n = 379$) or died of heart failure ($n = 40$). Compared with no regular chocolate intake, the multivariable-adjusted rate ratio of heart failure was 0.74 for women consuming one to three servings of chocolate per month, 0.68 for those consuming one to two servings per week, 1.09 for those consuming three to six servings per week and 1.23 for those consuming one or more servings per day ($P = 0.0005$ for quadratic trend). The investigators concluded that moderate habitual chocolate intake in women was associated with a lower rate of hospitalisation or death due to heart failure, but the protective association was not observed with intake of one or more servings per day.

8.5 Cocoa effects on vascular endothelial function and platelet activity

There is strong and consistent evidence that dark chocolate and cocoa have beneficial short-term effects on endothelial function in healthy volunteers free of disease, patients with CVD risk factors and subjects with coronary artery disease (CAD). Optimal vascular endothelial function can improve the overall integrity of the cardiovascular system. The severity of endothelial dysfunction is correlated with initial and reoccurring cardiovascular events (Vita and Keaney, Jr. 2002; Widlansky et al. 2003). CVD risk factors such as obesity, dyslipidaemia, diabetes, hypertension, smoking and age are all independently associated with reduced endothelium-dependent vasodilation (Steinberg et al. 1996). After adjustment for known risk factors for CAD, impaired epicardial coronary vasoreactivity to a cold pressor test was associated with the risk of developing cardiovascular events ($P = 0.040$) (Schindler et al. 2003).

Mechanisms by which cocoa may mediate vascular endothelial function are shown in Figure 8.3. In healthy adults, daily intake of high flavan-3-ol dark chocolate for 2 weeks resulted in improved endothelium-dependent flow-mediated dilation (FMD) of the brachial artery compared to low-flavan-3-ol chocolate (Engler et al. 2004). In a 3-hour randomised, single-blind, sham-controlled, crossover study with young healthy volunteers, dark chocolate led to a significant increase in resting and hyperaemic

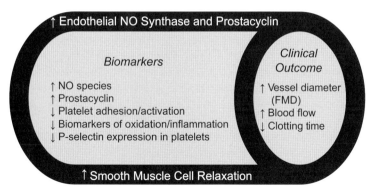

Figure 8.3 Mechanisms by which cocoa mediates vascular endothelial function.

brachial artery diameter throughout the study and FMD increased significantly at 1 hour (Vlachopoulos *et al.* 2005). They also reported a decrease in the aortic augmentation index, indicating a decrease in wave reflections, an independent marker of CVD risk. The carotid–femoral pulse velocity was also measured as a marker of aortic stiffness and CVD risk, but this was not significantly altered by the dark chocolate. A later cross-sectional study by the same group, however, showed significant improvements in aortic stiffness, wave reflections, and peripheral and central haemodynamics in healthy subjects that habitually consumed cocoa-containing products (Vlachopoulos *et al.* 2007). A landmark study demonstrated that the main flavan-3-ol in chocolate, (−)-epicatechin, was responsible for the vascular endothelial effects of cocoa consumption (Schroeter *et al.* 2006). This trial showed that purified (−)-epicatechin and a cocoa product containing the same dosage of (−)-epicatechin had similar effects on FMD.

Patients with CVD and other health risk factors also appear to benefit from increased endothelial function after dark chocolate consumption. In smokers, 40 g of dark chocolate improved FMD in comparison to baseline (Hermann *et al.* 2006). Two hours after dark chocolate ingestion, shear stress-dependent platelet function also was reduced, whereas no significant effect was seen in the white chocolate group. The effect of dark chocolate on FMD lasted about 8 hours (Hermann *et al.* 2006). Grassi *et al.* (2005a) showed that dark chocolate improved FMD in a crossover trial of males with hypertension that were randomised to receive 100 g of dark chocolate or flavan-3-ol-free isocaloric white chocolate for a period of 7 days on each treatment. Improvements in vascular function have been seen in medicated diabetics (Balzer *et al.* 2008), postmenopausal women (Wang-Polagruto *et al.* 2006), smokers (Hermann *et al.* 2006), heart transplant patients (Flammer *et al.* 2007), as well as overweight and obese adults (Faridi *et al.* 2008). Davison *et al.* (2008) investigated the effects of the addition of high or low flavan-3-ol cocoa to regular exercise over a 12-week period in overweight and obese adults. Although the high flavan-3-ol intervention did not enhance the effects of exercise on body composition or fat oxidation during exercise, the high flavan-3-ol cocoa group had increased FMD acutely (2 hours post-dose) and chronically (over 12 weeks). An epidemiological study showed decreased risk of pre-eclampsia in pregnancy in women that reported consuming five or more servings of chocolate per week. The cord blood concentration of theobromine was negatively associated with pre-eclampsia, supporting the validity of the self-reported intake data (Triche et al. 2008). Whether the reduced risk of pre-eclampsia is due to the vasodilatory effects of cocoa deserves further consideration.

Nitric oxide (NO) release from the endothelium is an important mediator of vasomotor responses and vasodilation. The vasodilatory effects of flavan-3-ols in cocoa appear to be mediated by NO (Fisher et al. 2003). Pulse wave amplitude (an indicator of vasodilation) was measured on the finger in 27 healthy people with a plethysmograph, before and after 5 days of consumption of flavan-3-ol-rich cocoa (821 mg of flavan-3-ols per day). N-nitro-L-arginine methyl ester (L-NAME) was infused intravenously on day 1, before cocoa, and on day 5, after an acute ingestion of cocoa. Four days of flavan-3-ol-rich cocoa consumption induced consistent peripheral vasodilation. On day 5, pulse wave amplitude exhibited a large additional acute response to cocoa ($P = 0.01$), which was completely reversed by the administration of L-NAME, a specific inhibitor of NO synthase (Fisher et al. 2003). In a crossover study with cocoa that was rich or poor in flavan-3-ols, the flavan-3-ol-rich cocoa increased plasma levels of circulating NO species from 22 to 36 nmol/L after consumption. FMD also increased from 3.4% to 6.3% (Heiss et al. 2003). A study comparing responses in younger and older individuals suggested that the NO-dependent vascular effects of flavan-3-ol-rich cocoa may be greater among older people, in whom endothelial function is more disturbed (Fisher and Hollenberg 2006).

Platelets play an important role in the formation of infarction and stroke. NO, along with prostacyclin, controls platelet activation and function. Rein et al. were the first to show that cocoa consumption altered ADP- or epinephrine-stimulated primary haemostasis platelet microparticle formation. The results were compared to a caffeine (positive) and water (negative) control beverages (Rein et al. 2000b). Early work demonstrated a decrease in the plasma leukotriene/prostacyclin ratio ∼2 hours after chocolate consumption. The changes provide a mechanistic explanation for reductions in platelet activity (Schramm et al. 2001). The same group then demonstrated that the reductions in primary haemostasis correlated to the change in the prostacyclin/leukotriene ratio, indicating an overall antithrombotic eicosanoid response. The administration of 234-mg cocoa flavan-3-ols in a capsule resulted in lower P-selectin expression and lower ADP-induced platelet aggregation when compared to a placebo (Murphy et al. 2003). Pearson et al. compared the ex vivo acute effects of flavan-3-ol-rich cocoa (897 mg flavan-3-ols), aspirin (81 mg) or a combination of both treatments in a crossover study on inhibition of platelet function. The effects induced by cocoa were qualitatively similar to aspirin, but less profound (Pearson et al. 2002). Although the vast majority of studies report statistically significant improvements in vascular function, Farouque et al. (2006) reported that vascular function was unchanged after a 6-week treatment period with flavan-3-ol-rich cocoa.

While the beneficial vascular effects of cocoa and dark chocolate appear to be well proven, there is less convincing evidence that cocoa can lower blood pressure (see Section 8.6). Although vascular resistance is correlated with hypertension, vasodilators are not always anti-hypertensive (Fries 1960). While peripheral vasodilators can lead to short-term blood pressure reductions, the kidney is the dominant controller of long-term blood pressure (Cowley 1992). Unless the renal pressure threshold for maintaining sodium homeostasis is set at a lower level, long-term blood pressure will not change significantly. Thus, cocoa's vasodilatory effects will result only in sustained reductions in blood pressure if the renal threshold for pressure natriuresis is reset to a lower level.

8.6 Cocoa and hypertension

Hypertension is a prevalent condition affecting approximately 65 million individuals in the United States and approximately 1 billion people worldwide (i.e. one-fourth of the adult population) (Kearney *et al.* 2005). Hypertension is an established risk factor for CVD, and hypertensive individuals are at greater risk of developing type 2 diabetes. There have been 13 intervention studies that measured the effects of cocoa or chocolate on blood pressure (Figure 8.4). Collectively, these reports have not reached a consensus on whether cocoa/chocolate is able to lower blood pressure. The varying effects observed in studies may be due to blood pressure measurement devices and techniques, cocoa product composition and intervention paradigms (Egan *et al.* 2010).

There are inherent difficulties in developing appropriate placebo products for foods as many biologically active dietary components are key components of the flavour and texture of food. Most of the flavan-3-ols in cocoa are dark coloured and impart the

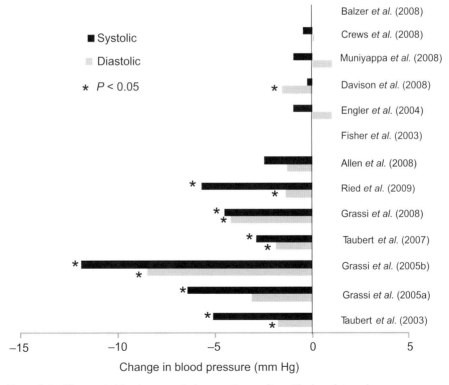

Figure 8.4 Changes in blood pressure in intervention studies with chocolate and cocoa.

characteristic bitter flavour and astringency of dark chocolate. Studies that support nutritional interventions are often impossible to double blind as subjects are able to perceive the removal/addition of many components from foods. It is important to note that dietary-induced changes in weight, plasma lipid concentrations and even blood pressure from randomised open-label crossover designs have been considered sufficient to change dietary recommendations (Chobanian *et al.* 2003; Azadbakht *et al.* 2005). Thus, some researchers have used an open-label study design to study cocoa's effects on blood pressure, while others have blinded the cocoa or chocolate to the best of their ability.

In six of seven open-label studies, cocoa significantly lowered both the systolic and diastolic blood pressure. Although changes were modest, the mean decrease in diastolic ranged from 3 to 12 mm Hg; mean decrease in systolic ranged from 2 to 9 mm Hg. The decrease in systolic was generally greater than the reduction in diastolic. These reductions were in the range of those observed with dietary changes such as the 'Dietary Approach to Stop Hypertension' (DASH) diet (Azadbakht *et al.* 2005).

Most randomised double-blind studies, on the other hand, have not reported measurable effects of cocoa or chocolate consumption on blood pressure (Fisher *et al.* 2003; Engler *et al.* 2004; Fisher and Hollenberg 2006; Allen *et al.* 2008; Balzer *et al.* 2008; Crews *et al.* 2008; Davison *et al.* 2008; Faridi *et al.* 2008; Muniyappa *et al.* 2008). A recent double-blind study demonstrated that cocoa attenuated exercise-induced increases in blood pressure (Berry *et al.* 2010).

One potential explanation for the lack of measurable effect in some of the randomised double-blind trials is the lack of appropriate instrumentation for blood pressure measurement. While the open-label studies used instruments that received either an A or B rating by the British Hypertension Society Protocol (Chobanian *et al.* 2003), three of these studies used instruments that received a 'C' rating and one study did not describe the instrument used to measure blood pressure. Two of the studies reported using appropriate blood pressure measurement methodology (Crews *et al.* 2008; Muniyappa *et al.* 2008). Most of the open-label studies had more accurate instrumentation and methodology for blood pressure measurement (Taubert *et al.* 2003; Grassi *et al.* 2005a, 2005b; Taubert *et al.* 2007a, 2007b; Grassi *et al.* 2008; Ried *et al.* 2009).

Another possible reason for the lack of measurable effects in the randomised trials is the composition of the chocolate or cocoa product used in the study. The variable effect of dark chocolate/cocoa may reflect differences in the content of flavan-3-ols or possibly other phytochemicals that lower blood pressure and others that raise blood pressure, for example, fivefold differences in the epicatechin/catechin ratio observed between some chocolate/cocoa samples (Miller *et al.* 2009). Although most of the trials report the total cocoa content of the product and the amounts of monomers and some oligomers, none have reported the amounts of specific enantiomers of the monomers in the samples used in the trials.

In vitro studies have not provided convincing evidence that a particular component is responsible for the blood pressure effects, as these studies have not accounted for flavan-3-ol metabolism and have not studied flavan-3-ol concentrations that exist *in vivo*, which are on the order of 1 μM or less. Flavan-3-ols are also present in the circulatory system almost exclusively as conjugated metabolites after consumption from most foods (see Chapter 5). These metabolites are not readily available, so meaningful *in vitro* studies are difficult to conduct. For example, angiotensin converting enzyme (ACE) inhibitors are effective anti-hypertensive agents, and one study reported that chocolate and certain flavan-3-ols inhibited ACE activity *in vitro* (Ottaviani *et al.* 2006). The lowest concentrations studied were at least an order of magnitude higher than would be expected to occur *in vivo* after chocolate consumption, and only parent unmetabolised flavan-3-ols were investigated.

A recent study demonstrated blood pressure-lowering effects of (−)-epicatechin in an animal model of hypertension induced by NO synthase inhibition. The addition of (−)-epicatechin to the animal diet modulated the blood pressure in a dose-dependent manner and also resulted in increases in plasma NO concentrations (Jaggers *et al.* 2010). While (−)-epicatechin may contribute to or even be responsible for the blood pressure-lowering effects of chocolate consumption, there is no apparent correlation between the dosages of (−)-epicatechin and whether the particular chocolate used lowers blood pressure. From the available data, one might speculate that the at least one of component(s) involved in the changes in blood pressure varies greatly between chocolate samples.

One highly variable component of cocoa/chocolate is (−)-catechin (Cooper *et al.* 2007a). This flavan-3-ol is not present in most other foods, but is thought to be formed from (−)-epicatechin during processing. Although (+)-catechin is not thought to have significant vascular effects, the two enantiomers of catechin showed opposite effects on glycogen metabolism in isolated rat hepatocytes. Since (−)-epicatechin reduced blood pressure in an animal model, it appears plausible that (−)-catechin could counteract this effect. The stereochemical properties of monomeric flavan-3-ols influenced vasodilation in humans fed purified forms of the enantiomers of both catechin and epicatechin. (−)-Epicatechin was found to be the only stereoisomer capable of mediating a significant arterial dilation response in rats, but none of the components had a measurable effect on blood pressure (Ottaviani *et al.* 2011). There is no information on the contribution of the individual stereoisomers to cocoa's effect on blood pressure in humans, but this should be studied further. If differences in product composition are responsible for the lack of blood pressure effects observed in some studies, cocoa processing techniques could be optimised to ensure that market products contain the requisite doses of specific components needed to achieve the desired biological effect.

8.7 Antioxidant and anti-inflammatory effects of cocoa

Many of the health benefits associated with chocolate and other flavan-3-ol-rich foods have been associated with their activity as antioxidants and anti-inflammatory agents (Sies *et al.* 2005). Oxidative stress and associated inflammation are related to the cause and progression of a wide variety of disease processes, including CVD, some forms of cancer and neurocognitive diseases. A summary of clinical intervention trials and epidemiological studies that reported changes in biomarkers associated with oxidation and inflammation status is provided in Table 8.1.

A biomarker of oxidation that has received significant attention is the oxidation of the low-density lipoprotein cholesterol (LDL-C) particle. The oxidation of the LDL-C particle is a critical step in the progression of atherosclerosis. In 1996, Waterhouse

Table 8.1 Changes in markers of oxidation or inflammation status in humans after cocoa product consumption

Citation	Population	Duration	Antioxidant marker	Inflammatory marker
Rein et al. (2000)	Healthy	2 h	↑TRAP, ↓TBARS	NM
Wang et al. (2000)	Healthy	2 h	↑TRAP, ↓TBARS NS: F2-isoprostanes	NM
Schramm et al. (2001)	Healthy	2 h	NM	↓Prostacyclin/leukotriene
Wan et al. (2001)	Healthy	4 wk	↓LDL-C oxidation, ↑ORAC	NS: thromboxane B2, 6-keto-prostaglandin F1α
Osakabe et al. (2001)	Healthy	2 wk	↓LDL-C oxidation NS: α-tocopherol, ascorbic acid, ß-carotene	NM
Mathur et al. (2002)	Healthy	6 wk	↓LDL-C oxidation NS: ORAC or F2-isoprostanes	NS: cytokines, CRP
Holt et al. (2002)	Healthy	2 h	NM	↓Prostacyclin/leukotriene
Serafini et al. (2003)	Healthy	1 h	↑FRAP	NM
Murphy et al. (2003)	Healthy	28 d	↑Ascorbic acid NS: TBARS, F2-isoprostanes, tocopherols	NM
Engler et al. (2004)	Healthy	2 wk	NS: F2-isoprostanes, ORAC, LDL-C oxidation	NM
Wiswedel et al. (2004)	Healthy	2 or 4 h (cocoa + exercise)	↓F2-isoprostanes NS:TBARS, TAC, MDA, ascorbic acid	NM
Grassi et al. (2005, 2010)	Hypertensive	14 d	NM	NS: CRP, ICAM-1
Vlachopoulos et al. (2005)	Healthy	3 h	NS: TAC, MDA	NM
Fraga et al. (2005)	Healthy athletes	14 d	↑ORAC ↓8-isoprostane, ↓LDL-C oxidation	NM
Farouque et al. (2006)	CVD patients	6 wk	NS: LDL-C oxidation	NS: P,C-selectins, I-CAM, V-CAM
Kurlandsky and Stote (2006)	Healthy females	6 wk	NM	↓ I-CAM

Table 8.1 (Continued)

Citation	Population	Duration	Antioxidant marker	Inflammatory marker
Wang-Polagruto et al. (2006)	Hypercholesterolaemic, post-menopausal females	6 wk	NM	↓VCAM-1 NS: P,C-selectins
Heinrich et al. (2006)	Healthy females	12 wk	NM	UV-induced erythema
Hermann et al. (2006)	Smokers	2 h	↑TAC	NM
Baba et al. 2007	Healthy males	4 wk	↓LCL-C oxidation	NM
Flammer et al. (2007)	Heart transplant patients	2 h	↓F2-isprostanes, ↑TRAP, ↑FRAP	NM
Taubert et al. (2007)	Hypertensive	18 wk	NS: F2-isoprostanes	NM
di Giuseppe et al. (2008)	Healthy chocolate consumers	1 serving/3 d	NM	↓CRP
Crews et al. (2008)	Healthy elderly	6 wk	NM	NS: CRP
Hamed et al. (2008)	Healthy	1 wk	NM	↓CRP women only
Allen et al. (2008)	Hypercholesterolaemic	8 wk	NM	NS: CRP, ICAM-1, sCD40L
Monagas et al. (2009)	CVD patients	4 wk	NM	↓P-selectin, I-CAM-1; NS: V-CAM-1, MCP-1, IL-6, CRP

NS, not significant; NM, not measured.

et al. reported that flavonoid-rich extracts of chocolatesamples inhibited the oxidation of LCL-C *in vitro*. This was followed by numerous reports that demonstrated reduced markers oxidation of LDL-C in healthy individuals after cocoa or chocolate consumption (Osakabe et al. 2001; Wan et al. 2001; Fraga et al. 2005; Baba et al. 2007). One study with healthy individuals was unable to show significant effects (Engler et al. 2004), and significant effects were not observed in CVD patients (Farouque et al. 2006).

Other studies have focused on the 'oxidation status' or 'free radical scavenging capacity' of plasma as a marker of antioxidant capacity. Common measures are the oxygen radical absorbance capacity (ORAC), total radical-trapping antioxidant parameter (TRAP), ferric-reducing antioxidant power (FRAP) and total antioxidant status (TAS). The formation of thiobarbituric acid reactive substances (TBARS), such as malondialdehyde (MDA), is thought to be inversely correlated to plasma antioxidant capacity. A critical review of these methods has been published by Huang et al. (2005). Rein et al. (2000a) were the first to show that consumption of cocoa increased the plasma TRAP value and lowered plasma TBARS, and this was followed by the numerous studies listed in Table 7.1. While the majority of studies showed that consumption of cocoa had positive effects on markers of plasma antioxidant activity, four studies reported effects that were not significant.

Pro-inflammatory cytokines such as tumour necrosis factor α (TNF-α) and interleukin 6 (IL-6) are transcription factors expressed by the vascular endothelium, macrophages and lymphocytes. Mao *et al.* (2000) reported cocoa-mediated inhibition of human cytokine transcription and secretion. While few other studies have focused on transcription of inflammatory cytokines, other investigations have centred on the downstream expression of intercellular adhesion molecules that are mobilised during an inflammatory response. These include P- and C-selectins along with intercellular adhesion molecule 1 (ICAM-1) and vascular cell adhesion molecule 1(VCAM-1). C-reactive protein (CRP) is synthesised in the liver and released into the bloodstream in response to inflammation. There are clear discrepancies among studies as to whether consumption of cocoa has any significant effects on the expression of these downstream biomarkers of inflammation.

The plasma prostacyclin/leukotriene ratio is considered a measure of the anti-inflammatory to pro-inflammatory eicosanoid balance. Schramm *et al.* (2001) demonstrated a significant reduction in the leukotriene/prostacyclin ratio in human plasma and aortic endothelial cells and were among the first to suggest that cocoa mediates some of its effects by modulation of endothelial cell eicosanoid synthesis. The F2 isoprostanes are prostaglandin-like compounds formed *in vivo* by a non-enzymatic mechanism involving the free radical-initiated oxidation of arachidonic acid that have been considered markers of oxidative injury. Milk chocolate produced beneficial effects on blood pressure, plasma cholesterol and the level of F2 isoprostanes in young exercising males (Wiswedel *et al.* 2004). Schewe *et al.* (2002) demonstrated that (−)-epicatechin inhibited mammalian 5-lipoxygenase activity *in vitro* but at concentrations greater than would be expected to occur after cocoa consumption (i.e. $>20\ \mu M$). To our knowledge, the inhibition of 5-lipoxygenase activity has not been shown to occur *in vivo* after cocoa consumption. Whether the conflicting results in the direction of changes in oxidation and inflammation biomarkers among the human studies listed below are due to the type of cocoa intervention, treatment duration or study population is not clear. The intimate relationship between mechanisms involved in oxidation and inflammation lend credence to the opinion that a simple antioxidant mechanism is not likely, but that effects are achieved by alteration of one or more specific pathways that reduce pathogenic events that lead to a chronic disease state (Cooper *et al.* 2007b).

8.8 Effects of cocoa consumption on lipid and lipoprotein metabolism

High plasma triglycerides and LDL-C and low HDL-C are significant risk factors for CAD. The INTERHEART study reported that those with abnormal concentrations of blood lipids have three times the risk of a heart attack compared to those with normal concentrations (Yusuf *et al.* 2004). The fat in chocolate comes from cocoa butter and consists of approximately equal amounts of oleic acid (a monounsaturated fat) and stearic and plamitic acids (both saturated fats). The consumption of excess fats, especially saturated fatty acids, has raised concerns that consumption of large amounts of cocoa/chocolate may have negative effects on blood lipids.

An early study showed that substitution of a milk chocolate bar for a high-carbohydrate snack did not adversely affect the LDL-C response in individuals on a recommended American Heart Association Step 1 Diet. The lack of change in LDL concentrations was despite an increase in total fat and saturated fatty acid content of the diet (Kris-Etherton *et al.* 1994). It has been suggested that the particular fatty acids in cocoa have a neutral effect on blood cholesterol levels (Kris-Etherton *et al.* 1993, 2005) and that even if these fatty acids altered blood lipid concentrations, cocoa/chocolate contributes to a small proportion of saturated fat intake in the overall diet and should not significantly alter blood lipid concentrations when eaten in moderation (Kris-Etherton *et al.* 2000).

Interestingly, some cocoa/chocolate intervention studies have shown positive effects on blood lipid and lipoprotein concentrations. Many of the studies used white chocolate as a control (which presumably contains similar amounts of fatty acids as the treatment group), which indicates that beneficial effects are due to the flavonoid components of the cocoa/chocolate. Grassi *et al.* (2005b) observed favourable effects on serum LDL-C concentrations in subjects consuming dark chocolate in comparison to white chocolate. Kurlandsky and Stote (2006) showed that a 6-week treatment period with 41 g per day of chocolate along with a low-fat diet reduced serum triglycerides by 21%, while the control group that consumed the low-fat diet only had a reduction of 11%.

An increase in HDL-C concentrations was observed in subjects that supplemented their 'standard American diet' for 4 weeks with cocoa and chocolate in comparison to those who did not (Wan *et al.* 2001). Monagas *et al.* (2009) also observed an increase in HDL-C concentrations in older volunteers with CVD risk factors during a 4-week period. Baba *et al.* (2007) showed a reduction in LDL-C and an increase in HDL-C in hypercholesterolaemic individuals after three different levels of cocoa powder. Cocoa bars enriched with phytosterols were shown to lower both total and LDL-C (Polagruto *et al.* 2006). Most recently, a 1-week treatment period with a large quantity (100 g per day) of dark chocolate decreased LDL-C by 6%, while HDL-C concentrations rose by 9% (Hamed *et al.* 2008).

The limitation of most of the intervention studies performed to date is that they had a restricted number of participants and were conducted over long periods (6 weeks maximum). Nevertheless, a recent meta-analysis of eight randomised controlled trials, which included a total of 215 subjects, showed that cocoa consumption did significantly reduce blood cholesterol levels and concluded the likelihood of a beneficial effect on the overall lipid and lipoprotein profile (Jia *et al.* 2010).

8.9 Cocoa effects on insulin sensitivity

Insulin resistance is part of the cluster of risk factors in the metabolic syndrome. When hepatic, muscle and fat cells fail to respond adequately, blood sugar concentrations rise. Insulin resistance increased risk of type 2 diabetes mellitus. Although chocolate and cocoa-containing foods contain large quantities of sugars, studies on the glycaemic index of foods indicate that chocolate bars elicit relatively low levels of postprandial glycaemia compared with the consumption of similar amounts of carbohydrate in starchy foods, such as bread or potatoes (Foster-Powell *et al.* 2002). The lower glycaemic index of chocolate bars has been partially attributed to its content of sucrose, which elicits less of a glycaemic response than glucose or fructose.

Brand-Miller *et al.* (2003) studied six pairs of common foods (i.e. 12 foods in total) that may use chocolate as an adjunctive flavour (e.g. vanilla pudding *vs.* chocolate pudding and sweetened strawberry-flavoured milk *vs.* sweetened chocolate-flavoured milk). They found that the mean insulin/glucose area under the concentration *vs.* time curve (AUC) ratio (a measure of insulinaemia relative to glycaemia) was 30% greater for the chocolate-flavoured version of each pair of foods. However, none of the individual food pairs differed significantly. The lack of significant differences observed for the specific pairs of foods may have been due to the small number of individuals enrolled in the study ($n = 11$) in consideration of the variability of insulin and glycaemic responses among individuals. This was the first study to demonstrate that chocolate (cocoa powder) has a specific insulinotropic effect, irrespective of food source or the overall macronutrient composition of the food. Grassi *et al.* (2005a) performed oral glucose tolerance tests in a single-blind crossover trial using males and females with hypertension who were randomised to receive 100 g of dark chocolate or flavan-3-ol-free isocaloric white chocolate for a period of 7 days on each treatment. They showed that the dark chocolate decreased the homeostasis model assessment of insulin resistance (HOMA-IR; $P < 0.0001$), quantitative insulin sensitivity check index (QUICKI) and insulin sensitivity index (ISI) calculated from the oral glucose tolerance test values. Davison *et al.* (2008) showed that addition of high flavan-3-ol cocoa to regular exercise reduced the HOMA2 (a similar marker of insulin resistance) in comparison to low flavan-3-ol cocoa.

Insulin sensitivity is partly dependent on insulin-mediated NO release (Stuhlinger *et al.* 2002; Naruse *et al.* 2006). Muniyappa and co-workers assessed insulin sensitivity using the hyperinsulinaemic–isoglycaemic glucose clamp and the QUICKI after cocoa consumption in individuals with essential hypertension. They found that although cocoa consumption for 2 weeks enhances insulin-mediated vasodilatation, they did not observe any significant differences in insulin sensitivity determined by glucose clamp or QUICKI. Similarly, values for fasting plasma glucose and insulin were unchanged by cocoa treatment. Although a well-designed crossover study with state-of-the-art methods, the lack of significant effects may have been due to the small sample size ($n = 9$, cocoa treatment first) compared with placebo treatment first ($n = 11$) (Muniyappa *et al.* 2008). The effects of cocoa or its components on glucose homeostasis in individuals with hypertension or other risk factors associated with the metabolic syndrome merits further study.

8.10 Cocoa effects on cerebral blood flow and neurocognitive functioning

The peripheral antioxidant, anti-inflammatory and vascular effects of cocoa that appear to promote cardiovascular health may also provide neuroprotection and inhibit neurocognitive decline (Joseph *et al.* 1999). The beneficial neurocognitive effects of products such as blueberries and *Ginkgo biloba* extract are thought to be attributed to their high concentrations of flavonoids (Mix 2002; Willis *et al.* 2005). There are a limited number of studies that examined potential neurocognitive effects of cocoa, but there are reports in both young cognitively intact people and older individuals that appear to have suffered some form of cognitive decline.

The decline in neurocognitive function in ageing is thought to be accompanied by a decrease in cerebral blood flow (Dinges 2006). Impaired cerebral perfusion is also associated with increased risk of stroke, dementia and cognitive decline. Nagahama *et al.* (2003) utilised single-photon emission tomography to evaluate the relationship between regional cerebral blood flow and cognitive decline in Alzheimer's disease patients. Their findings revealed that reduced cerebral perfusion in the right posterodorsal, anterior and superior prefrontal cortices and the inferior parietal cortex correlated significantly with rapid deterioration in the Mini-Mental State Examination. Reduced cerebral blood flow in these regions also predicted a more rapidly progressive cognitive decline in the patients with this cerebral pattern. Assessment of the effects of cocoa on cerebral perfusion have been undertaken by numerous scientists as a potential agent in the prevention of pathological conditions that are associated with decreased brain perfusion. Francis *et al.* (2006) used a technique known as blood oxygenation level-dependent (BOLD) functional magnetic resonance imaging (fMRI) to measure changes in middle cerebral arterial oxygenation and blood flow in young, healthy participants in response to a task-switching cognitive challenge. In this study, subjects were studied at baseline and then again after drinking a cocoa beverage for 5 days. They showed for the first time that cocoa increased cerebral blood flow and suggested that cocoa or its active components could be useful in prevention or treatment of dementia or ischaemic stroke. Interestingly, there were no differences in the functional performance in the task-switching challenge between the two treatment groups. The authors hypothesised that the lack of measurable changes were due to the high level of functioning in the young, healthy participants, who performed near optimally in the task-switching challenge. The same research group also investigated the acute and short-term effects of cocoa consumption on cerebral blood flow in older, healthy volunteers.

Transcranial Doppler ultrasound technology was used to measure mean blood flow velocity in the middle cerebral artery of subjects who were given daily supplementation for 2 weeks of either flavan-3-ol-rich or flavan-3-ol-poor cocoa. The study demonstrated that 2 weeks of daily flavan-3-ol-rich cocoa consumption, providing 900 mg of flavan-3-ols per day, resulted in a significant increase in peak cerebral blood flow response in the middle cerebral artery. Favourable responses as evidenced by increased cerebral blood flow were also demonstrated after 1 week of daily flavan-3-ol-rich cocoa consumption, though a statistically significant difference was not demonstrated. Given the demonstrated increase in cerebral blood flow associated with the dietary intake of flavan-3-ol-rich cocoa, the authors suggest a promising role for regular cocoa flavan-3-ol consumption in the treatment of cerebrovascular ischaemic syndromes, including dementias and stroke (Sorond *et al.* 2008). More recently, the same research group designed a study to compare velocity changes to flow changes as measured by using two specific MRI techniques known as perfusion MRI and arterial spin labelling, using flavan-3-ol-rich cocoa to induce cerebral blood flow changes in healthy volunteers.

All 20 subjects underwent daily standardised cocoa consumption for 7–14 days, and cerebral blood flow velocity was evaluated with transcranial Doppler ultrasound at baseline and following cocoa consumption. Fourteen of the subjects were evaluated at baseline and following cocoa supplementation with perfusion MRI with gadolinium. The remaining six of the subjects underwent arterial spin labelling and MRI at baseline and following the cocoa supplementation. The results of this study demonstrated that changes in the flow velocity in the middle cerebral artery associated with drinking cocoa were highly correlated with changes in cerebral blood flow using perfusion MRI and arterial spin labelling. Additionally, they report that the data validate the use of transcranial Doppler ultrasound as a representative measure of the assessment of cerebral blood flow (Sorond *et al.* 2010).

Despite the apparent increase in cerebral blood flow, significant changes in actual cognitive performance due to cocoa consumption have not been documented in humans, young or old. Crews *et al.* (2008) failed to document any significant differences in performance on memory and intelligence tests with elderly ($\geq$60-year-old) men and women in a crossover study after a 6-week treatment period with a dark chocolate bar and cocoa beverage or matched controls. Interestingly, this study also failed to provide any evidence of any peripheral vascular or inflammatory effects. Whether there is a true absence of changes in cognitive performance in people with documented peripheral vascular changes in response to cocoa treatment remains to be determined.

The cognitive effects of flavonoid-rich chocolate, wine and tea were evaluated in 2031 people 70–74 years old, who consumed one or a combination of chocolate, wine and tea (Nurk *et al.* 2009). Each participant underwent a cognitive test battery that included the Kendrick Object Learning Test, Trail Making Test, part A (TMT-A), modified versions of the Digit Symbol Test, Block Design, Mini-Mental State Examination and Controlled Oral Word Association Test. The results of this study demonstrated that those who consumed chocolate, wine or tea had significantly better mean test scores and lower prevalence of poor cognitive performance than those that did not consume the products. Study subjects who consumed all three of the flavonoid-rich products had the best test scores and the lowest risk for poor test performance. A dose-dependent relationship was demonstrated between the intake of food and the cognitive performance. The authors conclude that a diet rich in flavonoid foods is associated with better performance in several cognitive abilities in the elderly and that this relationship exists in a dose-dependent manner.

Animal models support beneficial effects of cocoa on cognitive function and suggest changes in cell signalling and inflammation in neurons. One study in a rodent model showed that cocoa prevented the inflammatory responses in trigeminal ganglion neurons. Mitogen-activated kinase (MAP) phosphatases MKP-1 and MKP-3 were elevated in neurons. The stimulatory effects of peripheral inflammation on neuronal expression of the MAPK p38 and extracellular signal-regulated kinases (ERK) were significantly repressed in response to cocoa. Cocoa also suppressed basal neuronal expression of calcitonin gene-related peptide (CGRP) and stimulated levels of the inducible form of nitric oxide synthase (iNOS). The authors suggested that because these proteins are implicated in the underlying pathology of migraine and temporomandibular joint disorder (TMJ), cocoa could be beneficial in their treatment (Cady and Durham 2010).

Cognitive performance in rodents stressed by heat exposure was improved after the administration of a cocoa extract. Animals treated with cocoa performed better in discriminating between an active and inactive lever in a light extinction paradigm and had decreased escape latencies in the Morris water maze (Rozan *et al.* 2007). Bisson *et al.* (2008) assessed cognitive performance in rats following a 1-year treatment period with a cocoa. They demonstrated that the cocoa extract improved cognitive performances in light extinction and water maze paradigms, increased lifespan and preserved high urinary free dopamine level.

The anxiolytic effects of cocoa were evaluated using the elevated T-maze test as an animal model of anxiety (Yamada *et al.* 2009). Two separate arms of the study were carried out to evaluate the effects of cocoa on both short- and long-term administration of cocoa. The first study involved the administration of cocoa to rats immediately prior to performance of the elevated T-maze. They found that the cocoa treatment reduced conditional fear-relating behaviour as evidenced by markedly reduced delayed avoidance latency. A longer term study, however, failed to show an increase in avoidance latency despite changes in brain serotonin turnover. The authors suggested that the cocoa may influence the mood control system, such as dopamine receptors. Although the study of the effects of cocoa on cognitive function is in its infancy and there are still few studies in humans, the extant data appear to support beneficial effects by multiple mechanisms.

8.11 Potential negative health effects of cocoa consumption

8.11.1 Obesity

Fat and sugar are major components of chocolate, and provide significant calories that need to be taken into account when assessing possible risks and benefits of recommending chocolate consumption for health purposes. Chocolate, especially of the milk variety, contains high amounts of sugar and calories and thus has possible implications for weight if eaten in large amounts. Cocoa itself is not high in sugar and could be processed in ways to maximise its potential benefits and reduce the amounts of sugar and calories.

8.11.2 Testicular health

An ecological study showed that the incidence of testicular cancers in 18 countries on five continents was correlated with the average per-capita consumption of cocoa (kg per capita per year) ($r = 0.859$; $P < 0.001$). An analogous significant correlation was also observed between early cocoa consumption and the prevalence rates of hypospadias ($r = 0.760$; $P < 0.001$). Although ecological studies limit causal inference, animal and human intervention studies have reported that cocoa and its main stimulant theobromine exert toxic effects on the testis, inducing testicular atrophy accompanied by aspermatogenesis or oligospermatogenesis, and high doses of cocoa may impair sperm quality (Tarka 1981; Gans 1984; Funabashi *et al.* 2000). More studies in human males need to be performed to clarify the potential negative effects of cocoa products on testicular health.

8.11.3 Acne

It is widely reported in the popular literature and believed by younger individuals that chocolate is a primary cause of acne lesions. A landmark study by Fulton *et al.* (1969) found no effect of a chocolate bar on the number or severity of acne lesions in teenagers when compared with a placebo bar. Although the methodology in this paper has recently been challenged (Mackie and Mackie 2007), there still appears to be no scientific evidence that chocolate causes acne.

8.11.4 Dental caries

Cocoa *per se* has not been associated with dental caries but most cocoa-containing products contain high amounts of sugars, which are the primary cause of the caries. One study showed that children that received 1 pint per day of milk with cocoa and sugar had a greater increase in caries after a 2-year period in comparison to groups that received milk only or milk with cocoa and an artificial sweetener (Dunning and Hodge 1971). The study suggests that associations between cocoa product consumption are likely to be due to the addition of sugar rather than any specific effect of the cocoa. Clearly, these risks can be reduced by brushing the teeth post-consumption, but this is not always common practice in school-age children.

8.12 Effects of consumption of cocoa with milk or other foods

Milk is often consumed with or as a part of a cocoa-containing food. Milk chocolate contains milk powder, liquid milk or condensed milk. Also a cocoa beverage may be made with either water or milk. Milk chocolate generally contains less cocoa liquor than dark chocolate and therefore generally contains less polyphenols; however, there are likely to be instances where a specific brand of milk chocolate will have used more cocoa than another brand of dark chocolate (Cooper *et al.* 2007a; Miller *et al.* 2009). In addition, since flavan-3-ols can degrade during certain processing techniques, a chocolate may contain 70% cocoa solids but have the same, or even a lower, content of flavan-3-ols as a milk chocolate that contains less cocoa.

Most studies have investigated dark chocolate or cocoa dissolved in water to avoid the possibility that milk might interfere with absorption or the biological activity under investigation. There are conflicting results on whether milk or other food matrix factors alter the bioavailability or metabolism of the flavan-3-ols in chocolate. Serafini *et al.* (2003) reported that the AUC of epicatechin metabolites was lower for milk chocolate in comparison to dark chocolate with identical epicatechin content. This group also demonstrated that the addition of milk to the cocoa beverage blunted the significant increase in total antioxidant capacity. Their hypothesis was that milk proteins bind to cocoa polyphenols, which in turn prevents their absorption in the gastrointestinal tract. Most subsequent studies have not found a reduction in epicatechin bioavailability when cocoa was consumed with milk (Schramm *et al.* 2003; Schroeter *et al.* 2003; Keogh *et al.* 2007; Roura *et al.* 2007, 2008; Neilson *et al.* 2009). Experimental differences, such as giving cocoa powder in a drink comparing either water or milk as a matrix, rather than as a solid confection, may be one explanation for the different findings.

Schramm *et al.* (2003) found that the addition of a carbohydrate source (sucrose or bread) increased the maximum plasma concentrations of epicatechin metabolites. Another study, however, compared the effects of cocoa with and without sugar on FMD. Although they both significantly improved FMD over placebo, the sugar-free cocoa resulted in the greatest improvement. The authors suggested that sugar content may attenuate the effects of cocoa on FMD and sugar-free preparations may augment them (Faridi *et al.* 2008). More recently, Mullen *et al.* (2009) showed that milk decreased the urinary excretion of flavan-3-ols from cocoa but did not have a effect on the plasma pharmacokinetics. This group and others also showed differences in the amounts of specific phase II metabolites that may in turn alter the resulting biological activity (Roura *et al.* 2008; Mullen *et al.* 2009). Whether the modest changes in flavonoid absorption/metabolism/excretion translate to significant differences in most of the biological effects of chocolate or cocoa is still not clear.

8.13 Conclusions

There is a convincing body of evidence indicating that cocoa and its components have the potential to provide significant health benefits. Cocoa has beneficial effects on vascular endothelial function, platelet activation and aggregation, and lipoprotein oxidation, and consumption has been correlated with reduced incidence of CVD. Studies have reported increases in biomarkers of insulin sensitivity, which may reduce the risk of developing type 2 diabetes. Cocoa and its components may interfere with the pathogenesis of a variety of other degenerative diseases by acting as *in vivo* antioxidants and possibly reducing inflammation. There have been conflicting results in studies that have investigated cocoa's effects on blood pressure. It is important to understand why some studies observed significant decreases and others have not. Cocoa production and/or processing techniques could potentially be optimised for specific components if a variable phytochemical component mediates cocoa's blood pressure effects.

The widely held beliefs that cocoa and chocolate have deleterious effects on human health have not been scientifically substantiated. There is little or no evidence that cocoa or chocolate significantly contributes to obesity, and cocoa may actually have beneficial effects on LDL-C and HDL-C concentrations. At present, it is not known how much cocoa is truly needed to provide a real benefit and exactly which products are most effective. Clearly, the amount of cocoa in a particular product, such as chocolate, is an important determinant of its potential biological activity. Long-term clinical interventions to assess the risks and benefits of regularly consuming various amounts of cocoa products are needed. However, for those individuals that do enjoy the flavours of cocoa, replacing less healthy snacks with dark chocolate or other products that contain significant amounts of cocoa's active constituents would appear to be very rational, considering what is currently known about cocoa's effects on human health.

References

Allen, R.R., Carson, L., Kwik-Uribe, C. et al. (2008) Daily consumption of a dark chocolate containing flavanols and added sterol esters affects cardiovascular risk factors in a normotensive population with elevated cholesterol. *J. Nutr.*, **138**, 725–731.

Arts, I.C., Hollman, P.C. and Kromhout, D. (1999) Chocolate as a source of tea flavonoids. *Lancet*, **354**, 488.

Azadbakht, L., Mirmiran, P., Esmaillzadeh, A. et al. (2005) Beneficial effects of a dietary approaches to stop hypertension eating plan on features of the metabolic syndrome. *Diabetes Care*, **28**, 2823–2831.

Baba, S., Natsume, M., Yasuda, A. et al. (2007) Plasma LDL and HDL cholesterol and oxidized LDL concentrations are altered in normo- and hypercholesterolemic humans after intake of different levels of cocoa powder. *J. Nutr.*, **137**, 1436–1441.

Balzer, J., Rassaf, T., Heiss, C. et al. (2008) Sustained benefits in vascular function through flavanol-containing cocoa in medicated diabetic patients: a double-masked, randomized, controlled trial. *J. Am. Coll. Cardiol.*, **51**, 2141–2149.

Bayard, V., Chamorro, F., Motta, J. et al. (2007) Does flavanol intake influence mortality from nitric oxide-dependent processes? Ischemic heart disease, stroke, diabetes mellitus, and cancer in panama. *Int. J. Med. Sci.*, **4**, 53–58.

Berry, N., Davison, K., Coates, A. et al. (2010) Impact of cocoa flavanol consumption on blood pressure responsiveness to exercise. *Br. J. Nutr.*, **103**, 1480–1484.

Bisson, J.F., Nejdi, A., Rozan, P. et al. (2008) Effects of long-term administration of a cocoa polyphenolic extract (acticoa powder) on cognitive performances in aged rats. *Br. J. Nutr.*, **100**, 94–101.

Brand-Miller, J., Holt, S., de Jong, V. et al. (2003) Cocoa powder increases postprandial insulinemia in lean young adults. *J. Nutr.*, **133**, 3149–3152.

Buijsse, B., Feskens, E.J.M., Kok, F.J. et al. (2006) Cocoa intake, blood pressure, and cardiovascular mortality: the Zutphen Elderly Study. *Arch. Int. Med.*, **166**, 411–417.

Buijsse, B., Weikert, C., Drogan, D. et al. (2010) Chocolate consumption in relation to blood pressure and risk of cardiovascular disease in German adults. *Eur. Heart J.*, **31**, 1616–1623.

Cady, R. and Durham, P. (2010) Cocoa-enriched diets enhance expression of phosphatases and decrease expression of inflammatory molecules in trigeminal ganglion neurons. *Brain Res.*, **1323**, 18–32.

Chobanian, A.V., Bakris, G.L., Black, H.R. et al. (2003) Seventh report of the joint national committee on prevention, detection, evaluation, and treatment of high blood pressure. *JAMA* **19**, 2560–2572.

Chun, O.K., Chung, S.J. and Song W.O. (2007) Estimated dietary flavonoid intake and major food sources of U.S. adults. *J. Nutr.*, **137**, 1244–1252.

Cooper, K.A., Campos-Gimenez, E., Alvarez, D.J. et al. (2007a) Rapid reversed phase ultraperformance liquid chromatography analysis of the major cocoa polyphenols and interrelationships of their concentrations in chocolate. *J. Agric. Food Chem.*, **55**, 2841–2847.

Cooper, K.A., Donovan, J.L., Waterhouse, A.L. et al. (2007b) Cocoa and health: a decade of research. *Br. J. Nutr.*, **99**, 1–11.

Cowley, A.W. (1992) Long-term control of arterial blood pressure. *Physiol. Rev.*, **72**, 231–300.

Crews, W.D., Jr., Harrison, D.W. and Wright, J.W. (2008) A double-blind, placebo-controlled, randomized trial of the effects of dark chocolate and cocoa on variables associated with neuropsychological functioning and cardiovascular health: clinical findings from a sample of healthy, cognitively intact older adults. *Am. J. Clin. Nutr.*, **87**, 872–880.

Davison, K., Coates, A.M., Buckley, J.D. *et al.* (2008) Effect of cocoa flavanols and exercise on cardiometabolic risk factors in overweight and obese subjects. *Int. J. Obes*, **32**, 1289–1296.

di Giuseppe, R., di Castelnuovo, A., Centritto, F. *et al.* (2008) Regular consumption of dark chocolate is associated with low serum concentrations of c-reactive protein in a healthy Italian population. *J. Nutr.*, **138**, 1939–1945.

Dillinger, T.L., Barriga, P., Escarcega, S. *et al.* (2000) Food of the gods: cure for humanity? A cultural history of the medicinal and ritual use of chocolate. *J. Nutr.*, **130**, S2057–S2072.

Dinges, D. (2006) Cocoa flavanols, cerebral blood flow, cognition, and health: going forward. *J. Cardiovasc. Pharmacol.*, **47**, S221–S223.

Donovan, J.L. and DeVane, C.L. (2001) A primer on caffeine pharmacology and its drug interactions in clinical psychopharmacology. *Psychopharm. Bull.*, **35**, 30–48.

Dunning, J.M. and Hodge, A.T. (1971) Influence of cocoa and sugar in milk on dental caries incidence. *J. Dent Res.*, **50**, 854–859.

Egan, B.M., Laken, M.A., Donovan, J.L. *et al.* (2010) Does dark chocolate have a role in the prevention and management of hypertension?: commentary on the evidence. *Hypertension*, **55**, 1289–1295.

Engler, M.B., Engler, M.M., Chen, C.Y. *et al.* (2004) Flavonoid-rich dark chocolate improves endothelial function and increases plasma epicatechin concentrations in healthy adults. *J. Am. Coll. Nutr.*, **23**, 197–204.

Faridi, Z., Njike, V.Y., Dutta, S. *et al.* (2008) Acute dark chocolate and cocoa ingestion and endothelial function: a randomized controlled crossover trial. *Am. J. Clin. Nutr.*, **88**, 58–63.

Farouque, H., Leung, M., Hope, S. *et al.* (2006) Acute and chronic effects of flavanol-rich cocoa on vascular function in subjects with coronary artery disease: a randomized double-blind placebo-controlled study. *Clin. Sci.*, **111**, 71–80.

Fisher, N.D. and Hollenberg, N.K. (2006) Ageing and vascular responses to flavanol-rich cocoa. *J. Hypertens.*, **24**, 1575–1580.

Fisher, N.D.L., Hughes, M., Gerhard-Herman, M. *et al.* (2003) Flavanol-rich cocoa induces nitric-oxide-dependent vasodilation in healthy humans. *J. Hyperten.*, **21**, 2281–2286.

Flammer, A.J., Hermann, F., Sudano, I. *et al.* (2007) Dark chocolate improves coronary vasomotion and reduces platelet reactivity. *Circulation*, **116**, 2376–2382.

Foster-Powell, K., Holt, S.H. and Brand-Miller, J.C. (2002) International table of glycemic index and glycemic load values. *Am. J. Clin. Nutr.*, **76**, 5–56.

Fraga, C.G., Actis-Goretta, L., Ottaviani, J.I. *et al.* (2005) Regular consumption of a flavanol-rich chocolate can improve oxidant stress in young soccer players. *Clin. Dev. Immunol.*, **12**, 11–17.

Francis, S., Head, K., Morris, P. *et al.* (2006) The effect of flavanol-rich cocoa on the fmri response to a cognitive task in healthy young people. *J. Cardiovasc. Pharmacol.*, **47**, S215–S220.

Fries, E.D. (1960) Hemodynamics of hypertension. *Physiol Rev.*, **40**, 27–54.

Fulton, J.E., Plewig, G., Kligman, A.M. (1969) Effect of chocolate on acne vulgaris. *JAMA*, **210**, 2071–2074.

Funabashi, H., Fujioka, M,. Kohchi, M. *et al.* (2000) Collaborative work to evaluate toxicity on male reproductive organs by repeated dose studies in rats: effects of 2-and 4-week administration of theobromine on the testis. *J. Toxicol. Sci.*, **25**, 211–221.

Gans, J. (1984) Comparative toxicities of dietary caffeine and theobromine in the rat. *Food Chem. Toxicol.*, **22**, 365–369.

Gotti, R., Furlanetto, S., Pinzauti, S. *et al.* (2006) Analysis of catechins in *Theobroma cacao* beans by cyclodextrin-modified micellar electrokinetic chromatography. *J. Chromatogr. A*, **1112**, 345–352.

Grassi, D., Lippi, C., Necozione, S. *et al.* (2005a) Short-term administration of dark chocolate is followed by a significant increase in insulin sensitivity and a decrease in blood pressure in healthy persons. *Am. J. Clin. Nutr.*, **81**, 611–614.

Grassi, D., Necozione, S., Lippi, C. *et al.* (2005b) Cocoa reduces blood pressure and insulin resistance and improves endothelium-dependent vasodilation in hypertensives. *Hypertension*, **46**, 398–405.

Grassi, D., Desideri, G., Necozione, S. *et al.* (2008) Blood pressure is reduced and insulin sensitivity increased in glucose-intolerant, hypertensive subjects after 15 days of consuming high-polyphenol dark chocolate. *J. Nutr.*, **138**, 1671–1676.

Hamed, M., Gambert, S., Bliden, K. *et al.* (2008) Dark chocolate effect on platelet activity, c-reactive protein and lipid profile: a pilot study. *South. Med. J.*, **101**, 1203–1208.

Heinrich, U., Neukam, K., Tronnier, H. *et al.* (2006) Long-term ingestion of high flavanol cocoa provides photoprotection against UV-induced erythema and improves skin condition in women. *J. Nutr.*, **136**, 1565–1569.

Heiss, C., Dejam, A., Kleinbongard, P. *et al.* (2003) Vascular effects of cocoa rich in flavan-3-ols. *JAMA*, **290**, 1030–1031.

Hermann, F., Spieker, L.E., Ruschitzka, F. *et al.* (2006) Dark chocolate improves endothelial and platelet function. *Heart*, **92**, 119–120.

Holt, R.R., Schramm, D.D., Keen, C.L. *et al.* (2002) Chocolate consumption and platelet function. *JAMA*, **287**, 2212–2213.

Hu, F.B., Stampfer, M.J., Manson, J.E. *et al.* (1999) Dietary saturated fats and their food sources in relation to the risk of coronary heart disease in women. *Am. J. Clin. Nutr.*, **70**, 1001–1008.

Hu, F.B., Stampfer, M.J. and Willett, W.C. (2000) Reply to P.M. Kris-Etherton *et al. Am. J. Clin. Nutr.*, **72**, 1059–1060.

Huang, D., Ou, B. and Prior, R.L. (2005) The chemistry behind antioxidant capacity assays. *J. Agric. Food Chem.*, **53**, 1841–1856.

Hurst, W., Tarka, S., Powis, T. *et al.* (2002) Archaeology: cacao usage by the earliest Maya civilization. *Nature*, **418**, 289–290.

Jaggers, G., Celep, G., Galleano, M. *et al.* (2010) Blood pressure lowering effect of dietary (−)-epicatechin in L-NAME induced hypertension in rats. *FASEB J*, **24**(meeting abstracts), 722.21.

Janszky, I., Mukamal, K., Ljung, R. *et al.* (2009) Chocolate consumption and mortality following a first acute myocardial infarction: the Stockholm Heart Epidemiology Program. *J. Intern. Med.*, **266**, 248–257.

Jia, L., Liu, X., Bai, Y.Y. *et al.* (2010) Short-term effect of cocoa product consumption on lipid profile: a meta-analysis of randomized controlled trials. *Am. J. Clin. Nutr.*, **92**, 218–225.

Joseph, J.A., Shukitt-Hale, B., Denisova, N.A. *et al.* (1999) Reversals of age-related declines in neuronal signal transduction, cognitive, and motor behavioral deficits with blueberry, spinach, or strawberry dietary supplementation. *J. Neurosci.*, **19**, 8114–8121.

Judelson, D.A., Griel, A.E., Miller, D. *et al.* (2010) Effects of theobromine, a caffeine-like substance found in cocoa and chocolate, on mood and vigilance. *FASEB J*, **24** (meeting abstracts), 209.5.

Kearney, P.M., Whelton, M., Reynolds, K. *et al.* (2005) Global burden of hypertension: analysis of worldwide data. *Lancet*, **365**, 217–223.

Keogh, J.B., McInerney, J. and Clifton, P.M. (2007) The effect of milk protein on the bioavailability of cocoa polyphenols. *J. Food Sci.*, **72**, S230–S233.

Kris-Etherton, P.M., Derr, J., Mitchell, D.C. et al. (1993) The role of fatty acid saturation on plasma lipids, lipoproteins, and apolipoproteins: I. Effects of whole food diets high in cocoa butter, olive oil, soybean oil, dairy butter, and milk chocolate on the plasma lipids of young men. *Metabolism*, **42**, 121–129.

Kris-Etherton, P.M., Derr, J.A., Mustad, V.A. et al. (1994) Effects of a milk chocolate bar per day substituted for a high-carbohydrate snack in young men on an ncep/aha step 1 diet. *Am. J. Clin. Nutr.*, **60**, S1037S–S1042.

Kris-Etherton, P.M., Pelkman, C.L., Zhao, G. et al. (2000) No evidence for a link between consumption of chocolate and coronary heart disease. *Am. J. Clin. Nutr.*, **72**, 1059.

Kris-Etherton, P.M., Griel, A.E., Psota, T.L. et al. (2005) Dietary stearic acid and risk of cardiovascular disease: intake, sources, digestion, and absorption. *Lipids*, **40**, 1193–1200.

Kurlandsky, S.B. and Stote, K.S. (2006) Cardioprotective effects of chocolate and almond consumption in healthy women. *Nutr. Res.*, **26**, 509–516.

Lamuela-Raventos, R., Romero-Perez, A., Andres-Lacueva, C. et al. (2005) Review: health effects of cocoa flavonoids. *Food Sci. Technol. Int.*, **11**, 159–176.

Mackie, B.S. and Mackie, L.E. (2007) Chocolate and acne. *Austral. J. Dermatol.*, **15**, 103–109.

Mao, T., van de Water, J., Keen, C.L. et al. (2000) Cocoa procyanidins and human cytokine transcription and secretion. *J. Nutr.*, **130**, S2093–S2099.

Mathur, S., Devaraj, S., Grundy, S. et al. (2002) Cocoa products decrease low density lipoprotein oxidative susceptibility but do not affect biomarkers of inflammation in humans. *J. Nutr.*, **132**, 3663–3667.

Miller, K.B., Hurst, W.J., Flannigan, N. et al. (2009) Survey of commercially available chocolate- and cocoa-containing products in the United States. 2. Comparison of flavan-3-ol content with nonfat cocoa solids, total polyphenols, and percent cacao. *J. Agric. Food Chem.*, **57**, 9169–9180.

Mink, P.J., Scrafford, C.G., Barraj, L.M. et al. (2007) Flavonoid intake and cardiovascular disease mortality: a prospective study in postmenopausal women. *Am. J. Clin. Nutr.*, **85**, 895–909.

Mix, J. (2002) A double blind, placebo controlled, randomized trial of *Ginkgo biloba* extract egb 761® in a sample of cognitively intact older adults: neuropsychological findings. *Human Psychopharm. Clin. Exper.*, **17**, 267–277.

Monagas, M., Khan, N., Andres-Lacueva, C. et al. (2009) Effect of cocoa powder on the modulation of inflammatory biomarkers in patients at high risk of cardiovascular disease. *Am. J. Clin. Nutr.*, **90**, 1144–1150.

Mostofsky, E., Levitan, E.B., Wolk, A. et al. (2010) Chocolate intake and incidence of heart failure: a population-based prospective study of middle-aged and elderly women. *Circ. Heart Fail.*, **3**, 612–616.

Mullen, W., Borges, G., Donovan, J.L. et al. (2009) Milk decreases urinary excretion but not plasma pharmacokinetics of cocoa flavan-3-ol metabolites in humans. *Am. J. Clin. Nutr.*, **89**, 1784–1791.

Muniyappa, R., Hall, G., Kolodziej, T.L. et al. (2008) Cocoa consumption for 2 weeks enhances insulin-mediated vasodilatation without improving blood pressure or insulin resistance in essential hypertension. *Am. J. Clin. Nutr.*, **88**, 1685–1696.

Murphy, K.J., Chronopoulos, A.K., Singh, I. et al. (2003) Dietary flavanols and procyanidin oligomers from cocoa (*Theobroma cacao*) inhibit platelet function. *Am. J. Clin. Nutr.*, **77**, 1466–1473.

Nagahama, Y., Nabatame, H., Okina, T. et al. (2003) Cerebral correlates of the progression rate of the cognitive decline in probable Alzheimer's disease. *Eur. Neurol.*, **50**, 1–9.

Naruse, K., Rask-Madsen, C., Takahara, N. et al. (2006) Activation of vascular protein kinase c- inhibits akt-dependent endothelial nitric oxide synthase function in obesity-associated insulin resistance. *Diabetes*, **55**, 691–698.

Neilson, A.P., George, J.C., Janle, E.M. *et al.* (2009) Influence of chocolate matrix composition on cocoa flavan-3-ol bioaccessibility in vitro and bioavailability in humans. *J. Agric. Food Chem.*, **57**, 9418–9426.

Nurk, E., Refsum, H., Drevon, C. *et al.* (2009) Intake of flavonoid-rich wine, tea, and chocolate by elderly men and women is associated with better cognitive test performance. *J. Nutr.*, **139**, 120–127.

Osakabe, N., Baba, S., Yasuda, A. *et al.* (2001) Daily cocoa intake reduces the susceptibility of low-density lipoprotein to oxidation as demonstrated in healthy human volunteers. *Free Rad. Res.*, **34**, 93–99.

Ottaviani, J.I., Actis-Goretta, L., Villordo, J.J. *et al.* (2006) Procyanidin structure defines the extent and specificity of angiotensin I converting enzyme inhibition. *Biochimie*, **88**, 359–365.

Ottaviani, J.I., Momma, T., Heiss, C. *et al.* (2011) The stereochemical configuration of flavanols influences the level and metabolism of flavanols in humans and their biological activity in vivo. *Free Rad. Biol. Med.*, **50**, 237–244.

Pearson, D.A., Paglieroni, T.G., Rein, D. *et al.* (2002) The effects of flavanol-rich cocoa and aspirin on *ex vivo* platelet function. *Thromb. Res.*, **106**, 191–197.

Perkel, J. (2007) Research profile: a molecular picture of chocoholics. *J. Proteome Res.*, **6**, 4105–4105.

Polagruto, J.A., Hackman, R.M., Braun, M.M. *et al.* (2006) Cocoa flavanol-enriched snack bars containing phytosterols effectively lower total and low-density lipoprotein cholesterol levels. *J. Am. Diet. Assoc.*, **106**, 1804–1813.

Rein, D., Lotito, S., Holt, R.R. *et al.* (2000a) Epicatechin in human plasma: in vivo determination and effect of chocolate consumption on plasma oxidation status. *J. Nutr.*, **130**, S2109–S2114.

Rein, D., Paglieroni, T.G., Pearson, D.A. *et al.* (2000b) Cocoa and wine polyphenols modulate platelet activation and function. *J. Nutr.*, **130**, S2120S–S2126.

Ried, K., Frank, O. and Stocks, N. (2009) Dark chocolate or tomato extract for prehypertension: a randomised controlled trial. *BMC Comp. Altern. Med.*, **9**, 22.

Roura, E., Andrés-Lacueva, C., Estruch, R. *et al.* (2007) Milk does not affect the bioavailability of cocoa powder flavonoid in healthy human. *Ann. Nutr. Metab.*, **51**, 493–498.

Roura, E., Andrés-Lacueva, C., Estruch, R. *et al.* (2008) The effects of milk as a food matrix for polyphenols on the excretion profile of cocoa (−)-epicatechin metabolites in healthy human subjects. *Br. J. Nutr.*, **100**, 846–851.

Rozan, P., Hidalgo, S., Nejdi, A. *et al.* (2007) Preventive antioxidant effects of cocoa polyphenolic extract on free radical production and cognitive performances after heat exposure in wistar rats. *J. Food Sci.*, **72**, S203–S206.

Schewe, T., Kuhn, H. and Sies, H. (2002) Flavonoids of cocoa inhibit recombinant human 5-lipoxygenase. *J. Nutr.*, **132**, 1825–1829.

Schindler, T.H., Hornig, B., Buser, P.T. *et al.* (2003) Prognostic value of abnormal vasoreactivity of epicardial coronary arteries to sympathetic stimulation in patients with normal coronary angiograms. *Arterioscl. Thromb. Vasc. Biol.*, **23**, 495–501.

Schramm, D.D., Wang, J.F., Holt, R.R. *et al.* (2001) Chocolate procyanidins decrease the leukotriene-prostacyclin ratio in humans and human aortic endothelial cells. *Am. J. Clin. Nutr.*, **73**, 36–40.

Schramm, D.D., Karim, M., Schrader, H.R. *et al.* (2003) Food effects on the absorption and pharmacokinetics of cocoa flavanols. *Life Sci.*, **73**, 857–869.

Schroeter, H., Holt, R.R., Orozco, T.J. *et al.* (2003) Nutrition milk and absorption of dietary flavanols. *Nature*, **426**, 787–788.

Schroeter, H., Heiss, C., Balzer, J. *et al.* (2006) (−)-Epicatechin mediates beneficial effects of flavanol-rich cocoa on vascular function in humans. *PNAS USA*, **103**, 1024–1029.

Serafini, M., Bugianesi, R., Maiani, G. et al. (2003) Plasma antioxidants from chocolate. *Nature*, **424**, 1013.

Sies, H., Schewe, T., Heiss, C. et al. (2005) Cocoa polyphenols and inflammatory mediators. *Am. J. Clin. Nutr.*, **81**, S304–S312.

Sorond, F., Lipsitz, L., Hollenberg, N. et al. (2008) Cerebral blood flow response to flavanol-rich cocoa in healthy elderly humans. *Neuropsych. Dis. Treat.*, **4**, 433–440.

Sorond, F.A., Hollenberg, N.K., Panych, L.P. et al. (2010) Brain blood flow and velocity: correlations between magnetic resonance imaging and transcranial doppler sonography. *J. Ultrasound Med.*, **29**, 1017–1022.

Steinberg, H.O., Chaker, H., Leaming, R. et al. (1996) Obesity/insulin resistance is associated with endothelial dysfunction. *J. Clin. Invest.*, **97**, 2601–2610.

Stuhlinger, M.C., Abbasi, F., Chu, J.W. et al. (2002) Relationship between insulin resistance and an endogenous nitric oxide synthase inhibitor. *JAMA*, **287**, 1420–1426.

Tarka, S. (1981) Effects of continuous administration of dietary theobromine on rat testicular weight and morphology. *Toxicol. Appl. Pharmacol.*, **58**, 76–82.

Taubert, D., Berkels, R., Roesen, R. et al. (2003) Chocolate and blood pressure in elderly individuals with isolated systolic hypertension. *JAMA*, **290**, 1029–1030.

Taubert, D., Roesen, R., Lehmann, C. et al. (2007a) Effects of low habitual cocoa intake on blood pressure and bioactive nitric oxide: a randomized controlled trial. *JAMA*, **298**, 49–60.

Taubert, D., Roesen, R. and Schomig, E. (2007b) Effect of cocoa and tea intake on blood pressure: a meta-analysis. *Arch. Int. Med.*, **167**, 626–634.

Tomas-Barberán, F., Cienfuegos-Jovellanos, E., Marín, A. et al. (2007) A new process to develop a cocoa powder with higher flavonoid monomer content and enhanced bioavailability in healthy humans. *J. Agric. Food Chem.*, **55**, 3926–3935.

Triche, E., Grosso, L., Belanger, K. et al. (2008) Chocolate consumption in pregnancy and reduced likelihood of preeclampsia. *Epidemiology*, **19**, 459–464.

Vinson, J., Proch, J., Bose, P. et al. (2006) Chocolate is a powerful *ex vivo* and *in vivo* antioxidant, an antiatherosclerotic agent in an animal model, and a significant contributor to antioxidants in the European and American diets. *J. Agric. Food Chem.*, **54**, 8071–8076.

Vita, J.A. and Keaney, J.F., Jr. (2002) Endothelial function: a barometer for cardiovascular risk? *Circulation*, **106**, 640–642.

Vlachopoulos, C., Aznaouridis, K., Alexopoulos, N. et al. (2005) Effect of dark chocolate on arterial function in healthy individuals. *Am. J. Hypertens.*, **18**, 785–791.

Vlachopoulos, C.V., Alexopoulos, N.A., Aznaouridis, K.A. et al. (2007) Relation of habitual cocoa consumption to aortic stiffness and wave reflections, and to central hemodynamics in healthy individuals. *Am. J. Cardiol.*, **99**, 1473–1475.

Wan, Y., Vinson, J.A., Etherton, T.D. et al. (2001) Effects of cocoa powder and dark chocolate on LDL oxidative susceptibility and prostaglandin concentrations in humans. *Am. J. Clin. Nutr.*, **74**, 596–602.

Wang, J.F., Schramm, D.D., Holt, R.R. et al. (2000) A dose-response effect from chocolate consumption on plasma epicatechin and oxidative damage. *J. Nutr.*, **130**, S2115–S2119.

Wang-Polagruto, J., Villablanca, A., Polagruto, J. et al. (2006) Chronic consumption of flavanol-rich cocoa improves endothelial function and decreases vascular cell adhesion molecule in hypercholesterolemic postmenopausal women. *J. Cardiovasc. Pharm.*, **47**, S177–S186.

Waterhouse, A.L., Shirley, J.R. and Donovan, J.L. (1996) Antioxidants in chocolate. *Lancet*, **348**, 834.

Widlansky, M.E., Gokce, N., Keaney, J.F., Jr., et al. (2003) The clinical implications of endothelial dysfunction. *J. Am. Coll. Cardiol.*, **42**, 1149–1160.

Willis, L., Bickford, P., Zaman, V. et al. (2005) Blueberry extract enhances survival of intraocular hippocampal transplants. *Cell Transplant.*, **14**, 213–223.

Wiswedel, I., Hirsch, D., Kropf, S. *et al.* (2004) Flavanol-rich cocoa drink lowers plasma f2-isoprostane concentrations in humans. *Free Rad. Biol. Med.*, **37**, 411–421.

Wollgast, J. and Anklam, E. (2000) Review on polyphenols in *Theobroma cacao*: changes in composition during the manufacture of chocolate and methodology for identification and quantification. *Food Res. Intern.*, **33**, 423–447.

Yamada, T., Yamada, Y., Okano, Y. *et al.* (2009) Anxiolytic effects of short- and long-term administration of cacao mass on rat elevated t-maze test. *J. Nutr. Biochem.*, **20**, 948–955.

Yusuf, S., Hawken, S., Ôunpuu, S. *et al.* (2004) Effect of potentially modifiable risk factors associated with myocardial infarction in 52 countries (The Interheart Study): case-control study. *Lancet*, **364**, 937–952.

Index

Note: Page numbers with italicised '*f*'s and '*t*'s refer to figures and tables, respectively.

acne, 238
adrenaline, 35
aminophilline, 36
anastatica tea (*Anastatica hirerochuntica*), 63–5, 65–6*f*. *See also* tea (*Camellia sinensis*); tisanes
anastatin, 66*f*
angiotensin-converting enzyme inhibitors (ACEI), 123
anthocyanins, 123, 197
antioxidants
    cocoa/cocoa products as, 229–32
    coffee as, 170–80
apigenin, 63*f*
apigenin-6,8-*C*-glucoside, 67*f*
apigenin-6-*C*-glucoside, 60*f*, 67*f*
apigenin-6-*C*-glucosyl-*C*-8-arabinoside, 67*f*
apigenin-6-glucoside, 67*f*
Arabica coffee beans, 33, 143
aspalathin, 59, 60*f*, 82–3, 83*t*
Aspalathus linearis, 59
aspermatogenesis, 237
assamaicin, 46*f*, 48
atractyligenin, 151
Aztecs, 5–6

beer, 15–16
bisabolol, 62, 62*f*
black tea. *See also* green tea
    bioavailability studies, 77–9
    caffeine content, 33
    health benefits
        bone health, 107
        cancer, 100–105
        cardiovascular disease, 105–6
        cognition, 106
        diabetes, 106–7
        neurodegenerative disease, 106
    history, 12
    human intervention studies, 101*t*

    polyphenols in, 55*t*
    production of, 48
bladder cancer, 107
blood pressure, lowering, 122–4
bone health, 107
Boston Tea Party, 21
breast cancer, 105, 110–11

cacahuatl, 7
cacao (*Theobroma cacao*), 193
    origin of, 5–8
    pods, 220
    seed, 26
cafestol, 151–2, 152*f*, 183
caffeine, 169. *See also* purine alkaloids
    biochemical and biological actions, 34–7
    biological properties, 35*t*
    biosynthesis of, 27, 29
    catabolism in plants, 30*f*
    in cocoa, 194
    effect on human health, 33–9
        dependence, 39
        mental performance enhancement, 37
        physical performance enhancement, 37–8
        pregnancy, 39
        tolerance, 39
        toxicity, 38–9
        withdrawal, 39
    FDA classification of, 34
    half-life of, 31
    metabolism by humans, 31–3
    as performance-enhancing drug, 38
    plant sources, 26
    structure of, 2*f*, 194*f*
    synthetic, 25
caffeine poisoning, 38
caffeinism, 38
caffeoyl-L-glutamic acid, 196

Teas, Cocoa and Coffee: Plant Secondary Metabolites and Health, First Edition.
Edited by Professor Alan Crozier, Professor Hiroshi Ashihara and Professor F. Tomás Barberán.
© 2012 Blackwell Publishing Ltd. Published 2012 by Blackwell Publishing Ltd.

caffeoylquinic acid, 57*f*, 63*f*, 145*f*, 157*t*
caffeoylquinic lactones, 150, 157*t*
caffeoylquinides, 149*f*
Camellia, 26, 45–6
Camellia sinensis. *See* tea (*Camellia sinensis*)
cancer
  black tea for, 100–105
  green tea for, 109–12
  oolong tea for, 107
  risk of, and coffee consumption, 186–8
captopril, 123
cardiovascular disease
  black tea for, 105–6
  cocoa and cocoa products for, 222–7
  green tea for, 112–13
  oolong tea for, 107–8
  risk of, and coffee consumption, 184–6
caries, 238
Castaing, Jonathan, 18
catechin, 46*f*, 56*f*, 60*f*, 195*f*, 221–2
catecholamines, 35
cerebral blood flow, 234–7
cha, 3, 9
Ch'a Ching, 3, 10
chai, 9
chamazulene, 62, 62*f*
chamomile (*Chamomilla recutita* L.). *See also* tea (*Camellia sinensis*); tisanes
  health benefits of, 119–20
  human intervention studies, 121*t*
  phytochemicals in, 62
chaube, 13
chay, 9
chicory-coffee, 21
China Drink, 18
Chinese green tea, 48
chlorogenic acids
  bioavailability studies, 155–6
  effect of roasting on, 149–53
  in fruits and vegetables, 155*t*
  in green coffee beans, 144–9
  in ileal fluid, 160*t*
  intake of, and coffee consumption, 154–5
  metabolism of, 163*t*
  metabolites, 161*t*
  pharmacokinetic parameters of derivatives and metabolites, 158*t*
  studies in humans with/without functioning colon, 156–64
  urinary excretion of metabolites, 161*t*
  in yerba maté tea, 54
chocolatl, 7
chocolate, 220
  and the Catholic aristocracy, 14–15
  dark, 220, 238
  future direction, 22
  health claims, 19–22
  history, 14–15
  matrix effects on bioavailability of (epi)catechins, 204–5
  milk, 220, 238
  public advertisement, 18
  use in New Spain, 7–8
chocolate houses, 16–17

chrysoeriol, 60*f*
cinnamoyl-amino acid conjugates, 146
cinnamoyl-L-aspartic acid, 196
cinnamtannin, 195*f*
Clostridium botulinum, 120
clovamide, 196
Coca-Cola, 34
cocoa
  consumption with milk, 238–9
  discovery of, 5–8
  ecological studies, 222–4
  epidemiological studies, 222–4
  health effects, negative
    acne, 238
    dental caries, 238
    obesity, 237
    testicular health, 237–8
  health effects, positive, 224–37
    anti-inflammatory effects, 229–32
    antioxidant effects, 229–32
    cardiovascular disease, 222–7
    cerebral blood flow, 234–7
    hypertension, 227–9
    insulin sensitivity, 233–4
    lipid/lipoprotein metabolism, 232–3
    neurocognitive functioning, 234–7
    on platelet activity, 224–7
    on vascular endothelial function, 224–7
  history, 14–18
  myth and legend, 5–8
  origin of, 2
  phytochemicals in, 193–213, 220–21
    anthocyanins, 197
    caffeine, 194
    flavan-3-ols, 194–6
    flavonols, 197
    phenolic acid derivatives, 196
    purine alkaloids, 194
    stilbene, 198
    theobromine, 194
    variation in, 221
  polyphenols in, 220–21
  worldwide consumption, 222
cocoa beans, 26–7
  flavanoid content of, 221
  processing of, 193–4
  roasting, 194
  worldwide production of, 222
cocoa liquor, 194, 220
cocoa powder, production of, 194
cocoa products, 220
Coffea arabica, 26, 169
Coffea canephora, 169
Coffea dewevrei, 143
Coffea liberica, 143
Coffea pseudozanguebariae, 143
Coffea racemosa, 143
Coffea species, 26
coffee, 162
  antioxidant status, 170–79
    cohort studies, 179
    effect of consumption, 179
    epidemiological studies, 179
    intervention studies, 171–8*t*, 179–80

chlorogenic acids in, 169
consumption in the US, 33
decaffeinated, 29–31
discovery of, 4–5
effect of roasting on phytochemical content, 149–53
etymology, 13
Europe's first experience of, 13–14
future direction, 22
health claims, 19–22
health risks
    cancer, 186–8
    cardiovascular disease, 183–6
    diabetes, 180–83
    inflammation, 186
history, 13–14, 15–18
human intervention studies, 171–8t
major producers, 143
myth and legend, 4–5
origin of, 1–2
phytochemicals in, 144–9
plant sources, 26
public advertisement, 18
coffee beans
    blending, 144
    harvesting, 144
    major producers, 143
    roasting, 144
coffee houses, 16–18
coffee sniffers, 21
cognition, 106, 113–14
colorectal cancer, 109–10
Columbus, Christopher, 7
conching, 220
consecutive reaction monitoring (CRM), 68
coumaroylquinic acid, 145f, 157t
C-reactive protein, 185
cupuaçu (*Theobroma grandiflorum*), 26–7
cut-tea-curl process, 48
cyclic AMP-phosphodiesterase, 35
cyclic diketopiperazines, 153
Cyclopia genistoides, 59
Cyclopia subternata, 59, 61–2
cytosolic β-glucosidase (CBg), 66

dark chocolate, 220, 238
decaffeinated coffee, 29–31, 171t, 177–8t
decaffeinated tea, 29–31
dicaffeoylquinic acid, 145f
dicaffeoylquinic lactones, 150
dehydroisokahweol, 151f
dehydrokahweol, 151f
dehydrotheasinensin, 51
dental caries, 238
dental health, green tea for, 115
deoxyclovamide, 196
dermatological health, green tea for, 115–16
devils brew. *See* coffee
diabetes
    black tea for, 106–7
    green tea for, 114
    oolong tea for, 108
    risk for, and coffee consumption, 180–83
dicaffeoylquinic acid, 157t

dihydroferulic acid, 156f
3′,4′-dihydroxybenzoic acid, 85f
3′,4′-dihydroxyphenylacetic acid, 85f
3′,4′-dimethoxycinnamic acid, 148f
1,2-dimethylpyridinium, 150
dimethylxanthines, 31
Dirx, Nikolas, 20
diterpenes, 151
dopamine, 35
dry coffee, 5
Dutch East India Company, 10, 20
Dutching, 194

East India Company, 10, 20
emodin, 58f
endometrial cancer, 111
energy drinks, 34, 39
epiafzelechin, 46f
epicatechin, 46f, 53, 56f, 195f, 200–207
(-)-epicatechin, degradation in large intestine, 207–10
epigallocatechin, 46f, 56f
(-)-epigallocatechin-3-O-gallate, 77
epitheaflavic acid, 50f
eriocitrin, 61
erythema, 116
espresso coffee
    chlorogenic acid in, 155t
    CQA isomers in, 154t
estragole, 64f, 85–7

falcarindiol, 64f
falcarinol, 64f
fenchone, 63, 64f
fennel (*Foeniculum vulgare*), 63. *See also* tea (*Camellia sinensis*); tisanes
    bioavailability studies, 85–7
    phytochemicals in, 63, 64f
ferulic acid, 156f, 162
feruloylglycine, 162
feruloylquinic acid, 145f, 157t
feruloylquinic lactones, 150
ficus tea, 66. *See also* tisanes
filter coffee, 172t, 176t
filter decaffeinated coffee, 177t
First Opium War, 20
flavan-3-ols, 194–6. *See also* phytochemicals in cocoa
    bioavailability of, 198–212
    concentration of (epi)catechin metabolites in plasma, 201f
    degradation of (epi)catechin in large intestine, 207–10
    in green tea, 69–76
    matrix effects, 204–7
    monomers, 198–210
    procyanidins, 210–12
    stability of, 76–7
    in tea leaf, 48
flavonoids
    daily intake, 222
    in rooibos tea, 60f, 60t
flavonols, 197
flavanones, 59
Fortune, Robert, 11

gallocatechin, 46*f*, 53, 56*f*
Garraway, Thomas, 19
gastric cancer, 109
German chamomile (*Matricaria reticulata*), 62
Ginkgo biloba, 235
green coffee extract, 172*t*
green tea. *See also* black tea
   caffeine in, 33
   Chinese, 48
   dose effects, 76
   flavan-3-ols in, 48, 69–76
   health benefits
      cancer, 109–12
      cardiovascular disease, 112–13
      cognition, 113–14
      dental health, 115
      dermatological health, 115–16
      diabetes, 114
      neurodegenerative disease, 113–14
      obesity, 114–15
   human intervention studies, 102–4*t*
   inhibition of adrenaline metabolism, 77
   in vitro faecal incubation studies, 72–5
   Japanese, 48
   phytochemicals in, 47*t*
   polyphenols in, 47*t*, 48–9, 55*t*
   stability of flavan-3-ols, 76–7
   studies with healthy subjects, 68–72
   studies with ileostomists, 72–5
   transformation products in, 49*f*
   types of, 48
Greenwich Tea Party, 21
Grindel, Martin, 10–11

Hadji, Omar, 5
haematological cancer, 110
herbal teas. *See* tisanes
hibiscus (*Hibiscus sabdariffa L.*), 62–3, 120. *See also* tea (*Camellia sinensis*); tisanes
   bioavailability studies, 85
   blood pressure-lowering effect of, 122–4
   cholesterol-lowering effect of, 122
   comparison with angiotensin converting enzyme inhibitors, 123
   as folk remedy, 122
   health benefits of, 120–26
   hepatoprotective properties of, 122
   human intervention studies, 125*t*
   phytochemicals in, 62–3
high-density lipoprotein, 184
high-density lipoprotein cholesterol (HDL-C), 185, 232–3
Hipocrates/Galen humerol system, 19
hippuric acid, 156*f*
honeybush (*Cyclopia intermedia*), 59–62, 128
   anticancer properties of, 129–30
   bioavailability studies, 84–5
   in folk medicine, 129
   health benefits, 128–30
   modulation of mutagenesis, 129
   phytochemicals in, 59–62
   stimulation of milk production with, 128
hsun, 3
hydrogen peroxide, 53

3'-hydroxyhippuric acid, 156*f*
hydroxyhydroquinone-free, 173–4*t*
1-hydroxypyrene glucuronide (1-OHPG), 120
hyperglycaemia, 180
hypertension, cocoa for, 227–9
hypospadias, 237

ibuprofen, 62*f*
ileostomists, 72–5, 156–64
impaired fasting glucose, 180
impaired glucose tolerance, 180
inflammation
   effect of cocoa consumption on, 229–32
   effect of coffee consumption on, 186
instant arabica coffee, 173*t*
instant coffee, 171*t*, 173*t*
   chlorogenic acids in, 157*t*
insulin sensitivity, 233–4
International Olympic Committee, 38
isoferulic acid, 156*f*
isokahweol, 151*f*
isomangiferin, 61
isosilybin, 66*f*
isovitexin, 59
itadori (*Polygonum cuspidatum*), 58. *See also* tea (*Camellia sinensis*); tisanes
   bioavailability studies, 80–81
   health benefits of, 118
   phytochemicals in, 58

Japanese green tea, 48
Japanese knotweed. *See* itadori (*Polygonum cuspidatum*)

kahveh, 13
kahweol, 151, 151*f*, 183
Kair Bey (governor of Mecca), 13
Kaldi (goat herder), discovery of coffee by, 4–5
King Gustav III, 21
Koprili (Ottoman Grand Vizier), 13

lactase phlorizin hydrolase (LPH), 66
leukotriene, 232
Levant, 13
liberine, structure of, 26*f*
Linnaeus, Carl, 11
Linschoten, Jan Hugo van, 9
lipids, 232–3
lipoproteins, 232–3
lisinopril, 123–4
Lloyds of London, 17–18
low-density lipoprotein cholesterol (LDL-C), 229–30, 232–3
low-density lipoprotein (LDLs), 185
luteolin, 60*f*
luteolin-8-*C*-diglucoside, 67*f*

Madrid Codex, 6
Maillard products, 153
mangiferin, 61
matesaponins, 56
matricin, 62, 62*f*
Mayans, 5–6
melanoidins, 153

mental performance, effect of caffeine on, 37
Mesoamerica, 5
3-methoxy-4-hydroxybenzoic acid, 156f
1-methyl-2-pyridone-5-carboxamide, 150
1-methyl-4-pyridone-5-carboxamide, 150
methycafestol, 151
methylkahweol, 151
methylliberine, structure of, 26f
1-methylpyridinium, 150
methyluric acids, 25
methylxanthines, 25
milk, effect on polyphenol absorption, 79
milk chocolate, 220, 238
mocca, 178t
Mocha region, 4
Mohammedans, 13

N-caffeoyl-3-O-hydroxytyrosine, 196
neotheaflavin, 53
neurocognitive functioning, 234–7
neurodegenerative disease, 106, 113–14
nibs, 220
nicotinic acid, 150
Nikolas Tulp (pseudonym of Nikolas Dirx), 20
nitric oxide, 226
N-methyltransferases, 27
N-nitro-L-arginine methyl ester (L-NAME), 226
noradrenaline, 35
norothyriol, 85f
nothofagin, 59, 60f
N-p-coumaroyl-tyrosine, 196
nutraceuticals, 21

obesity, 108–9, 114–15, 237
oesophageal cancer, 109
oligospermatogenesis, 237
oolong tea
  health benefits
    cancer, 107
    cardiovascular disease, 107–8
    diabetes, 108
    obesity, 108–9
  human intervention studies, 101–2t
oolongtheanin, 49f
opium, 20
oral contraceptives, and caffeine metabolism, 31
oral glucose tolerance test, 180, 182
orientin, 59
orobol, 61f
ovarian cancer, 101, 105, 111

p-anisaldehyde, 64f
paraxanthine, 37
  structure of, 26f
Parkinson's disease, 106
Pen ts'ao, 3
phenolic acid derivatives, 196
phenylindans, 149f
physical performance, effect of caffeine on, 37–8
phytochemicals in cocoa, 193–213, 220–21
  anthocyanins, 197
  caffeine, 194

flavan-3-ols, 194–6
flavonols, 197
phenolic acid derivatives, 196
purine alkaloids, 194
stilbene, 198
theobromine, 194
variation in, 221
phytochemicals in coffee, 144–55
  coffee beverages, 154–5
  green coffee beans, 145–9
  roasted coffee beans, 149–53
phytochemicals in teas, 45–88
  absorption, 66–8
  anastatica tea, 63–5, 65f
  bioavailability, 66–8
  Camellia teas, 45–54
  chamomile tea, 62
  distribution, 66–8
  excretion, 66–8
  fennel tea, 63, 64f
  ficus tea, 66
  hibiscus tea, 62–3
  honeybush tea, 59–62
  itadori tea, 58
  metabolism, 66–8
  rooibos tea, 59
  yerba maté tea, 54–7
Pole Clement VIII, 13–14
polyphenol oxidase, 12
polyphenon E, 76
pregnancy, caffeine in, 39
proanthocyanidins, 194, 221
procyanidin glycosides, 195f
procyanidins, 195f, 210–12, 221
prostacyclin, 232
prostate cancer, 111
pu-erh tea, 49, 54
puerin, 49f
purine alkaloids, 25–40. See also phytochemicals in cocoa
  biosynthesis of, 27
  caffeine consumption and health, 33–9
  caffeine metabolism by humans, 31–3
  in cocoa, 194
  degradation of, 27–9
  occurrence of, 26–7
  plant sources, 26–7
  structures of, 26f
  toxicity in animals, 40
Pythagoras, 13–14

Qing Dynasty, 20
quercetin, 60f
quercetin-3-O-rhamnoside, 63f
quercetin-3-O-rutinoside, 57f, 63f
quercetin-O-glycosides, 197

Ramusio, Giambattista, 9
Rauwolf, Leonhard, 13
rectal cancer, 105
Red Bull, 39
red bush tea, 59
red tea. See rooibos (Aspalathus linearis)
resveratrol, 80–81, 82f, 82t

roasted coffee beans, 5
    aroma, 152–3
    phytochemicals in, 149–53
robusta coffee (*Coffee canephora*), 33, 143
rooibos (*Aspalathus linearis*), 59. *See also* tea (*Camellia sinensis*); tisanes
    antiallergenic effect of, 127–8
    bioavailability studies, 81–4
    in folk medicine, 127
    health benefits, 126–8
    human intervention studies, 127*t*
    phytochemicals in, 59
rosmarinic acid, 64*f*

salsoline, 194*f*
sambubiosides, 85
saponins, 56
scolymoside, 61
selected ion monitoring (SIM), 68
serotonin, 150
Shen Nung (Chinese emperor), 3
silybin, 66*f*
sinapic acid, 148*f*, 157
sinus tachycardia, 36
skin care, green tea for, 115–16
slave trade, and coffee trade, 4
sleep, effect of caffeine on, 31, 36
smoking, and caffeine metabolism, 31
soft drink, 34
stilbene, 58*t*, 198
Stockholm Heart Epidemiology Study, 224
strictinin, 46*f*
sugar, 16

tea (*Camellia sinensis*), 45–54. *See also* tisanes
    black tea, 100–107
    caffeine in, 26, 33
    consumption in the UK, 33
    decaffeinated, 29–31
    discovery of, 3
    etymology, 9
    future direction, 22
    global trade/history, 9–12
    green tea, 109–16
    health claims, 19–22
    history, 15–18
    myth and legend, 3–4
    origin of, 1
    phytochemicals in, 45–54
    public advertisement, 18
Tea Act of 1773, 21
tea houses, 16–17
telangiectasias, 116
testicular cancer, 237
Thea bohea, 11
Thea sinensis, 11
Thea viridis, 11
theacitrins, 50*f*
theacrine, 25
    structure of, 26*f*
theadibenztropolone A, 50*f*
theaflagallin, 50*f*
theaflagallins, 51
theaflavate A, 50*f*

theaflavin gallates, 50
theaflavins, 50, 50*f*, 53, 78
theaflavonin, 50*f*
theanaphthoquinones, 51
theanine, 46*f*
thearubigins, 51–2, 77–8
theasinensins, 49*f*
theobromine, 194. *See also* phytochemicals in cocoa
    plant sources, 26–7
    structure of, 2*f*, 26*f*, 194*f*
    toxicity in animals, 40
theogallinin, 50*f*
theophylline, 36
    structure of, 2*f*, 26*f*
tisanes. *See also* tea (*Camellia sinensis*)
    anastatica tea, 63–5
    chamomile tea, 62, 119–20, 121*t*
    fennel tea, 63, 64*f*, 85–7
    ficus tea, 66
    hibiscus tea, 62–3, 85, 120–26
    honeybush tea, 59–62, 84–5, 128–30
    itadori tea, 58, 80–81, 118
    oolong tea, 107–9
    rooibos tea, 59, 81–4, 126–8
    yerba maté tea, 54–7, 116–18
toxicity, caffeine, 38–9
trans-anethole, 63, 64*f*
trans-resveratrol, 58*f*, 80–81
trans-resveratrol-3-*O*-glucoside, 58*f*
Treatise on Warm Beer, 10
tricetanidin, 50*f*, 51
trigonelline, 150–51
2′,4′,6′-trihydroxybenzoic acid, 85*f*
3′,4′,5′-trihydroxybenzoic acid, 85*f*
3′,4′,5′-trimethoxycinnamic acid, 148*f*
t'u, 3, 9
tumour necrosis factor (TNF), 232

ultrasonography, prenatal, 64*f*
umbelliferone, 63*f*
unfiltered coffee, 177–8*t*

Van Houten, Coenraad Johannes, 15
ventricular extrasystoles, 36
Vesling, Johann, 13
vitexin, 59

weak coffee, 176*t*
wet coffee, 5
white tea, 54
Wickham, John, 10
William of Orange, 16
Women's Petition Against Coffee, 19

xanthine, 27–8
xanthosine, 27, 28*f*

yerba maté (*Ilex paraguariensis*). *See also* tea (*Camellia sinensis*); tisanes
    health benefits of, 116–18
    phytochemicals in, 54–7
    purine alkaloids in, 27
yu, 3

Zutphen Elderly Study, 223

Food Science and Technology

GENERAL FOOD SCIENCE & TECHNOLOGY, ENGINEERING AND PROCESSING

Title	Author	ISBN
Organic Production and Food Quality: A Down to Earth Analysis	Blair	9780813812175
Handbook of Vegetables and Vegetable Processing	Sinha	9780813815411
Nonthermal Processing Technologies for Food	Zhang	9780813816685
Thermal Procesing of Foods: Control and Automation	Sandeep	9780813810072
Innovative Food Processing Technologies	Knoerzer	9780813817545
Handbook of Lean Manufacturing in the Food Industry	Dudbridge	9781405183673
Intelligent Agrifood Networks and Chains	Bourlakis	9781405182997
Practical Food Rheology	Norton	9781405199780
Food Flavour Technology, 2nd edition	Taylor	9781405185431
Food Mixing: Principles and Applications	Cullen	9781405177542
Confectionery and Chocolate Engineering	Mohos	9781405194709
Industrial Chocolate Manufacture and Use, 4th edition	Beckett	9781405139496
Chocolate Science and Technology	Afoakwa	9781405199063
Essentials of Thermal Processing	Tucker	9781405190589
Calorimetry in Food Processing: Analysis and Design of Food Systems	Kaletunç	9780813814834
Fruit and Vegetable Phytochemicals	de la Rosa	9780813803203
Water Properties in Food, Health, Pharma and Biological Systems	Reid	9780813812731
Food Science and Technology (textbook)	Campbell-Platt	9780632064212
IFIS Dictionary of Food Science and Technology, 2nd edition	IFIS	9781405187404
Drying Technologies in Food Processing	Chen	9781405157636
Biotechnology in Flavor Production	Havkin-Frenkel	9781405156493
Frozen Food Science and Technology	Evans	9781405154789
Sustainability in the Food Industry	Baldwin	9780813808468
Kosher Food Production, 2nd edition	Blech	9780813820934

FUNCTIONAL FOODS, NUTRACEUTICALS & HEALTH

Title	Author	ISBN
Functional Foods, Nutraceuticals and Degenerative Disease Prevention	Paliyath	9780813824536
Nondigestible Carbohydrates and Digestive Health	Paeschke	9780813817620
Bioactive Proteins and Peptides as Functional Foods and Nutraceuticals	Mine	9780813813110
Probiotics and Health Claims	Kneifel	9781405194914
Functional Food Product Development	Smith	9781405178761
Nutraceuticals, Glycemic Health and Type 2 Diabetes	Pasupuleti	9780813829333
Nutrigenomics and Proteomics in Health and Disease	Mine	9780813800332
Prebiotics and Probiotics Handbook, 2nd edition	Jardine	9781905224524
Whey Processing, Functionality and Health Benefits	Onwulata	9780813809038
Weight Control and Slimming Ingredients in Food Technology	Cho	9780813813233

INGREDIENTS

Title	Author	ISBN
Hydrocolloids in Food Processing	Laaman	9780813820767
Natural Food Flavors and Colorants	Attokaran	9780813821108
Handbook of Vanilla Science and Technology	Havkin-Frenkel	9781405193252
Enzymes in Food Technology, 2nd edition	Whitehurst	9781405183666
Food Stabilisers, Thickeners and Gelling Agents	Imeson	9781405132671
Glucose Syrups – Technology and Applications	Hull	9781405175562
Dictionary of Flavors, 2nd edition	De Rovira	9780813821351
Vegetable Oils in Food Technology, 2nd edition	Gunstone	9781444332681
Oils and Fats in the Food Industry	Gunstone	9781405171212
Fish Oils	Rossell	9781905224630
Food Colours Handbook	Emerton	9781905224449
Sweeteners Handbook	Wilson	9781905224425
Sweeteners and Sugar Alternatives in Food Technology	Mitchell	9781405134347

FOOD SAFETY, QUALITY AND MICROBIOLOGY

Title	Author	ISBN
Food Safety for the 21st Century	Wallace	9781405189118
The Microbiology of Safe Food, 2nd edition	Forsythe	9781405140058
Analysis of Endocrine Disrupting Compounds in Food	Nollet	9780813818160
Microbial Safety of Fresh Produce	Fan	9780813804163
Biotechnology of Lactic Acid Bacteria: Novel Applications	Mozzi	9780813815831
HACCP and ISO 22000 – Application to Foods of Animal Origin	Arvanitoyannis	9781405153669
Food Microbiology: An Introduction, 2nd edition	Montville	9781405189132
Management of Food Allergens	Coutts	9781405167581
Campylobacter	Bell	9781405156288
Bioactive Compounds in Foods	Gilbert	9781405158756
Color Atlas of Postharvest Quality of Fruits and Vegetables	Nunes	9780813817521
Microbiological Safety of Food in Health Care Settings	Lund	9781405122207
Food Biodeterioration and Preservation	Tucker	9781405154178
Phycotoxins	Botana	9780813827001
Advances in Food Diagnostics	Nollet	9780813822211
Advances in Thermal and Non-Thermal Food Preservation	Tewari	9780813829685

For further details and ordering information, please visit www.wiley.com/go/food

Food Science and Technology from Wiley-Blackwell

SENSORY SCIENCE, CONSUMER RESEARCH & NEW PRODUCT DEVELOPMENT

Title	Author	ISBN
Sensory Evaluation: A Practical Handbook	Kemp	9781405162104
Statistical Methods for Food Science	Bower	9781405167642
Concept Research in Food Product Design and Development	Moskowitz	9780813824246
Sensory and Consumer Research in Food Product Design and Development	Moskowitz	9780813816326
Sensory Discrimination Tests and Measurements	Bi	9780813811116
Accelerating New Food Product Design and Development	Beckley	9780813808093
Handbook of Organic and Fair Trade Food Marketing	Wright	9781405150583
Multivariate and Probabilistic Analyses of Sensory Science Problems	Meullenet	9780813801780

FOOD LAWS & REGULATIONS

Title	Author	ISBN
The BRC Global Standard for Food Safety: A Guide to a Successful Audit	Kill	9781405157964
Food Labeling Compliance Review, 4th edition	Summers	9780813821818
Guide to Food Laws and Regulations	Curtis	9780813819464
Regulation of Functional Foods and Nutraceuticals	Hasler	9780813811772

DAIRY FOODS

Title	Author	ISBN
Dairy Ingredients for Food Processing	Chandan	9780813817460
Processed Cheeses and Analogues	Tamime	9781405186421
Technology of Cheesemaking, 2nd edition	Law	9781405182980
Dairy Fats and Related Products	Tamime	9781405150903
Bioactive Components in Milk and Dairy Products	Park	9780813819822
Milk Processing and Quality Management	Tamime	9781405145305
Dairy Powders and Concentrated Products	Tamime	9781405157643
Cleaning-in-Place: Dairy, Food and Beverage Operations	Tamime	9781405155038
Advanced Dairy Science and Technology	Britz	9781405136181
Dairy Processing and Quality Assurance	Chandan	9780813827568
Structure of Dairy Products	Tamime	9781405129756
Brined Cheeses	Tamime	9781405124607
Fermented Milks	Tamime	9780632064588
Manufacturing Yogurt and Fermented Milks	Chandan	9780813823041
Handbook of Milk of Non-Bovine Mammals	Park	9780813820514
Probiotic Dairy Products	Tamime	9781405121248

SEAFOOD, MEAT AND POULTRY

Title	Author	ISBN
Handbook of Seafood Quality, Safety and Health Applications	Alasalvar	9781405180702
Fish Canning Handbook	Bratt	9781405180993
Fish Processing – Sustainability and New Opportunities	Hall	9781405190473
Fishery Products: Quality, safety and authenticity	Rehbein	9781405141628
Thermal Processing for Ready-to-Eat Meat Products	Knipe	9780813801483
Handbook of Meat Processing	Toldra	9780813821825
Handbook of Meat, Poultry and Seafood Quality	Nollet	9780813824468

BAKERY & CEREALS

Title	Author	ISBN
Whole Grains and Health	Marquart	9780813807775
Gluten-Free Food Science and Technology	Gallagher	9781405159159
Baked Products – Science, Technology and Practice	Cauvain	9781405127028
Bakery Products: Science and Technology	Hui	9780813801872
Bakery Food Manufacture and Quality, 2nd edition	Cauvain	9781405176132

BEVERAGES & FERMENTED FOODS/BEVERAGES

Title	Author	ISBN
Technology of Bottled Water, 3rd edition	Dege	9781405199322
Wine Flavour Chemistry, 2nd edition	Bakker	9781444330427
Wine Quality: Tasting and Selection	Grainger	9781405113663
Beverage Industry Microfiltration	Starbard	9780813812717
Handbook of Fermented Meat and Poultry	Toldra	9780813814773
Microbiology and Technology of Fermented Foods	Hutkins	9780813800189
Carbonated Soft Drinks	Steen	9781405134354
Brewing Yeast and Fermentation	Boulton	9781405152686
Food, Fermentation and Micro-organisms	Bamforth	9780632059874
Wine Production	Grainger	9781405113656
Chemistry and Technology of Soft Drinks and Fruit Juices, 2nd edition	Ashurst	9781405122863

PACKAGING

Title	Author	ISBN
Food and Beverage Packaging Technology, 2nd edition	Coles	9781405189101
Food Packaging Engineering	Morris	9780813814797
Modified Atmosphere Packaging for Fresh-Cut Fruits and Vegetables	Brody	9780813812748
Packaging Research in Food Product Design and Development	Moskowitz	9780813812229
Packaging for Nonthermal Processing of Food	Han	9780813819440
Packaging Closures and Sealing Systems	Theobald	9781841273372
Modified Atmospheric Processing and Packaging of Fish	Otwell	9780813807683
Paper and Paperboard Packaging Technology	Kirwan	9781405125031

For further details and ordering information, please visit www.wiley.com/go/food